Mathematics

Teacher's Guide

CONTENTS

Revision Editor: Alan Christopherson, M.S.

Alpha Omega Publications ®

300 North McKemy Avenue, Chandler, Arizona 85226-2618

MATHEMATICS

Curriculum Overview
Grades K–12

Kindergarten

1-40	41-80	81-120	121-160
Directions-right, left, high, low, etc. **Comparisons**-big, little, alike, different **Matching** **Cardinal Numbers**-to 9 **Colors**-red, blue, green, yellow, brown ,purple **Shapes**-circle, square, rectangle, triangle **Number Order** **Before and After** **Ordinal Numbers**-to 9th **Problem Solving**	**Directions**-right, left, high, low, etc. **Comparisons**-big, little, alike, different **Matching** **Cardinal Numbers**-to 12 **Colors**-orange **Shapes**-circle, square, rectangle, triangle **Number Order** **Before and After** **Ordinal Numbers**-to 9th **Problem Solving** **Number Words**-to nine **Addition**-to 9	**Directions**-right, left, high, low ,etc. **Comparisons**-big, little, alike, different **Matching** **Cardinal Numbers**-to 19 **Colors**-black, white **Shapes**-circle, square, rectangle, triangle **Number Order** **Before and After** **Ordinal Numbers**-to 9th **Problem Solving** **Number Words**-to nine **Addition**-multiples of 10 **Subtraction**-to 9 **Place Value** **Time/Calendar**	**Directions**-right, left, high, low, etc. **Comparisons**-big, little, alike, different **Matching** **Cardinal Numbers**-to 100 **Colors**-pink **Shapes**-circle, square, rectangle, triangle **Number Order** **Before and After** **Ordinal Numbers**-to 9th **Problem Solving** **Number Words**-to nine **Addition**-to 10 and multiples of 10 **Subtraction**-to 10 **Place Value** **Time/Calendar** **Money** **Skip Counting**-2's, 5's, 10's **Greater/ Less than**

Mathematics LIFEPAC Overview

	Grade 1	Grade 2	Grade 3
LIFEPAC 1	NUMBERS TO 99 • Number order, skip-count • Add, subtract to 9 • Story problems • Measurements, shapes	NUMBERS TO 100 • Numbers and words to 100 • Operation symbols +, –, =, >, < • Add, subtract, story problems • Place value, fact families	NUMBERS TO 999 • Digits, place value to 999 • Add, subtract, time • LInear measurements, dozen • Operation symbols +, –, =, ≠, >, <
LIFEPAC 2	NUMBERS TO 99 • Add, subtract to 10 • Number words • Place value, shapes • Patterns, sequencing, estimation	NUMBERS TO 200 • Numbers and words to 200 • Add, subtract, even and odd • Skip-count 2's, 5's, 10's, shapes • Ordinal numbers, fractions, money	NUMBERS TO 999 • Fact families, patterns, fractions • Add, subtract - carry, borrow • Skip count 2's, 5's, 10's • Money, shapes, lines, even, odd
LIFEPAC 3	NUMBERS TO 100 • Number sentences, • Fractions, oral directions • Story problems • Time, symbols =, ≠	NUMBERS TO 200 • Add w/ carry to 10's place • Subtract, standard measurements • Flat shapes, money, AM/PM • Rounding to 10's place	NUMBERS TO 999 • Add 3 numbers w/ carry • Coins, weight, volume, AM/PM • Fractions, oral instructions • Skip count 3's, subtract w/ borrow
LIFEPAC 4	NUMBERS TO 100 • Add to 18, place value • Skip-count, even and odd • Money • Shapes, measurement	NUMBERS TO 999 • Numbers and words to 999 • Add, subtract, place value • Calendar, making change • Measurements, solid shapes	NUMBERS TO 9,999 • Place value to 9,999 • Rounding to 10's, estimation • Add and subtract fractions • Roman numerals, 1/4 inch
LIFEPAC 5	NUMBERS TO 100 • Add 3 numbers - 1 digit • Ordinal numbers, fractions • Time, number line • Estimation, charts	NUMBERS TO 999 • Data and bar graphs, shapes • Add, subtract to 100's • Skip-count 3's, place value to 100's • Add fractions, temperature	NUMBERS TO 9,999 • Number sentences, temperature • Rounding to 100's, estimation • Perimeter, square inch • Bar graph, symmetry, even/odd rules
LIFEPAC 6	NUMBERS TO 100 • Number words to 99 • Add 2 numbers - 2 digit • Symbols >, < • Fractions, shapes	NUMBERS TO 999 • Measurements, perimeter • Time, money • Subtract w/ borrow from 10's place • Add, subtract fractions	NUMBERS TO 9,999 • Add, subtract to 9,999 • Multiples, times facts for 2 • Area, equivalent fractions, money • Line graph, segments, angles
LIFEPAC 7	NUMBERS TO 200 • Number order, place value • Subtract to 12 • Operation signs • Estimation, graphs, time	NUMBERS TO 999 • Add w/ carry to 100's place • Fractions as words • Number order in books • Rounding and estimation	NUMBERS TO 9,999 • Times facts for 5, missing numbers • Mixed numbers - add, subtract • Subtract with 0's in minuend • Circle graph, probability
LIFEPAC 8	NUMBERS TO 200 • Addition, subtract to 18 • Group counting • Fractions, shapes • Time, measurements	NUMBERS TO 999 • Add, subtract, measurements • Group count, 'think' answers • Convert coins, length, width • Directions-N, S, E, W	NUMBERS TO 9,999 • Times facts for 3, 10 - multiples of 4 • Convert units of measurement • Decimals, directions, length, width • Picture graph, missing addend
LIFEPAC 9	NUMBERS TO 200 • Add 3 numbers - 2 digit • Fact families • Sensible answers • Subtract 2 numbers - 2 digit	NUMBERS TO 999 • Area and square measurement • Add 3 numbers - 20 digit w/ carry • Add coins and convert to cents • Fractions, quarter-inch	NUMBERS TO 9,999 • Add, subtract whole numbers, fractions, mixed numbers • Standard measurements, metrics • Operation symbols, times facts for 4
LIFEPAC 10	NUMBERS TO 200 • Add, subtract, place value • Directions - N, S, E, W • Fractions • Patterns	NUMBERS TO 999 • Rules for even and odd • Round numbers to 100's place • Time - digital, sensible answers • Add 3 numbers - 3 digit	NUMBERS TO 9,999 • Add, subtract, times facts 2,3,4,5,10 • Rounding to 1,000's, estimation • Probability, equations, parentheses • Perimeter, area

	Grade 4	Grade 5	Grade 6
LIFEPAC 1	WHOLE NUMBERS & FRACTIONS • Naming whole numbers • Naming Fractions • Sequencing patterns • Numbers to 1,000	WHOLE NUMBERS & FRACTIONS • Operations & symbols • Fraction language • Grouping, patterns, sequencing • Rounding & estimation	FRACTIONS & DECIMALS • Number to billions' place • Add & subtract fractions • Add & subtract decimals • Read and write Fractions
LIFEPAC 2	WHOLE NUMBERS & FRACTIONS • Operation symbols • Multiplication - 1 digit multiplier • Fractions - addition & subtraction • Numbers to 10,000	WHOLE NUMBERS & FRACTIONS • Multiplication & division • Fractions - +, –, simplify • Plane & solid shapes • Symbol language	FINDING COMMON DENOMINATORS • Prime factors • Fractions with unlike denominators • Exponential notation • Add & subtract mixed numbers
LIFEPAC 3	WHOLE NUMBERS & FRACTIONS • Multiplication with carrying • Rounding & estimation • Sequencing fractions • Numbers to 100,000	WHOLE NUMBERS & FRACTIONS • Short division • Lowest common multiple • Perimeter & area • Properties of addition	MULTIPLYING MIXED NUMBERS • Multiply mixed numbers • Divide decimals • Bar and line graphs • Converting fractions & decimals
LIFEPAC 4	LINES & SHAPES • Plane & solid shapes • Lines & line segments • Addition & subtraction • Multiplication with carrying	WHOLE NUMBERS • Lines - shapes - circles • Symmetric - congruent - similar • Decimal place value • Properties of multiplication	DIVIDING MIXED NUMBERS • Divide mixed numbers • Area and perimeter • Standard measurements
LIFEPAC 5	WHOLE NUMBERS • Division - 1 digit divisor • Families of facts • Standard measurements • Number grouping	WHOLE NUMBERS & FRACTIONS • Multiply & divide by 10, 100, 1,000 • Standard measurements • Rate problems • Whole number & fraction operations	METRIC MEASURE • Metric measures • Plane & solid shapes • Multi-operation problems • Roman Numerals
LIFEPAC 6	WHOLE NUMBERS & FRACTIONS • Division - 1 digit with remainder • Factors & multiples • Fractions - improper & mixed • Equivalent fractions	FRACTIONS & DECIMALS • Multiplication of fractions • Reading decimal numbers • Adding & subtracting decimals • Multiplication - decimals	LCM & GCF • LCM, GCF • Fraction and decimal equivalents • Percent • Variables, functions & formulas
LIFEPAC 7	WHOLE NUMBERS & FRACTIONS • Multiplication - 2 digit multiplier • Simplifying fractions • Averages • Decimals in money problems	WHOLE NUMBERS & FRACTIONS • Division - 2-digit divisor • Metric units • Multiplication - mixed numbers • Multiplication - decimals	INTEGERS, RATIO & PROPORTION • Positive and negative integers • Ratio & proportion • Fractions, decimals & percents • Statistics
LIFEPAC 8	WHOLE NUMBERS & FRACTIONS • Division 1 digit divisor • Fractions - unlike denominators • Metric units • Whole numbers - +, –, x, ÷	WHOLE NUMBERS • Calculators & whole numbers • Calculators & decimals • Estimation • Prime factors	PROBABILITY & GRAPHING • Probability • Graphs • Metric and standard units • Square root
LIFEPAC 9	DECIMALS & FRACTIONS • Reading and writing decimals • Mixed numbers - +, – • Cross multiplication • Estimation	FRACTIONS & DECIMALS • Division - fractions • Division - decimals • Ratios & ordered pairs • Converting fractions to decimals	CALCULATORS & ESTIMATION • Calculators • Estimation • Geometric symbols & shapes • Missing number problems
LIFEPAC 10	PROBLEM SOLVING • Estimation & data gathering • Charts & Graphs • Review numbers to 100,000 • Whole numbers - +, –, x, ÷	PROBLEM SOLVING • Probability & data gathering • Charts & graphs • Review numbers to 100 million • Fractions & decimals - +, –, x, ÷	INTEGERS & OPERATIONS • Mental arithmetic • Fraction operations • Variables & properties • Number lines

	Grade 7	Grade 8	Grade 9
LIFEPAC 1	WHOLE NUMBERS • Number concepts • Addition • Subtraction • Applications	WHOLE NUMBERS • The set of whole numbers • Graphs • Operations with whole numbers • Applications with whole numbers	VARIABLES AND NUMBERS • Variables • Distributive Property • Definition of signed numbers • Signed number operations
LIFEPAC 2	MULTIPLICATION AND DIVISION • Basic facts • Procedures • Practice • Applications	NUMBERS AND FACTORS • Numbers and bases • Sets • Factors and multiples • Least common multiples	SOLVING EQUATIONS • Sentences and formulas • Properties • Solving equations • Solving inequalities
LIFEPAC 3	GEOMETRY • Segments, lines, and angles • Triangles • Quadrilaterals • Circles and hexagons	RATIONAL NUMBERS • Proper and improper fractions • Mixed numbers • Decimal fractions • Percent	PROBLEM ANALYSIS AND SOLUTION • Words and symbols • Simple verbal problems • Medium verbal problems • Challenging verbal problems
LIFEPAC 4	RATIONAL NUMBERS • Common fractions • Improper fractions • Mixed numbers • Decimal fractions	FRACTIONS AND ROUNDING • Common fraction addition • Common fraction subtraction • Decimal fractions • Rounding numbers	POLYNOMIALS • Addition of polynomials • Subtraction of polynomials • Multiplication of polynomials • Division of polynomials
LIFEPAC 5	SETS AND NUMBERS • Set concepts and operations • Early number systems • Decimal number system • Factors and multiples	FRACTIONS AND PERCENT • Multiplication of fractions • Division of fractions • Fractions as percents • Percent exercises	ALGEBRAIC FACTORS • Greatest common factor • Binomial factors • Complete factorization • Word problems
LIFEPAC 6	FRACTIONS • Like denominators • Unlike denominators • Decimal fractions • Equivalents	STATISTICS, GRAPHS, & PROBABILITY • Statistical measures • Types of graphs • Simple probability • And–Or statements	ALGEBRAIC FRACTIONS • Operations with fractions • Solving equations • Solving inequalities • Solving word problems
LIFEPAC 7	FRACTIONS • Common fractions • Decimal fractions • Percent • Word problems	INTEGERS • Basic concepts • Addition and subtraction • Multiplication and division • Expressions and sentences	RADICAL EXPRESSIONS • Rational and irrational numbers • Operations with radicals • Irrational roots • Radical equations
LIFEPAC 8	FORMULAS AND RATIOS • Writing formulas • A function machine • Equations • Ratios and proportions	FORMULAS AND GEOMETRY • Square root • Perimeter, circumference, and area • Rectangular solid • Cylinder, cone, and sphere	GRAPHING • Equations of two variables • Graphing lines • Graphing inequalities • Equations of lines
LIFEPAC 9	DATA, STATISTICS AND GRAPHS • Gathering and organizing data • Central tendency and dispersion • Graphs of statistics • Graphs of points	ALGEBRAIC EQUATIONS • Variables in formulas • Addition and subtraction • Multiplication and division • Problem solving	SYSTEMS • Graphical solution • Algebraic solutions • Determinants • Word problems
LIFEPAC 10	MATHEMATICS IN SPORTS • Whole numbers • Geometry, sets, and systems • Fractions • Formulas, ratios, and statistics	NUMBERS, FRACTIONS, ALGEBRA • Whole numbers and fractions • Fractions and percent • Statistics, graphs and probability • Integers and algebra	QUADRATIC EQUATIONS AND REVIEW • Solving quadratic equations • Equations and inequalities • Polynomials and factors • Radicals and graphing

	Grade 10	Grade 11	Grade 12
LIFEPAC 1	A MATHEMATICAL SYSTEM • Points, lines, and planes • Definition of definitions • Geometric terms • Postulates and theorems	SETS, STRUCTURE, AND FUNCTION • Properties and operations of sets • Axioms and applications • Relations and functions • Algebraic expressions	RELATIONS AND FUNCTIONS • Relations and functions • Rules of correspondence • Notation of functions • Types of functions
LIFEPAC 2	PROOFS • Logic • Reasoning • Two-column proof • Paragraph proof	NUMBERS, SENTENCES, & PROBLEMS • Order and absolute value • Sums and products • Algebraic sentences • Number and motion problems	SPECIAL FUNCTIONS • Linear functions • Second-degree functions • Polynomial functions • Other functions
LIFEPAC 3	ANGLES AND PARALLELS • Definitions and measurement • Relationships and theorems • Properties of parallels • Parallels and polygons	LINEAR EQUATIONS & INEQUALITIES • Graphs • Equations • Systems of equations • Inequalities	TRIGONOMETRIC FUNCTIONS • Definition • Evaluation of functions • Trigonometric tables • Special angles
LIFEPAC 4	CONGRUENCY • Congruent triangles • Corresponding parts • Inequalities • Quadrilaterals	POLYNOMIALS • Multiplying polynomials • Factoring • Operations with polynomials • Variations	CIRCULAR FUNCTIONS & GRAPHS • Circular functions & special angles • Graphs of sin and cos • Amplitude and period • Phase shifts
LIFEPAC 5	SIMILAR POLYGONS • Ratios and proportions • Definition of similarity • Similar polygons and triangles • Right triangle geometry	RADICAL EXPRESSIONS • Multiplying and dividing fractions • Adding and subtracting fractions • Equations with fractions • Applications of fractions	IDENTITIES AND FUNCTIONS • Reciprocal relations • Pythagorean relations • Trigonometric identities • Sum and difference formulas
LIFEPAC 6	CIRCLES • Circles and spheres • Tangents, arcs, and chords • Special angles in circles • Special segments in circles	REAL NUMBERS • Rational and irrational numbers • Laws of Radicals • Quadratic equations • Quadratic formula	TRIGONOMETRIC FUNCTIONS • Trigonometric functions • Law of cosines • Law of sines • Applied problems
LIFEPAC 7	CONSTRUCTION AND LOCUS • Basic constructions • Triangles and circles • Polygons • Locus meaning and use	QUADRATIC RELATIONS & SYSTEMS • Distance formulas • Conic sections • Systems of equations • Application of conic sections	TRIGONOMETRIC FUNCTIONS • Inverse functions • Graphing polar coordinates • Converting polar coordinates • Graphing polar equations
LIFEPAC 8	AREA AND VOLUME • Area of polygons • Area of circles • Surface area of solids • Volume of solids	EXPONENTIAL FUNCTIONS • Exponents • Exponential equations • Logarithmic functions • Matrices	QUADRATIC EQUATIONS • Conic sections • Circle and ellipse • Parabola and hyperbola • Transformations
LIFEPAC 9	COORDINATE GEOMETRY • Ordered pairs • Distance • Lines • Coordinate proofs	COUNTING PRINCIPLES • Progressions • Permutations • Combinations • Probability	PROBABILITY • Random experiments & probability • Permutations • Combinations • Applied problems
LIFEPAC 10	REVIEW • Proof and angles • Polygons and circles • Construction and measurement • Coordinate geometry	REVIEW • Integers and open sentences • Graphs and polynomials • Fractions and quadratics • Exponential functions	CALCULUS • Mathematical induction • Functions and limits • Slopes of functions • Review of 1200 mathematics

LIFEPAC

MANAGEMENT

STRUCTURE OF THE LIFEPAC CURRICULUM

The LIFEPAC curriculum is conveniently structured to provide one teacher handbook containing teacher support material with answer keys and ten student worktexts for each subject at grade levels two through twelve. The worktext format of the LIFEPACs allows the student to read the textual information and complete workbook activities all in the same booklet. The easy to follow LIFEPAC numbering system lists the grade as the first number(s) and the last two digits as the number of the series. For example, the Language Arts LIFEPAC at the 6th grade level, 5th book in the series would be LA 605.

Each LIFEPAC is divided into 3 to 5 sections and begins with an introduction or overview of the booklet as well as a series of specific learning objectives to give a purpose to the study of the LIFEPAC. The introduction and objectives are followed by a vocabulary section which may be found at the beginning of each section at the lower levels, at the beginning of the LIFEPAC in the middle grades, or in the glossary at the high school level. Vocabulary words are used to develop word recognition and should not be confused with the spelling words introduced later in the LIFEPAC. The student should learn all vocabulary words before working the LIFEPAC sections to improve comprehension, retention, and reading skills.

Each activity or written assignment has a number for easy identification, such as 1.1. The first number corresponds to the LIFEPAC section and the number to the right of the decimal is the number of the activity.

Teacher checkpoints, which are essential to maintain quality learning, are found at various locations throughout the LIFEPAC. The teacher should check 1) neatness of work and penmanship, 2) quality of understanding (tested with a short oral quiz), 3) thoroughness of answers (complete sentences and paragraphs, correct spelling, etc.), 4) completion of activities (no blank spaces), and 5) accuracy of answers as compared to the answer key (all answers correct).

The self test questions are also number coded for easy reference. For example, 2.015 means that this is the 15th question in the self test of Section II. The first number corresponds to the LIFEPAC section, the zero indicates that it is a self test question, and the number to the right of the zero the question number.

The LIFEPAC test is packaged at the centerfold of each LIFEPAC. It should be removed and put aside before giving the booklet to the student for study.

Answer and test keys have the same numbering system as the LIFEPACs and appear at the back of this handbook. The student may be given access to the answer keys (not the test keys) under teacher supervision so that he can score his own work.

A thorough study of the Curriculum Overview by the teacher before instruction begins is essential to the success of the student. The teacher should become familiar with expected skill mastery and understand how these grade level skills fit into the overall skill development of the curriculum. The teacher should also preview the objectives that appear at the beginning of each LIFEPAC for additional preparation and planning.

TEST SCORING and GRADING

Answer keys and test keys give examples of correct answers. They convey the idea, but the student may use many ways to express a correct answer. The teacher should check for the essence of the answer, not for the exact wording. Many questions are high level and require thinking and creativity on the part of the student. Each answer should be scored based on whether or not the main idea written by the student matches the model example. "Any Order" or "Either Order" in a key indicates that no particular order is necessary to be correct.

Most self tests and LIFEPAC tests at the lower elementary levels are scored at 1 point per answer; however, the upper levels may have a point system awarding 2 to 5 points for various answers or questions. Further, the total test points will vary; they may not always equal 100 points. They may be 78, 85, 100, 105, etc.

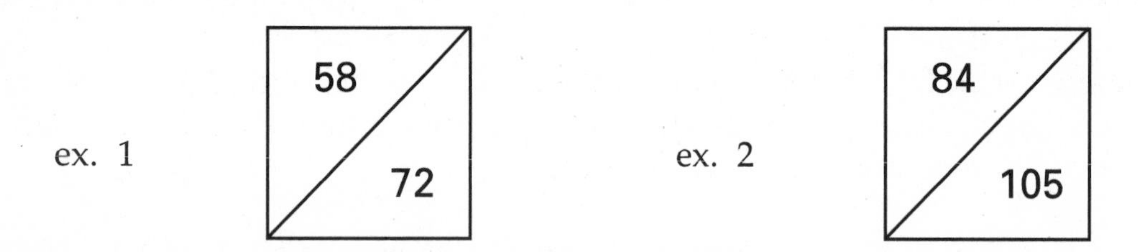

A score box similar to ex.1 above is located at the end of each self test and on the front of the LIFEPAC test. The bottom score, 72, represents the total number of points possible on the test. The upper score, 58, represents the number of points your student will need to receive an 80% or passing grade. If you wish to establish the exact percentage that your student has achieved, find the total points of his correct answers and divide it by the bottom number (in this case 72.) For example, if your student has a point total of 65, divide 65 by 72 for a grade of 90%. Referring to ex. 2, on a test with a total of 105 possible points, the student would have to receive a minimum of 84 correct points for an 80% or passing grade. If your student has received 93 points, simply divide the 93 by 105 for a percentage grade of 86%. Students who receive a score below 80% should review the LIFEPAC and retest using the appropriate Alternate Test found in the Teacher's Guide.

The following is a guideline to assign letter grades for completed LIFEPACs based on a maximum total score of 100 points.

LIFEPAC Test = 60% of the Total Score (or percent grade)
Self Test = 25% of the Total Score (average percent of self tests)
Reports = 10% or 10* points per LIFEPAC
Oral Work = 5% or 5* points per LIFEPAC
*Determined by the teacher's subjective evaluation of the student's daily work.

Example:

LIFEPAC Test Score	=	92%	92 x .60	=	55 points
Self Test Average	=	90%	90 x .25	=	23 points
Reports				=	8 points
Oral Work				=	4 points
TOTAL POINTS				=	90 points

Grade Scale based on point system:

100	–	94	=	A
93	–	86	=	B
85	–	77	=	C
76	–	70	=	D
Below		70	=	F

TEACHER HINTS and STUDYING TECHNIQUES

LIFEPAC Activities are written to check the level of understanding of the preceding text. The student may look back to the text as necessary to complete these activities; however, a student should never attempt to do the activities without reading (studying) the text first. Self tests and LIFEPAC tests are never open book tests.

Language arts activities (skill integration) often appear within other subject curriculum. The purpose is to give the student an opportunity to test his skill mastery outside of the context in which it was presented.

Writing complete answers (paragraphs) to some questions is an integral part of the LIFEPAC Curriculum in all subjects. This builds communication and organization skills, increases understanding and retention of ideas, and helps enforce good penmanship. Complete sentences should be encouraged for this type of activity. Obviously, single words or phrases do not meet the intent of the activity, since multiple lines are given for the response.

Review is essential to student success. Time invested in review where review is suggested will be time saved in correcting errors later. Self tests, unlike the section activities, are closed book. This procedure helps to identify weaknesses before they become too great to overcome. Certain objectives from self tests are cumulative and test previous sections; therefore, good preparation for a self test must include all material studied up to that testing point.

The following procedure checklist has been found to be successful in developing good study habits in the LIFEPAC curriculum.

1. Read the introduction and Table of Contents.
2. Read the objectives.
3. Recite and study the entire vocabulary (glossary) list.
4. Study each section as follows:
    a. Read the introduction and study the section objectives.
    b. Read all the text for the entire section, but answer none of the activities.
    c. Return to the beginning of the section and memorize each vocabulary word and definition.
    d. Reread the section, complete the activities, check the answers with the answer key, correct all errors, and have the teacher check.
    e. Read the self test but do not answer the questions.
    f. Go to the beginning of the first section and reread the text and answers to the activities up to the self test you have not yet done.
    g. Answer the questions to the self test without looking back.
    h. Have the self test checked by the teacher.
    i. Correct the self test and have the teacher check the corrections.
    j. Repeat steps a–i for each section.

5. Use the SQ3R* method to prepare for the LIFEPAC test.
6. Take the LIFEPAC test as a closed book test.
7. LIFEPAC tests are administered and scored under direct teacher supervision. Students who receive scores below 80% should review the LIFEPAC using the SQ3R* study method and take the Alternate Test located in the Teacher Handbook. The final test grade may be the grade on the Alternate Test or an average of the grades from the original LIFEPAC test and the Alternate Test.

   *SQ3R: Scan the whole LIFEPAC.
   Question yourself on the objectives.
   Read the whole LIFEPAC again.
   Recite through an oral examination.
   Review weak areas.

GOAL SETTING and SCHEDULES

Each school must develop its own schedule, because no single set of procedures will fit every situation. The following is an example of a daily schedule that includes the five LIFEPAC subjects as well as time slotted for special activities.

Possible Daily Schedule

8:15	–	8:25	Pledges, prayer, songs, devotions, etc.
8:25	–	9:10	Bible
9:10	–	9:55	Language Arts
9:55	–	10:15	Recess (juice break)
10:15	–	11:00	Mathematics
11:00	–	11:45	Social Studies
11:45	–	12:30	Lunch, recess, quiet time
12:30	–	1:15	Science
1:15	–		Drill, remedial work, enrichment*

*Enrichment: Computer time, physical education, field trips, fun reading, games and puzzles, family business, hobbies, resource persons, guests, crafts, creative work, electives, music appreciation, projects.

Basically, two factors need to be considered when assigning work to a student in the LIFEPAC curriculum.

The first is time. An average of 45 minutes should be devoted to each subject, each day. Remember, this is only an average. Because of extenuating circumstances a student may spend only 15 minutes on a subject one day and the next day spend 90 minutes on the same subject.

The second factor is the number of pages to be worked in each subject. A single LIFEPAC is designed to take 3 to 4 weeks to complete. Allowing about 3-4 days for LIFEPAC introduction, review, and tests, the student has approximately 15 days to complete the LIFEPAC pages. Simply take the number of pages in the LIFEPAC, divide it by 15 and you will have the number of pages that must be completed on a daily basis to keep the student on schedule. For example, a LIFEPAC containing 45 pages will require 3 completed pages per day. Again, this is only an average. While working a 45 page LIFEPAC, the student may complete only 1 page the first day if the text has a lot of activities or reports, but go on to complete 5 pages the next day.

Long range planning requires some organization. Because the traditional school year originates in the early fall of one year and continues to late spring of the following year, a calendar should be devised that covers this period of time. Approximate beginning and completion dates can be noted on the calendar as well as special occasions such as holidays, vacations and birthdays. Since each LIFEPAC takes 3-4 weeks or eighteen days to complete, it should take about 180 school days to finish a set of ten LIFEPACs. Starting at the beginning school date, mark off eighteen school days on the calendar and that will become the targeted completion date for the first LIFEPAC. Continue marking the calendar until you have established dates for the remaining nine LIFEPACs making adjustments for previously noted holidays and vacations. If all five subjects are being used, the ten established target dates should be the same for the LIFEPACs in each subject.

FORMS

The sample weekly lesson plan and student grading sheet forms are included in this section as teacher support materials and may be duplicated at the convenience of the teacher.

The student grading sheet is provided for those who desire to follow the suggested guidelines for assignment of letter grades found on page 3 of this section. The student's self test scores should be posted as percentage grades. When the LIFEPAC is completed the teacher should average the self test grades, multiply the average by .25 and post the points in the box marked self test points. The LIFEPAC percentage grade should be multiplied by .60 and posted. Next, the teacher should award and post points for written reports and oral work. A report may be any type of written work assigned to the student whether it is a LIFEPAC or additional learning activity. Oral work includes the student's ability to respond orally to questions which may or may not be related to LIFEPAC activities or any type of oral report assigned by the teacher. The points may then be totaled and a final grade entered along with the date that the LIFEPAC was completed.

The Student Record Book which was specifically designed for use with the Alpha Omega curriculum provides space to record weekly progress for one student over a nine week period as well as a place to post self test and LIFEPAC scores. The Student Record Books are available through the current Alpha Omega catalog; however, unlike the enclosed forms these books are not for duplication and should be purchased in sets of four to cover a full academic year.

WEEKLY LESSON PLANNER

Week of:

Subject	Subject	Subject	Subject
Monday			
Subject	Subject	Subject	Subject
Tuesday			
Subject	Subject	Subject	Subject
Wednesday			
Subject	Subject	Subject	Subject
Thursday			
Subject	Subject	Subject	Subject
Friday			

WEEKLY LESSON PLANNER

Week of:

	Subject	Subject	Subject	Subject
Monday				
	Subject	Subject	Subject	Subject
Tuesday				
	Subject	Subject	Subject	Subject
Wednesday				
	Subject	Subject	Subject	Subject
Thursday				
	Subject	Subject	Subject	Subject
Friday				

Student Name ______________________ Year ____________

Bible

LP #	Self Test Scores by Sections 1	2	3	4	5	Self Test Points	LIFEPAC Test	Oral Points	Report Points	Final Grade	Date
01											
02											
03											
04											
05											
06											
07											
08											
09											
10											

History & Geography

LP #	Self Test Scores by Sections 1	2	3	4	5	Self Test Points	LIFEPAC Test	Oral Points	Report Points	Final Grade	Date
01											
02											
03											
04											
05											
06											
07											
08											
09											
10											

Language Arts

LP #	Self Test Scores by Sections 1	2	3	4	5	Self Test Points	LIFEPAC Test	Oral Points	Report Points	Final Grade	Date
01											
02											
03											
04											
05											
06											
07											
08											
09											
10											

Student Name ______________________ Year __________

Mathematics

LP #	Self Test Scores by Sections 1	2	3	4	5	Self Test Points	LIFEPAC Test	Oral Points	Report Points	Final Grade	Date
01											
02											
03											
04											
05											
06											
07											
08											
09											
10											

Science

LP #	Self Test Scores by Sections 1	2	3	4	5	Self Test Points	LIFEPAC Test	Oral Points	Report Points	Final Grade	Date
01											
02											
03											
04											
05											
06											
07											
08											
09											
10											

Spelling/Electives

LP #	Self Test Scores by Sections 1	2	3	4	5	Self Test Points	LIFEPAC Test	Oral Points	Report Points	Final Grade	Date
01											
02											
03											
04											
05											
06											
07											
08											
09											
10											

TEACHER

NOTES

INSTRUCTIONS FOR EIGHTH GRADE MATHEMATICS

The LIFEPAC curriculum from grades two through twelve is structured so that the daily instructional material is written directly into the LIFEPACs. The student is encouraged to read and follow this instructional material in order to develop independent study habits. The teacher should introduce the LIFEPAC to the student, set a required completion schedule, complete teacher checks, be available for questions regarding both content and procedures, administer and grade tests, and develop additional learning activities as desired. Teachers working with several students may schedule their time so that students are assigned to a quiet work activity when it is necessary to spend instructional time with one particular student.

Mathematics is a subject that requires skill mastery. But skill mastery needs to be applied toward active student involvement. Measurements require measuring cups, rulers, empty containers. Boxes and other similar items help the study of solid shapes. Construction paper, beads, buttons, beans are readily available and can be used for counting, base ten, fractions, sets, grouping, and sequencing. Students should be presented with problem situtations and be given the opportunity to find their solutions.

Any workbook assignment that can be supported by a real world experience will enhance the student's ability for problem solving. There is an infinite challenge for the teacher to provide a meaningful environment for the study of mathematics. It is a subject that requires constant assessment of student progress. Do not leave the study of mathematics in the classroom.

The Teacher Notes section of the Teacher's Guide lists the required or suggested materials for the LIFEPACs and provides additional learning activities for the students. Additional learning activities provide opportunities for problem solving, encourage the student's interest in learning and may be used as a reward for good study habits.

I. MATERIALS NEEDED

Required:
none

Suggested:
none

II. ADDITIONAL LEARNING ACTIVITIES

Section I The Set of Whole Numbers

1. Discuss these questions with your class.
   a. What are the names of place valued beyond the trillions' place?
   b. What is the largest number?
2. Discuss the various types of graphs and the advantages of each type.
3. A small group of students can construct different types of graphs with the same data and compare the graphs.
4. Two or more students can research the base two number system, its place values, the conversion of numbers from base two to base ten, and the conversion of numbers from base ten to base two.
5. Research scientific notation for large (and small) numbers and explain the advantages of using scientific notation rather than writing out the numbers.
6. Make a pictograph showing how many males and how many females are in your class. Let 웃 represent one student.

Section II Operations with Whole Numbers

1. Discuss with the class the mathematical vocabulary used in addition, subtraction, multiplication, and division. Write addition, subtraction, multiplication, and division problems on the chalkboard and ask students to point to and label the different parts of the problems.
2. Ask the class how multiplication is related to addition.
3. Explain to the class how to multiply two-or-more digit numbers. Write several problems on the chalkboard as examples.
4. With a friend make enough times-table charts for the entire class. You can use light cardboard and a felt pen to make the charts.
5. Practice multiplying with a friend. Make flash cards with the problem on one side of the card and the answer on the other side of the card. Take turns quizzing each other.
6. Make a chart for the class illustrating the mathematical vocabulary you have learned.

Section III Applications with Whole Numbers

1. Discuss these questions with your class.
   a. What are the words that are clues in addition word problems? subtraction word problems? multiplication word problems? division word problems?
   b. What is geometry?
   c. What are some of the ways we use geometry every day?
2. With a friend make up your own super square. Ask the class to find the sum.
3. With a friend make a chart explaining how to use Egyptian multiplication.

4. Find the perimeter of square *ABCD* where each side of the square is 20 cm. Make up several other problems similar to this one. Make a poster showing how to solve this type of problem.

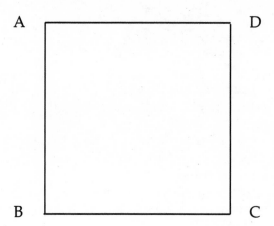

5. Expand on the terms used in addition, subtraction, multiplication, and division word problems. Find other words that are related to the four operations. Note: This activity can also be a class activity.

I. MATERIALS NEEDED

Required:
none

Suggested:
place value chart for display

II. ADDITIONAL LEARNING ACTIVITIES

Section I Numbers

1. Ask the class what the difference is between a numeral and a number.
2. Write on the chalkboard and discuss the meanings of the various code symbols used by the Egyptians, Greeks, and Romans.
3. Discuss base 10 with the students. Write large numbers on the chalkboard and have students write the numbers out in words.
4. With a friend make a poster showing the different code symbols used by various civilizations.
5. Make up several large numbers and have your friends write them out in words.
6. Research the numbers larger than quadrillion and the prefixes naming them. Share your findings with the class.
7. Research the Hindu-Arabic number system. Write a one-page paper on its history.

Section II Sets

1. Explain the terms *set, member*, and *element* to the class by using examples on the chalkboard.
2. Ask the class what the difference is between finite and infinite.
3. Ask each student to make up three problems that illustrate associative property of addition and three problems that illustrate the commutative property of addition. Have the students share their problems with the class. Then have the students decide if the problems are correct.
4. With a friend make a chart showing the associative and commutative properties of addition and multiplication.
5. With a friend visit a mathematics department at a local university or college. Observe the symbols and sets used there.
6. Make a poster showing several Venn diagrams. Demonstrate how each diagram can be written as a required set.
7. Write the definition for and illustrate each of the following terms.

commutative property of addition
set
member
element
finite set
infinite set
empty set
null set
union of sets
intersection of sets

Section III Factors and Multiples

1. Discuss these questions with your class.
   a. What is the difference between a prime number and composite number?
   b. With what type of problem is the phrase "lowest common denominator" used?
2. Invite a mathematician to speak to the class about factoring. Encourage the class to ask about practical applications of factoring and careers in mathematics.
3. Explain the Sieve of Eratosthenes to the class. Have class members practice finding prime numbers with the sieve on the chalkboard.
4. Explain the short cuts that may be used in factoring.
5. With a friend study the Fibonacci sequence. Explain it to the class. Demonstrate how to use it on the chalkboard.
6. With a friend find the least common multiple (LCM) of the following sets of numbers: 12, 14, 16, and 3, 5, 7.
7. With a friend research the different careers that involve mathematics.
8. Make a chart explaining and demonstrating Pascal's triangle.
9. Find five perfect numbers. A perfect number is a number that equals the sum of its factors not including itself.

I. MATERIALS NEEDED

Required:
none

Suggested:
none

II. ADDITIONAL LEARNING ACTIVITIES

Section I Common Fractions

1. Discuss these questions with your class.
   a. One definition of a rational number is "a number that can be expressed as a ratio of two whole numbers." Do the numerator or denominator have any restrictions placed on them?
   b. In which form is it easier to compare the order of rational numbers: as improper fractions or as mixed number fractions? Why?
2. Fractions can be reduced by factoring the numerators and denominators into their prime factors. For example, the fraction 24/36 has these factors: $24 = 2 \times 2 \times 2 \times 3$ and $36 = 2 \times 2 \times 3 \times 3$. The factors they have in common are $2 \times 2 \times 3$ or 12. The students may use this method on any of the problems on page 5 of the LIFEPAC (Problems 1.13 through 1.22).
3. With a friend make a chart defining and giving examples of the following terms: common fraction, proper fraction, numerator, denominator, equivalent fractions, factor, and improper fraction.
4. With a friend write five fractions equivalent to 2/3; 1/5; 4/9.
5. Make a number line using fractions and whole numbers. Use fractions with denominators smaller than 8.
6. Use the number lines you made in the preceding activity to show that $2\frac{3}{4} > 2\frac{2}{3}$; $5\frac{1}{4} > 5\frac{1}{6}$; $8\frac{2}{5} < 8\frac{1}{3}$.

Section II Decimal Fractions

1. Discuss these questions with your class.
   a. In general, which is easier: to change a percent to a fraction in lowest terms or to change a percent to a decimal? Why?
   b. Which is easier: to compare the order of rational numbers in fraction form or to compare the order of rational numbers in decimal form? Why?
   c. What does the term *equivalent fraction* mean? How does it differ from the equivalent fractions you studied in Section 1 of the LIFEPAC? (The two kinds of equivalent fractions are explained on pages 3 and 19 in the LIFEPAC.)
2. All rational numbers can be expressed as either terminating decimals or repeating decimals. Some repeating decimals repeat in a few digits and others repeat after many digits. Study the following list of repeating decimals with a friend and see if you can determine any relationships about the number of digits in the repeating part.

$\frac{1}{3} = 0.\overline{3}$

$\frac{1}{17} = 0.\overline{0588235294117647}$

$\frac{1}{6} = 0.1\overline{6}$

$\frac{1}{18} = 0.0\overline{5}$

$\frac{1}{7} = 0.\overline{142587}$	$\frac{1}{19} = 0.\overline{052631578947368421}$
$\frac{1}{9} = 0.\overline{1}$	$\frac{1}{21} = 0.\overline{047619}$
$\frac{1}{11} = 0.\overline{09}$	$\frac{1}{22} = 0.0\overline{45}$
$\frac{1}{12} = 0.08\overline{3}$	$\frac{1}{23} = 0.\overline{0434782608695652173913}$
$\frac{1}{13} = 0.\overline{076923}$	$\frac{1}{24} = 0.041\overline{6}$
$\frac{1}{14} = 0.0\overline{714285}$	$\frac{1}{26} = 0.0\overline{384615}$
$\frac{1}{15} = 0.0\overline{6}$	$\frac{1}{27} = 0.\overline{037}$

Hints: Consider the remainders after each division, and consider whether the denominator is prime of composite.

3. Fractions with denominators of 2, 4, 5, 8, 10, 16, 20, or 25 can be expressed as terminating decimals. Fractions with denominators of 3, 6, 7, 9, 11, 12, 13, 14, or 15 can be expressed only as repeating decimals. With a friend try to discover a rule concerning the denominator of a fraction that will tell you whether the fraction in decimal form will terminate or repeat, without actually converting the fraction to a decimal.
4. Make a poster defining and illustrating the following terms: decimal point, decimal fraction, and mixed decimal.
5. Change the following percentages to decimals: 20%, 25%, 50%, 58%, and 65%.

Section III Applications

1. Discuss these questions with your class.
   a. Can a unique (one and only one) solution be found to a proportion with two missing terms? (Hint: $2{:}x = x{:}8$)
   b. A special case of proportion may consist of more than two ratios. An example is 2:5 = 4:10 = 6:15 = 10:25. How many simple proportions can be made from this proportion?
2. Have the class name an irrational number that they have used previously. (Hint: Have them consider their work with circles.)
3. With a friend construct a metric ruler out of scrap wood. Use the ruler to explain to the class the metric powers of 10.
4. With a friend write a one-page report explaining the difference between rational numbers and irrational numbers. Use several examples in your report.
5. Use items with a beginning price of $100 to consider the following discount and price increases. In which instance would the item have a lesser price? What would be the difference, if any, between the final prices in each case?
   a. discounted 20%, then increased 20% *or* increased 20%, then discounted 20%
   b. discounted 20%, then increased 30% *or* increased 30%, then discounted 20%
   c. discounted 30%, then increased 20% *or* increased 20%, then discounted 30%
6. Special rules exist for adding and multiplying numbers that are written in scientific notation. Use simple examples of numbers written in scientific notation to see if you can discover the rules. Try also to discover the rules for subtracting and dividing numbers that are written in scientific notation.

I. MATERIALS NEEDED

Required:	Suggested:
calculators	none

II. ADDITIONAL LEARNING ACTIVITIES

Section I Addition of Common Fractions

1. Discuss these questions with your class.
   a. What is the difference between a proper and an improper fraction?
   b. What is an equivalent fraction?
2. Show the students how to find the least common multiple (LCM) of two numbers. Write several examples on the chalkboard.
3. Go to a grocery store with a friend. Find as many items as you can that are measured in fractions or decimals.
4. Read a recipe book with a friend. Practice doubling and tripling recipes.
5. Find out what the difference is between a least common multiple (LCM) and a lowest common denominator (LCD). Make a poster showing the difference.
6. Make up several improper fractions and change them to proper fractions. Have your teacher check your work.

Section II Subtraction of Common Fractions

1. Explain "borrowing" in subtraction of fractions. Work several examples on the chalkboard.
2. Show the students how to check subtraction problems by adding. Work several examples on the chalkboard.
3. Invite a contractor to class to talk about and demonstrate the importance of being able to add and subtract fractions on a job.
4. Visit a construction site with a friend. Observe the careful measuring with fractions that is done there.
5. With a friend make up several problems that involve subtraction of fractions. Quiz each other with your problems.
6. Go to the library and research careers that might involve computation with fractions. Share your information with the class.
7. Make up four word problems that involve subtraction of fractions. Solve your problems. Have your teacher check your work.

Section III Decimal Fractions

1. Demonstrate to the class that when decimal fractions are written as common fractions, the denominators are powers of 10.
2. Demonstrate addition and subtraction of decimal fractions on the chalkboard. Show the students how to check their work.
3. Go to the library with a friend. Ask the librarian to explain their decimal system for shelving books.
4. Make up several decimal problems with two or three friends. Take turns finding the answers to the problems with a calculator.
5. Ask your parents to let you help them balance their checkbook.
6. With your parents' permission, check the entries in their checkbook to see that the balance is correct. You may use a calculator if one is available.

Section IV Rounding and Review

1. Discuss and explain pages 32 and 33 in the LIFEPAC. Write numbers on the chalkboard and have students take turns rounding them.
2. Discuss and explain pages 35, 36, and 37 in the LIFEPAC. Write review problems on the chalkboard similar to the ones in the LIFEPAC. Have students come to the chalkboard and work the answers to the problems.
3. Study the review rules on pages 37 and 38 in the LIFEPAC with two or three friends. Quiz each other on the rules.
4. With a friend make up several review problems. Trade problems and check each other's work.
5. Make a chart illustrating the rules on pages 37 and 38 in the LIFEPAC. Make two examples for each rule.

I. MATERIALS NEEDED

Required:	Suggested:
none	none

II. ADDITIONAL LEARNING ACTIVITIES

Section I Multiplication

1. Give the class two examples of the necessity of being able to multiply fractions. Measurement with fractions is one example. Ask the class to give their own examples. Write the examples on the chalkboard.
2. Explain to the class how to use a function rule. Write a function rule with several examples on the chalkboard. Have the students work some problems using the function rule. Check their work to see that they understand how to use a function rule.
3. Explain to the class the rules for multiplication of common fractions, mixed numbers, and decimal fractions. Illustrate the rules with examples on the chalkboard.
4. With a friend write twenty examples of multiplication of fractions used in everyday life.
5. With a friend measure the windows in your classroom and figure how much drapery material you would need to make drapes that would hang from the ceiling to the floor.
6. Write a paper explaining the two different methods of multiplying mixed numbers. State which method you prefer and why.
7. Make a poster demonstrating how to use a function rule in the multiplication and addition of fractions.

Section II Division

1. Discuss these questions with your class.
   a. What is the reciprocal of a fraction?
   b. What is the product of a fraction and its reciprocal?
   c. If the answer to a division problem is not even, what do you do?
2. With a friend make a chart of your new vocabulary words for Section II. Define the words and give two examples for each definition.
3. With a friend find the assessed value of a $10,000 truck assessed at one-fourth of its value.
4. Solve this problem. If 45 nails cost $4.82, what is the cost per nail?
5. Make up four word problems using division of common fractions, mixed numbers, and decimal fractions. Let your teacher check your answers.

Section III Percent

1. Discuss these questions with your class.
   a. What is the meaning of the term *rate* as it applies to percent?
   b. What is the basic equation?
2. Demonstrate how to use the basic equation on the chalkboard with several examples.

3. With a friend find the answer to the following problem.
   a. For sales tax at 5%, between what amounts of purchase does the tax increase from 1¢ to 2¢ after rounding?
   b. Between what amounts of purchase does the tax increase from 6¢ to 7¢ after rounding?

   Answer:
   a. 5% of 29¢ = 1.45¢; rounded = 1¢
      5% of 30¢ = 1.5¢; rounded = 2¢
   b. 5% of $1.29 = 6.45¢; rounded = 6¢
      5% of $1.30 = 6.5¢; rounded = 7¢

4. With a friend find the answer to the following problem. If an apartment may be rented for $50 per week or $200 per month, how much per year can be saved by renting by the month?

   Answer:

   50 x 52 = $2,600 per year
   200 x 12 = $2,400 per year
   $ 200 per year can be saved by renting at $200 per month.

5. Figure the answer to the following problem.

   Joe has two choices of ways he can borrow $1,000. His employees' credit union charges 12% interest per year (1% per month) on the balance due (the amount owed). The credit union adds the interest to the balance each month, then subtracts the payment to get the new balance (the amount still owed). Payments on the $1,000 will be $90 per month for 11 months. The amount due, including the interest for the month, will be the payment for the last month.

   The bank charges 9% discount (called interest or carrying charges) or 9% of the $1,000.

   Find the difference between the bank's interest and the credit union's interest for the year.

   Answer:

   Figure the credit union's payment schedule a month at a time for all 12 months.

Credit Union Payment Schedule

	first month		second month
	$1,000.00		$920.00
Plus 1% interest:	10.00	Plus 1% interest:	9.20
Total:	$1,010.00	Total:	$929.20
Less payment:	90.00	Less Payment:	90.00
Balance:	$920.00	Balance:	$839.20

	third month		fourth month
	$839.20		$757.59
Plus 1% interest:	8.39	Plus 1% interest:	7.58
Total:	$847.59	Total:	$765.17
Less Payment:	90.00	Less payment:	90.00
Balance:	$757.59	Balance:	$675.17

fifth month	sixth month	seventh month	eighth month
$675.17	$591.92	$507.84	$422.92
+ 6.75	+ 5.92	+ 5.08	+ 4.23
$681.92	$597.84	$512.92	$427.15
- 90.00	- 90.00	- 90.00	- 90.00
$591.92	$507.84	$422.92	$337.15

ninth month	tenth month	eleventh month	twelfth month
$337.15	$250.52	$163.03	$74.66
+ 3.37	+ 2.51	+ 1.63	+ 0.75
$340.52	$253.03	$164.66	$75.41
- 90.00	- 90.00	- 90.00	
$250.52	$163.03	$74.66	

Add the interest for each month to find the total interest charged by the credit union for the year.

$10.00
9.20
8.39
7.58
6.75
5.92
5.08
4.23
3.37
2.51
1.63
0.75
$65.41

Now find the bank's interest. If the bank has a straight 9% charge, the interest will be 9% x $1,000, or $90. The difference between the bank's interest and the credit union's interest is $24.59, as shown.

$90.00
- 65.41
$24.59

The credit union would be cheaper than the bank for Joe to borrow $1,000.

Banks, credit unions, and other lending institutions are now required to quote to a person who is considering a loan or considering buying on payments, the true annual interest rate. If you figure the difference between the bank's interest and the credit union's interest based on the true annual interest rates of the bank and credit union, you will get a different figure than the previous one. You may want to find out more about true annual interest rates and how they work.

I. MATERIALS NEEDED

Required:
none

Suggested:
none

II. ADDITIONAL LEARNING ACTIVITIES

Section I Statistics

1. Discuss these questions with your class.
   a. What is the difference between the mean and the median?
   b. What does the term *mode* mean?
   c. What word do we use for the difference between the greatest value in a set of numbers and the least value in the set of numbers?
2. With a friend find the mean score, the median score, the modal score, and the spread of your test scores in Mathematics LIFEPACs 804 and 805.
3. With a friend chart the frequency distribution of your test scores in Mathematics LIFEPACs 801 through 805.
4. Find your median test score in English so far this year.
5. Write a one-page report explaining how to compute a batting average.

Section II Graphs

1. Invite a college student whose major is statistics to speak to the class on careers in statistics. Also, have them demonstrate how histograms, frequency polygons, bar graphs, and line-segment graphs are used in statistics.
2. Discuss the difference between a frequency polygon and a histogram. Use examples on the chalkboard.
3. Explain how to construct and use a line-segment graph.
4. With a friend contact the local weather bureau. Ask for the daily average temperatures for the past seven days. Construct a line-segment graph for the data.
5. With a friend talk to your school administrator or teacher and get some data on school costs or school enrollment. Construct a graph of the data.
6. Locate histograms, bar graphs, and line-segment graphs from newspaper, magazines, or other publications. Share them with the class.
7. Construct a bar graph charting the number of boys and girls in your class over the past four months.

Section III Probability

1. Discuss these questions with your class.
   a. What is probability?
   b. What is the difference between probability and statistics?
   c. What is the formula for finding the probability of an event?
2. With a friend turn to a page of a telephone book at random and select one column of telephone numbers. Find the average of the last digit only for one column of numbers. Why would you expect the average to be about 5?
3. With a friend turn to a page of a telephone book at random and find the average of the last two digits only for one column of numbers. Why would you expect the average to be about 50?

4. Solve this problem.

   Automobile license plates in the state of Dover consist of three digits and three letters (Example: 179-ABC). If license plates are issued at random and all digits and letters are used, what would be the probability of your receiving the license plate 123-JOE?

5. Research probability in the library. Write a one-page report on probability.

I. MATERIALS NEEDED

Required:
a large number line (marked with integers) at the front of the classroom; a number line can be purchased or made from strips of construction paper stapled end-to-end
a large chalkboard grid with a horizontal axis and a vertical axis; the grid can be hung from an existing chalkboard or placed on a tripod rack

Suggested:
none

II. ADDITIONAL LEARNING ACTIVITIES

Section I Introduction to Integers

1. Explain the basic concepts on pages 2 and 3 in the LIFEPAC with examples on the chalkboard.
2. Have the students make their own number lines from poster board. They can tape them on their desks or keep them inside their desks.
3. Explain page 4 of the LIFEPAC to the class. Some students will have trouble with the concept that a number such as -4 is an even number.
4. With a friend graph (-4, 3), (1, 4), and (3, -2) on a number-pair plane. Connect the points. Find the area of the figure with the appropriate formula.
5. With a friend visit an architect and find out how architects use graph paper to figure area.
6. Explain what a non-negative integer is; what a non-positive integer is. Write your explanations and check them with the LIFEPAC glossary.
7. Make a poster showing the formulas for the different geometric shapes. Show example problems next to each formula.

Section II Operations with Integers

1. Draw a number line on the chalkboard. Explain pages 38 and 39 in the LIFEPAC by working examples on the chalkboard number line.
2. Explain the short cut used on page 32 of the LIFEPAC (to think of a mental number line). Use a number line to demonstrate the short cut.
3. Invite a mathematician from a local college or university to speak to the class about operations with integers.
4. With a friend make a poster of the rules of addition. After each rule write an example to illustrate the rule.
5. With a friend make a poster of the rules of multiplication. After each rule write an example to illustrate the rule.
6. Go to the library and look up the meaning of the term *power* in different mathematics textbooks. Write a one-page report explaining the use of the term *power* in this LIFEPAC.
7. Find these powers: 4^3, 5^4, and $(-3)^6$.

Section III Applications of Integers

1. Discuss pages 66 and 67 in the LIFEPAC with the students. Work several examples on the chalkboard.
2. Discuss page 7 in the LIFEPAC with the students. Work several examples of sentences and number pairs on the chalkboard.
3. Give each student a sheet of graph paper. Write several sentences with two variables (such as $a + b = 5$) on the chalkboard. Have the students graph the solutions to each sentence on a separate area of their paper. Check their solutions.
4. Make a chart with a friend explaining the order of operation.
5. Write several sentences with two variables, graph the solutions, and trade them with those of a friend. Check each other's work.
6. Make a chart defining all of the new terms in Section III of the LIFEPAC.
7. Write several algebraic expressions of your own and evaluate them. Ask your teacher to check your work.

I. MATERIALS NEEDED

Required:
none

Suggested:
scissors and tape (for paper-folding)
rulers (with English and metric measurements)
calculators
a 50-ft. tape measure for group measuring or teacher demonstrations of large areas

II. ADDITIONAL LEARNING ACTIVITIES

Section I Square Root and Simple Formulas

1. Discuss the importance of being able to calculate area.
2. Discuss jobs that require calculations of area.
3. Demonstrate the use of $\pi = \frac{22}{7}$ as an alternate value of π.
4. With a friend make a poster explaining square root.
5. With a friend measure and calculate the cost of putting tile at $5.00 per square foot on your classroom floor.
6. Measure your bedroom at home. How much would the cost be to carpet your room at $7.18 per square foot?
7. Make a chart showing how the Pythagorean Theorem works.

Section II Area and Volume Formulas

1. Discuss with the class the importance of being able to calculate volume.
2. Invite a special guest from your church who does area or volume calculations in his work to talk about his work.
3. Take the class on a field trip to a construction site. Have the class look for and find one way that area and volume formulas are used.
4. With a friend measure and calculate the area of the school parking lot.
5. With a friend calculate the volume of a sink at school.
6. Ask the owner of an apartment building in your town for a sample floor plan; calculate the area of each of the rooms.
7. Draw your own chest of drawers or your parent's chest of drawers and calculate its capacity (volume).

Section III Area and Volume Formulas with π

1. Make (or have the students make) several construction paper pyramids, cylinders, cones, and spheres of various sizes. Give one to each student and have the students figure the surface area and volume of their figure.
2. Have the students draw a plan of the school or church complex.
3. With a friend make a poster illustrating cylinders, uses area and volume formulas with π.
4. Explain how you can find the volume of a ping-pong ball; then find the volume of a ping-pong ball.
5. Find the surface area of the sides and bottom of a glass in your home. Find the volume of the glass.

I. MATERIALS NEEDED

Required:
none

Suggested:
none

II. ADDITIONAL LEARNING ACTIVITIES

Section I The Variable

1. Work examples of the commutative and associative properties for addition and multiplication on the chalkboard.
2. Have the students come to the chalkboard and work problems involving variables (such as the ones on page 8 in the LIFEPAC). Guide them through the process. Some students will have problems transferring from numbers to variables. Check their work each day.
3. Demonstrate the use of variable and numbers combined (as on pages 8, 9, 13, 14, 15, and 16 in the LIFEPAC). Have the students make up problems and find the answers. Check their answers.
4. With a friend make a chart defining the important terms used in Section I of the LIFEPAC.
5. With two or three friends make up several problems using the distributive property. Give them to each other to solve.
6. Write five problems illustrating the associative property for addition. Have your teacher check your answers.
7. Write five problems illustrating the commutative property for multiplication. Have your teacher check your answers.

Section II The Equation

1. Demonstrate, using examples on the chalkboard, and open sentence, an equation, a solution set, the addition property of equality, and the subtraction property of equality.
2. Demonstrate, using examples on the chalkboard, the multiplication and division properties of equality.
3. Explain to the class how to solve and check equations (page 31 in the LIFEPAC). Work several examples on the chalkboard.
4. With a friend create and solve several equations involving the addition and subtraction properties of equality. Have your teacher check your work.
5. Make a poster with your friends illustrating and explaining the multiplication and division properties of equality.
6. Create and solve five equations that involve the division property of equality. Have your teacher check your answers.
7. Create and solve five equations that involve the four properties of equality and combination of terms. Have your teacher check your answers.

Section III Applications

1. Explain page 37 of the LIFEPAC to the class. Write several examples on the chalkboard similar to the models on pages 37 and 38.
2. Explain page 40 of the LIFEPAC to the class. Write several examples on the chalkboard similar to the models on page 40.
3. Have the student make up age problems to give to each other to solve. Check their work.
4. With members of your class, make a poster defining the vocabulary used in Section III of the LIFEPAC.
5. With a friend make up four number problems. Write them in the same form as that used on page 45 in the LIFEPAC (with a heading, equation, and solution). Have your teacher check your solutions.
6. Solve this problem. If you can travel 200 miles in 4 hours, how far can you travel at the same rate in 9 hours?
7. Make up several proportion problems and solve them. Have your teacher check your solutions.

I. MATERIALS NEEDED

Required:	Suggested:
none	none

II. ADDITIONAL LEARNING ACTIVITIES

Section I Numbers

1. Discuss these questions with your class.
   a. In how many ways can a number be written?
   b. How many different bases exist?
   c. Which number systems are based on ten?
   d. How do you change a decimal to a percent? a percent to a decimal?
2. Say a number and have the students write the number in Roman numerals. Check their answer together, discussing any questions they have.
3. Say a number and have a student tell the prime factors of the number (have each student think of the factors mentally).
4. This activity is for two to five students to do together. One student writes five to ten one- and two-digit numerals on a piece of paper and finds their sum. Then he repeats (slowly) the numbers to the other students, who add the numbers mentally as they are said. If necessary, the numbers may be repeated. When each of the students has the sum, they check their answer with the one the student has written on the piece of paper. Points may be kept, one point for each correct answer. After several turns the student writing the problems takes another student's place so that each student has a turn at writing and at mentally adding the numbers.
5. Each of three students writes any ten numbers (or any ten letters) as a set and labels the set. Each student tells the name of his set (but not the elements in the set) to a fourth student. The fourth student then says to the group to find the intersection (or union) of sets A and B, for example. The students whose sets are named find the intersection (or union) of their sets. The fourth student then names another set, and the appropriate students answer. After all possible combinations of sets are exhausted, the fourth student takes the place of one of the other students. The three students then think of three new sets and continue as before. Each student takes a turn at naming the intersections and unions of sets.
6. Ask a foreign language teacher or someone who speaks another language to write several numerals as words and to tell you their meanings in English (you write the meanings). Then write the words as numerals.

Section II Fractions

1. Do part (a) of this activity with the students before doing part (b); part (b) is more difficult.
   a. Have students multiply the following fractions mentally. The problems are arranged across so that the more difficult problems are at the end of the list. The answers will be in lowest terms. You may have the students write their answers and check them at the end of the activity or you may call on one student at a time to answer. Have them multiply as many as they can mentally. After a little practice they should be able to multiply all the fractions mentally.

$\frac{1}{2} \bullet \frac{1}{3}$ $\frac{1}{3} \bullet \frac{2}{3}$ $\frac{1}{4} \bullet \frac{1}{4}$ $\frac{1}{5} \bullet \frac{1}{2}$

$\frac{1}{7} \bullet \frac{3}{4}$ $\frac{1}{6} \bullet \frac{1}{5}$ $\frac{1}{9} \bullet \frac{2}{5}$ $\frac{1}{8} \bullet \frac{3}{5}$

$\frac{2}{3} \bullet \frac{4}{5}$ $\frac{1}{10} \bullet \frac{1}{6}$ $\frac{5}{6} \bullet \frac{5}{6}$ $\frac{2}{5} \bullet \frac{1}{7}$

$\frac{2}{7} \bullet \frac{2}{7}$ $\frac{3}{4} \bullet \frac{3}{7}$ $\frac{2}{9} \bullet \frac{4}{7}$ $\frac{3}{7} \bullet \frac{5}{7}$

$\frac{3}{5} \bullet \frac{6}{7}$ $\frac{3}{8} \bullet \frac{1}{8}$ $\frac{3}{10} \bullet \frac{3}{8}$ $\frac{5}{7} \bullet \frac{5}{8}$

$\frac{5}{8} \bullet \frac{7}{8}$ $\frac{4}{7} \bullet \frac{1}{9}$ $\frac{4}{9} \bullet \frac{2}{9}$ $\frac{4}{5} \bullet \frac{4}{9}$

$\frac{5}{9} \bullet \frac{5}{9}$ $\frac{8}{9} \bullet \frac{7}{9}$ $\frac{7}{9} \bullet \frac{8}{9}$ $\frac{3}{10} \bullet \frac{1}{10}$

$\frac{3}{10} \bullet \frac{3}{10}$ $\frac{9}{10} \bullet \frac{7}{10}$ $\frac{7}{10} \bullet \frac{9}{10}$

b. Have the students multiply the following fractions mentally. The problems are arranged across so that the more difficult problems are at the end of the list. The answers will need to be reduced. Students may need to use paper and pencil for some problems. After some practice though, the students should gain facility in mentally multiplying the fractions.

$\frac{2}{3} \bullet \frac{1}{2}$ $\quad \frac{3}{4} \bullet \frac{1}{3}$ $\quad \frac{5}{6} \bullet \frac{1}{5}$ $\quad \frac{7}{8} \bullet \frac{1}{7}$

$\frac{1}{4} \bullet \frac{4}{5}$ $\quad \frac{1}{5} \bullet \frac{5}{7}$ $\quad \frac{3}{7} \bullet \frac{7}{8}$ $\quad \frac{1}{7} \bullet \frac{7}{9}$

$\frac{1}{8} \bullet \frac{8}{9}$ $\quad \frac{1}{9} \bullet \frac{9}{10}$ $\quad \frac{2}{5} \bullet \frac{1}{4}$ $\quad \frac{5}{6} \bullet \frac{1}{10}$

$\frac{5}{7} \bullet \frac{3}{10}$ $\quad \frac{5}{8} \bullet \frac{4}{7}$ $\quad \frac{1}{3} \bullet \frac{6}{7}$ $\quad \frac{2}{7} \bullet \frac{1}{6}$

$\frac{3}{5} \bullet \frac{1}{9}$ $\quad \frac{4}{9} \bullet \frac{3}{5}$ $\quad \frac{5}{9} \bullet \frac{3}{8}$ $\quad \frac{2}{9} \bullet \frac{1}{8}$

$\frac{7}{8} \bullet \frac{2}{3}$ $\quad \frac{1}{10} \bullet \frac{2}{5}$ $\quad \frac{4}{5} \bullet \frac{5}{8}$ $\quad \frac{7}{9} \bullet \frac{3}{7}$

$\frac{7}{10} \bullet \frac{2}{7}$ $\quad \frac{1}{6} \bullet \frac{4}{9}$ $\quad \frac{6}{7} \bullet \frac{5}{9}$ $\quad \frac{2}{10} \bullet \frac{5}{6}$

$\frac{3}{8} \bullet \frac{2}{9}$ $\quad \frac{4}{7} \bullet \frac{7}{10}$

2. Say a decimal and have a student change it to a percent; say a percent and have a student change it to a decimal.
3. Ask the class how to find the average of a group of numbers.
4. Cut 2 red, 1 white, 3 green, and 5 blue squares from construction paper and put them in a small box with a lid. Write the color names and the number (how many) of each color on the chalkboard. Choose one student to come and stand by the box. Have the rest of the students figure the probability of the student drawing a red piece of paper; a white piece of paper; a green piece of paper; a blue piece of paper. Then have the student draw one piece of paper from the box without looking. Is the piece the student drew the color with the highest probability? the second highest probability? the third highest probability? the lowest probability? Have the student replace the piece of paper he drew and draw another piece. Consider the same questions. Repeat as many times as desired.
5. Have each student in the group write a fraction on the same piece of paper for another student to add or have one student write several fractions for another student to add.
6. Have each student in the group write a decimal on the same piece of paper for another student to add or have one student write several decimals for another student to add.

7. Find the average (mean) of the numbers 1, 2, 3, 4, 5, 6, 7, 8, and 9. Find the average (mean) of the numbers 1, 2, 3, 4, 5, 6, 7, 8, 9, 10, 11, 12, 13, 14, 15, 16, 17, 18, 19, and 20. Use the following short cut to mentally average groups of consecutive numbers that begin with 1 (such as the preceding ones). To average the numbers, add the first number and the last number and divide by 2. For example, to find the average of the numbers from 1 through 9, add 1 and 9 and divide by 2: ($\frac{1+9}{2} = \frac{10}{2} = 5$). With some practice you can find the average of large numbers such as from 1 to 99 ($\frac{1+94}{2} = \frac{100}{2} = 50$).

Section III Algebra

1. Review with the students the area and volume formulas and the area and volume formulas with π. A complete list of the formulas is on the back of the first page of Mathematics LIFEPAC 808. Two ways of reviewing the formulas are given.
   a. Tell the class a formula and have the students tell you the name of the formula.
      Example: Say "$\pi \bullet r^2$."
      The students respond with "the area of a circle."
   b. Tell the class the name of a formula and have the class say the formula itself.
      Example: Ask, "What is the formula for the circumference of a circle?"
      The students respond with "$C = 2\pi \bullet r$ or $C = \pi \bullet d$."
2. Have the class translate sentences to algebraic sentences and solve the sentences as you read them orally. Either write sentences and solve them beforehand or read the sentences in the LIFEPAC (page 90) and use the LIFEPAC Answer Key. Say the sentences slowly and direct the students to write each part as you say it to form the algebraic sentence. Emphasize key words such as *more than, quotient, sum,* and so on.
3. Have the class solve word problems such as the ones on pages 92 and 93 of the LIFEPAC using the three-step method (write the heading, the equation, and the solution) as you read the problems orally. Say the problems slowly and direct the students to write each part as you say it so they will have the information they need to solve the problem. Emphasize key words such as *product, consecutive* (integers), *times* (used as a product), as so on.
4. With a friend quiz each other with the area and volume formulas and with the area and volume formulas with π.
5. Find the following powers. You may use a calculator for this activity if one is available.

   $2^2,\ 2^3,\ 2^4,\ 2^5,\ 2^6,\ 2^7,\ 2^8,\ 2^9,\ 2^{10}$

   Find these powers.

   $4^2,\ 4^3,\ 4^4,\ 4^5,\ 4^6$

   Compare 2^4 to 4^2; 2^6 to 4^3; 2^8 to 4^4; and 2^{10} to 4^5.
   Do you see a pattern? Now find these powers.

   $16^2,\ 16^3$

   Compare 4^4 to 16^2 and 4^6 to 16^3. Do you see a pattern? What has to be true about the bases and the exponents for the pattern to exist? Do you think you would get the same pattern if you compared powers of 4 to powers of 8? Why? Test your answers to find out if you are correct. Compare other powers to each other (such as powers of 3 to powers of 9 and powers of 5 to powers of 25) and see what their patterns are.

6. Find the following powers. You may use a calculator for this activity if one is available.

$$2^2,\ 2^3,\ 2^4,\ 2^5$$

Find these powers.

$$4^2,\ 4^3,\ 4^4,\ 4^5$$

Now divide 2^2 by 4^2; 2^3 by 4^3; 2^4 by 4^4; and 2^5 by 4^5. Write the quotients as fractions.

Next find the following powers.

$$3^2,\ 3^3,\ 3^4,\ 3^5$$

Find these powers.

$$6^2,\ 6^3,\ 6^4,\ 6^5$$

Now divide 3^2 by 6^2; 3^3 by 6^3; 3^4 by 6^4; and 3^5 by 6^5. Write the quotients as fractions. Compare these fractions with the previous fractions (2^2 divided by 4^2, 2^3 divided by 4^3, and so on). Do you see a pattern? Do you think the pattern continues? What do you think 2^6 divided by 4^6 equals? Test your answer. What has to be true about the bases and the exponents for the pattern to exist? Divide other powers to see if you get the same pattern as you did with the powers in this activity.

ALTERNATE

TESTS & KEYS

Reproducible Tests
for use with the Mathematics
800 Teacher's Guide

Name ____________________

Match these items (each answer, 2 points).

1. ______ bar graph
2. ______ date
3. ______ numeral
4. ______ divisor
5. ______ quotient
6. ______ digit
7. ______ placeholder
8. ______ sum
9. ______ difference
10. ______ number line

a. the answer in addition
b. the amount left over
c. the number by which another number is divided
d. the facts
e. the answer in division
f. using points on a graph
g. any number less than ten
h. the answer in subtraction
i. using bars on a graph
j. a name for a number
k. zero used in place of a number

Complete the following items (each answer, 3 points).

11. The number 6,384 in words is ____________________________________

____ .

12. The number *twelve thousand, nine hundred forty-four* in numerals is ______________ .

13. The number 2,758 rounded to the nearest hundred is ______________ .

14. The digit in the thousands' place in the number 4,789,265 is ______________ .

15. A billion is the digit *one* followed by ___________ (how many) zeros.

Perform the following operations (each answer, 4 points).

16.
```
   81
   76
  931
  387
+  43
-----
```

17.
```
  53,253
- 32,445
--------
```

18.
```
  2,365
x   476
-------
```

19. $31\overline{)8{,}866}$

20. $25\overline{)4,614}$

21. $\begin{array}{r} 6,865 \\ \times\ 94 \\ \hline \end{array}$

22. $\begin{array}{r} 11,462 \\ -\ 6,565 \\ \hline \end{array}$

23. $\begin{array}{r} 2,810 \\ 4,156 \\ 8,751 \\ +\ 6,093 \\ \hline \end{array}$

Write the correct letter in the blank (each answer, 3 points).

24. In the problem, 38 + 41 = 79, the numbers 38 and 41 are called ________ .

a. addends b. terms c. sums d. divisors

25. In the problem, $8\overline{)56}$ with 7 written above, the 8 is called the ________ .

a. divisor b. dividend c. quotient d. remainder

26. A triangle has ________ .

a. three sides b. four sides c. five sides d. six sides

27. A trillion is a digit followed by ________ .

a. 6 zeros b. 9 zeros c. 12 zeros d. 15 zeros

28. The multiplicand in the problem, 15 x 431 = 6,465, is ________ .

a. 15 b. x c. 431 d. 6,465

29. The numbers that are shown as points on the number line are ________ .

a. 0, 4, 6, 10, and 12 b. 2 and 8 c. 2 d. 8

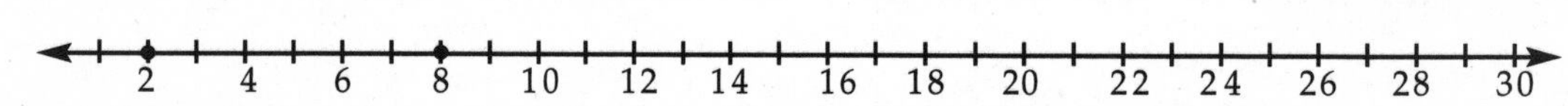

Work the following word problems (each answer, 4 points).

30. The speedometer read 47,916 at the beginning of a trip and 52,738 at the end of the trip. Find the total distance traveled.

Answer ________________________

31. Find out how much money the Jones family has totally, if Mrs. Jones has $71; Mr. Jones, $157; and Paul, $21; and Becky, $15.

Answer ________________________

32. If duck eggs cost $3 per dozen, how much do 17 dozen duck eggs cost?

Answer______________________________

Complete this activity (3 points).

33. Write the number 746 in expanded form. ______________________________

80
100

Date ______________________________

Score ______________________________

Name ____________________________

Match these items (each answer, 2 points).

1. ______ numeral
2. ______ LCD of $\frac{11}{24}$ and $\frac{15}{36}$
3. ______ prime numbers from 50 to 60
4. ______ GCF of 72 and 84
5. ______ 4 x 4 x 4 x 4 x 4
6. ______ complete factorization of 98
7. ______ test for divisibility of 3
8. ______ 11101_2
9. ______ 6 in ordered pair (6, 7)
10. ______ Commutative property
11. ______ proper subset of B = {1, 2, 3, 4, 5, 6}
12. ______ empty set

a. 12
b. 2×7^2
c. $(2 + 3) + 5 = 5 + (2 + 3)$
d. sums of digits is divisible by 3
e. element of the range
f. $(a + b) + c = a + (b + c)$
g. element of the domain
h. {2, 3, 4}
i. ϕ
j. 29
k. last 3 digits are evenly divisible by 3
l. 4^5
m. 72
n. 53 and 59
o. symbol for a number
p. 31
q. {51, 53, 57, 59}
r. 504
s. {5, 6, 7}

Answer these questions (each answer, 3 points).

13. Which of these number systems used pictures for numerals: Greek, Egyptian, or Roman?

14. For any real number a, $a + b = b + a$. What is the name of this property?

____

15. What is the sum of 3 and 2 in Mod 4? _______________

16. What is the intersection of set $A = \{2, 4, 6, 7, 9\}$ and set $B = \{2, 3, 4, 5\}$?

____

17. Is 8,473,296 divisible by 8? _______________

18. What is the base of 16? _______________

19. What is the least common multiple of 6 and 9? _______________

20. What is the sixth number in the Fibonacci sequence?
(Do not count the number *1* twice.) _______________

21. The difference of twin primes is always what number?

____

Convert to base 2 (each answer, 3 points).

22. 15 _______________

23. 25 _______________

24. 30 _______________

25. 35 _______________

Write in numbers (each answer, 3 points).

26. seven billion, seventy-seven _______________

27. seventy-seven million, seven thousand, seventy _______________

28. seven hundred thousand, seven hundred _______________

29. seventy thousand, seven hundred seven _______________

Given $M = \{5, 10, 15, 20, 25\}$, $N = \{5, 15, 25, 35, 45\}$, and $P = \{10, 20, 30, 40, 50\}$, write the required information (each answer, 3 points).

30. $M \cup P$ ______________________

31. $M \cap P$ ______________________

32. $M \cap N$ ______________________

33. $N \cap P$ ______________________

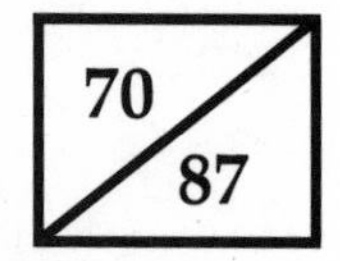

Date ______________________

Score ______________________

Name ____________________

Match these items (each answer, 2 points).

1.	______	factor	a. hundredths
2.	______	similar	b. a repeating decimal
3.	______	equivalent fractions	c. two equal ratios
4.	______	rational number	d. comparison by division
5.	______	denominator	e. numerator is less than denominator
6.	______	percent	f. numerator is more than denominator
7.	______	ratio	g. have equal numerical value
8.	______	proper fraction	h. upper part of a fraction
9.	______	proportion	i. lower part of a fraction
10.	______	numerator	j. same shape
			k. a multiplied number

Complete these items (each question, 3 points).

11. Seven-ninths is a fraction. Write it using numerals. ____________
12. Raise $\frac{8}{9}$ to higher terms with a denominator of 45. ____________
13. Reduce $\frac{65}{13}$ to a whole number. ____________
14. Change $7\frac{5}{6}$ to an improper fraction. ____________
15. Rewrite $\frac{411}{100}$ as a mixed number. ____________
16. Change $\frac{4}{5}$ to a decimal. ____________
17. Change 7% to a decimal. ____________
18. Write 30% as a fraction reduced to lowest terms. ____________
19. Write the ratio of 55 cm to 120 cm. ____________
20. Write 10,000 as a power of 10. ____________
21. Change 12 m to cm. ____________

22. Change 0.0023 to a percent. ____________

23. Change 369% to a decimal. ____________

24. Combine $\frac{3}{14}$ and $\frac{4}{14}$. ____________

25. On a map, 1 inch represents 150 miles. How many miles does 11 inches represent?

____

26. Write $\frac{1}{4}$ as a terminating decimal. ____________

27. Reduce $\frac{54}{90}$ to lowest terms. ____________

28. On the given number line locate $\frac{21}{5}$.

0 1 2 3 4 5 6

29. Take the following fraction, $\frac{7}{4}$, and write it <u>ten</u> equivalent ways using higher terms, decimals, and percents.

____

____

Write the correct letter on the line (each answer, 2 points).

30. The numerator of a fraction is the same as the part of a division problem called the _________ .

a. dividend
b. divisor
c. quotient
d. remainder

31. In comparing fractions the symbol < stands for _________ .

a. the same as
b. more than
c. less than
d. almost the same as

32. A decimal equal to $\frac{3}{8}$ is _________ .

a. 0.375
b. 0.25
c. 0.5
d. 0.125

33. The expression 9 out of 100 can be written _________ .

a. 0.9
b. 0.09
c. 90%
d. 9

34. The decimal 0.007 will become a percent that is _________ .

a. equal to 7
b. between 1 and 7
c. less than one
d. more than 100

35. The following choice is a rational number:_________ .

a. $\sqrt{2}$
b. 2.4494897. . .
c. $\sqrt{100}$
d. $\sqrt{5}$

36. On the number line the fraction with the smallest value is _________ .

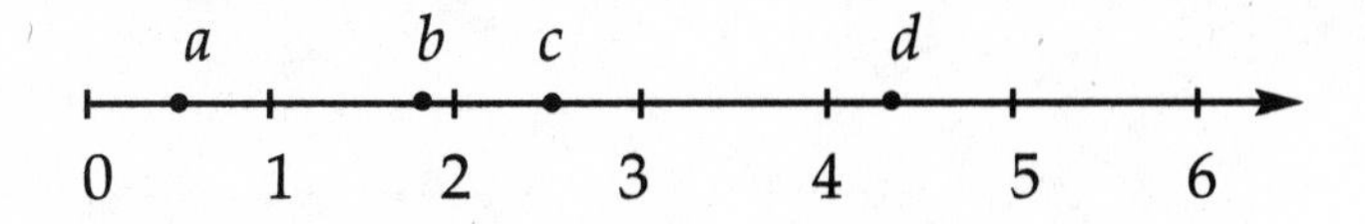

73/91

Date ______________________
Score ______________________

Math 804 Alternate Test

Name ____________________

Solve these word problems (each answer, 4 points).

1. After a board was cut into two pieces, the pieces measured $7\frac{3}{10}$ meters and $11\frac{9}{10}$ meters. How long was the original board?

 ____

2. Mary has $100.48 in her savings account and $109.42 in her checking account. Bob has $108.48 in his savings account and $104.29 in his checking account. How much more money does Bob have than Mary?

 ____

3. Mr. Jones' recipe calls for $1\frac{1}{4}$ cups of flour. He wants to double the recipe. How much flour will he need?

 ____

4. If Bill bought a baseball glove for $16.25, he should get ____________ change back from a twenty-dollar bill.

5. If one man can assemble a motor in $3\frac{1}{4}$ hours and another can do it in $2\frac{1}{2}$ hours, how much less time does the job take the second man?

 ____

Add or Subtract as indicated. Simplify answers that can be simplified (each answer, 4 points).

6. $\frac{1}{8} + \frac{1}{8}$

7. $6\frac{3}{4} + 14\frac{4}{7}$

8. $16\frac{3}{4} + 27\frac{7}{10}$

9. $\frac{5}{6} - \frac{1}{6}$

10. $7\frac{3}{8} - 2\frac{5}{8}$

11. $17\frac{3}{10} - 11\frac{5}{6}$

12. 17.56 + 21.7 + 5.23 = ____________________

13. 77.234 + 25.4 + 4,564.43 = ____________________

14. 625.34 + 66.66 + 243.517 = ____________________

15. $17\frac{5}{12} + 25\frac{4}{5} + 67\frac{3}{4} =$ __________

16. $452\frac{2}{3} + 25\frac{6}{7} - 467\frac{13}{14} =$ __________

17. $4{,}573.65 + 473.7 - 4{,}925.752 =$ __________

18. $64.75 + 25.7 - 65.9 =$ __________

19. $0.007 - 0.0035 =$ __________

20. $17.352 - 1.2376 =$ __________

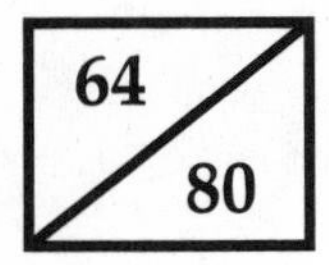

Date ____________________

Score ____________________

Math 805 Alternate Test

Name ____________________

Complete these items (each answer, 3 points).

1. To divide by a fraction, multiply by its ____________________ .

2. The best method for multiplying mixed numbers is to convert them to ____________________ and then multiply like other fractions.

Multiply or divide as indicated. Be sure that the common fraction answers are reduced to lowest terms (each answer, 3 points).

3.	$\frac{3}{8} \times \frac{6}{7} =$	________	9.	$3\frac{2}{3} \div 1\frac{1}{3} =$	________
4.	$\frac{2}{3} \times \frac{5}{6} =$	________	10.	$5\frac{1}{5} \div 2\frac{3}{4} =$	________
5.	$\frac{5}{8} \times \frac{1}{6} =$	________	11.	$519.7 \times 0.031 =$	________
6.	$\frac{4}{9} \times \frac{1}{5} =$	________	12.	$0.30 \times 0.03 =$	________
7.	$3\frac{1}{3} \div 2\frac{1}{3} =$	________	13.	$402.4 \times 0.24 =$	________
8.	$4\frac{1}{6} \div 10 =$	________	14.	$15 \times 0.051 =$	________

15. $0.065\overline{)0.1079}$

16. $15.1\overline{)83{,}080.2}$

17. $45.3\overline{)1{,}112.115}$

Work these percent problems. Write the answers on the lines (each answer 3 points).

18. What is the % change from 30 to 15? ________
19. What is the % change from 17 to 13.09? ________
20. 71% of 3.2 = what number? ________
21. 6.2 is what % of 12.5? ________
22. 0.05% of 62 = what number? ________
23. 32% of what number is 18.304? ________
24. 31.1% of what base is 0.18971? ________
25. 0.06 is what % of 32? ________

60/75

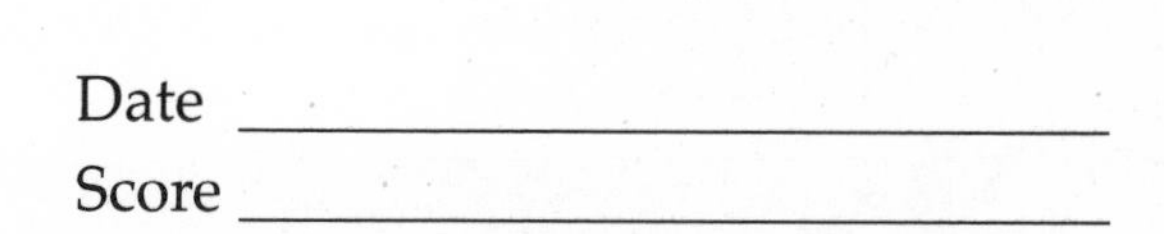

Name ________________________

Complete the ordered-pair number chart for the function rule $f(n) = 2n - 3$ (each answer, 3 points).

	n	$f(n)$
	2	1
	3	3
1.	4	____
2.	5	____
3.	6	____
4.	7	____
5.	8	____

Graph the following number pairs (each pair, 3 points).

6. (1, 0)
7. (0, 1)
8. (2, 6)
9. (4, 4)
10. (1, 7)

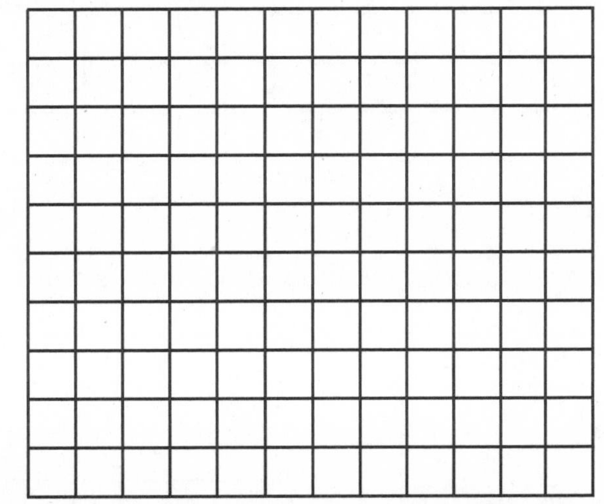

Complete the ordered-pair number chart (each answer, 3 points); write the function rule (3 points); graph the function (8 points).

	n	$f(n)$
	2	0
	4	1
	6	2
11.	8	____
12.	10	____
13.	12	____

14. ________________________________

15.

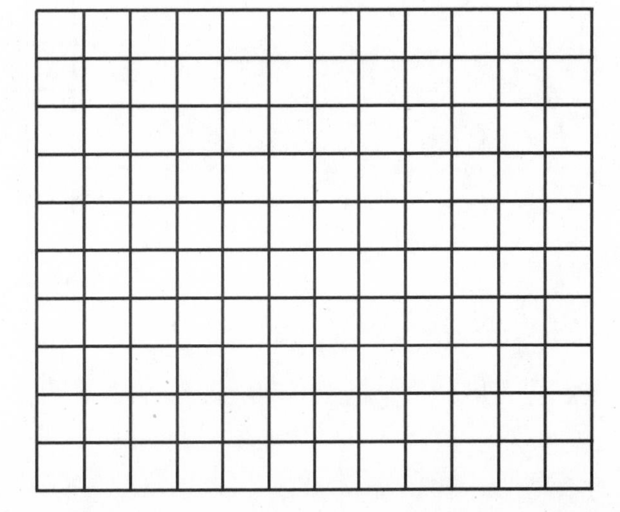

16. Construct a double bar graph from the following set of data (8 points).

Game	1	2	3	4	5
Strike-outs Yankees	6	9	8	4	10
Strike-outs White Sox	8	11	12	10	14

17. Construct a histogram from the following table (8 points).

Interval	Frequency
0- 6	10
7-12	20
13-18	40
19-24	20
25-30	10

Complete the following frequency chart (each answer, 3 points); construct a frequency polygon (8 points); find the mean, median, and mode (each answer, 3 points).

1, 12, 21, 3, 13, 5, 15, 25, 30, 28, 28, 31, 4, 19,
8, 21, 9, 29, 35, 42, 11, 20, 15, 25, 18, 23, 19, 39,
39, 26, 46, 15, 16, 21, 47, 35, 24, 30, 27, 38, 48,
20, 45, 31, 25, 21, 28, 39, 21, 41

	Interval	Frequency	I x F
18.	0- 9	_____	_____
19.	10-19	_____	_____
20.	20-29	_____	_____
21.	30-39	_____	_____
22.	40-49	_____	_____

Note: Use 5, 15, 25, 35, 45 as midpoints of the interval for I x F column.

23. mean = ____________________

24. median = ____________________

25. mode = ____________________

26.

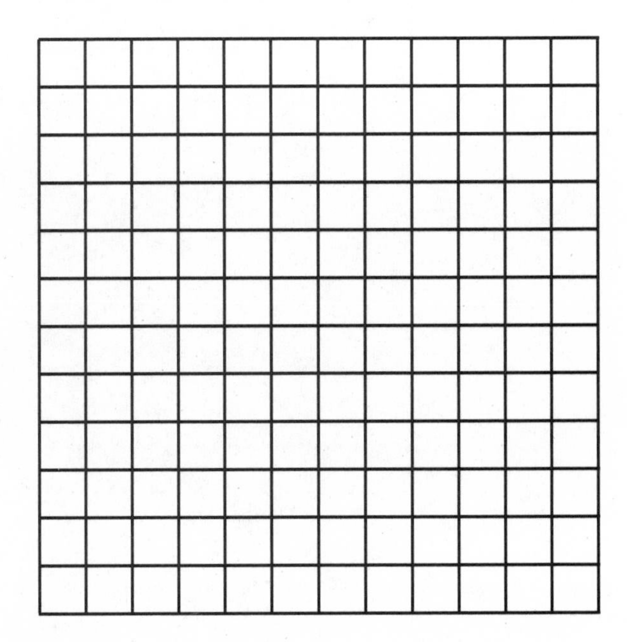

Math 806 Alternate Test

Write the probabilities (each answer, 3 points).

A box contains eight balls of like shape and size. Two are red, three are white, and three are blue. The balls are also numbered from 1 to 8. If the balls are drawn at random without replacement for each problem, find the following probabilities.

27. *P*(one red) = ______________
28. *P*(one white) = ______________
29. *P*(one blue) = ______________
30. *P*(number 6) = ______________
31. *P*(white and red) = ______________
32. *P*(white or blue) = ______________
33. *P*(1 ball > 5) = ______________
34. *P*(1 ball < 4) = ______________
35. *P*(prime numbers) = ______________
36. *P*(odd or even) = ______________
37. *P*(red and white and blue) = ______________
38. *P*(4 and 5 and 6) = ______________

119/149

Date ______________________

Score ______________________

Name ____________________

Write the opposite of each number (each answer, 2 points).

1. 14 ____________   2. -7 ____________

Complete each statement (each answer, 3 points).

3. The odd integers less than five are ________________ .

4. The nonpositive integers larger than negative four and less than two are ________________________________ .

Use < or = or > to make each statement true (each answer, 2 points).

5. $|7|$ ______ $|-7|$ 6. -30 ______ -20

Draw a number-line graph for each condition (each numbered item, 3 points).

7. -2, -1, 0, 4

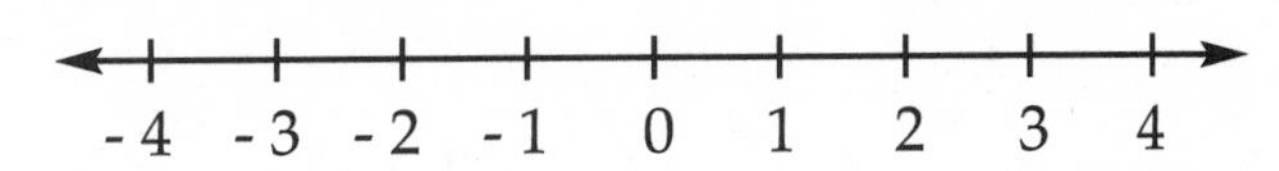

8. The integers having an absolute value less than three.

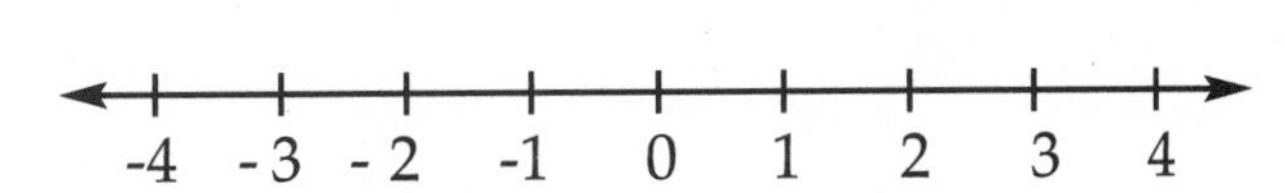

Write the number pair for each lettered point (each answer, 3 points).

9. *C* ____________

10. *D* ____________

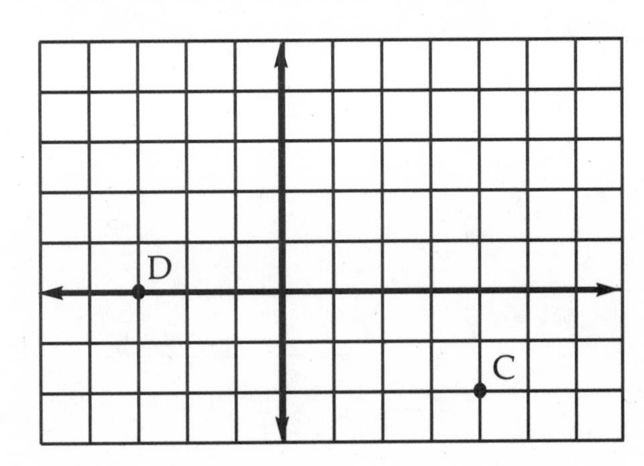

Perform the indicated operations (each answer, 3 points).

11. -10 + 7 ____________

12. 14 + (-5) ____________

13. 0 – (-11) ____________

14. $6 - 9$ ______________

15. $-4 \cdot (-2)$ ______________

16. $3 \cdot (-7)$ ______________

17. 8^2 ______________

18. $(-8)^{-2}$ ______________

19. $-3 \div 0$ ______________

20. $-27 \div 3$ ______________

Evaluate each algebraic expression if $a = 4$, $b = -1$ and $c = 2$ (each answer, 4 points).

21. $a + bc$ ______________

22. $ab - c$ ______________

Evaluate each algebraic expression if $f = -2$, $g = 0$, and $h = 3$ (each answer, 4 points).

23. $5h^3$ ______________

24. $f^2 + h^g$ ______________

Complete each table to find number pairs that make the given sentences true (each numbered item, 4 points).

25. $r + s = 4$ (,), (,), (,)

r	2	-3	
s			0

26. $y = -2x$ (,), (,), (,)

x	-1	0	7
y			

Graph the integral solutions to each sentence (each graph, 4 points).

27. $n = m + 3$ for $-5 \leq m \leq 1$

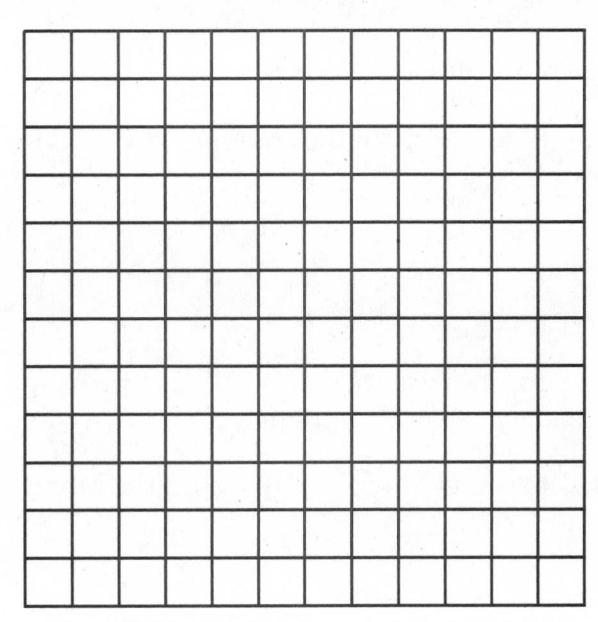

28. $b = a^2 - 3$ for $-3 \leq a \leq 2$

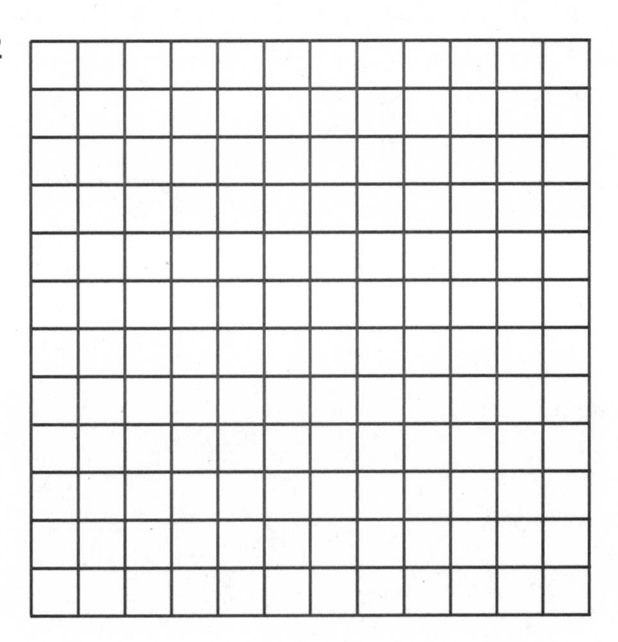

70/88

Date ______________________

Score ______________________

Math 808 Alternate Test

Name ______________________

Students may use the page of formulas for this test.

1. Find the square root, to tenths, by either method (4 points).

   $\sqrt{542.89}$

Solve these problems using appropriate formulas. When necessary, write the required information on the blank (each answer, 4 points).

2. A rectangular room measures 11 ft. by 15 ft. Find the area.

3. The room in Problem 2 has a door measuring 30 inches wide.
   a. How much baseboard that goes around the edge of the wall will be needed?
   b. Find the cost of the baseboard at $1.20 per foot.

4. Find the circumference of a circle with a 24-ft. diameter.

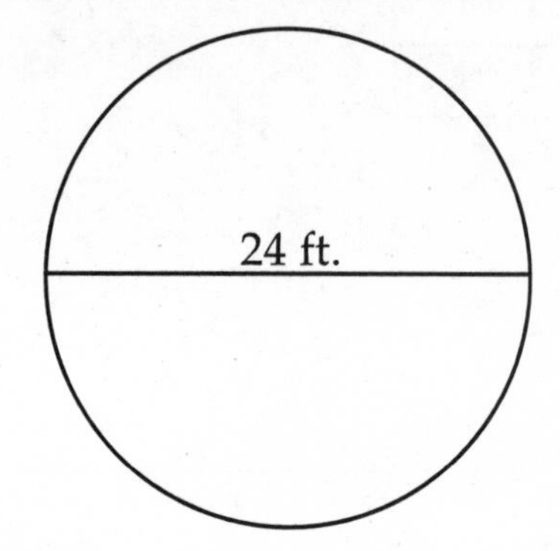

5. Find the area of the circle in Problem 4.

6. Find the area of the figure shown.

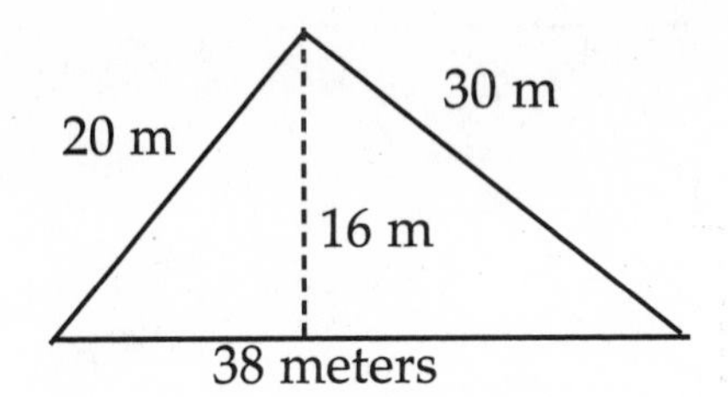

7. a. Name of figure: ____________________

   b. Find the area. ____________________

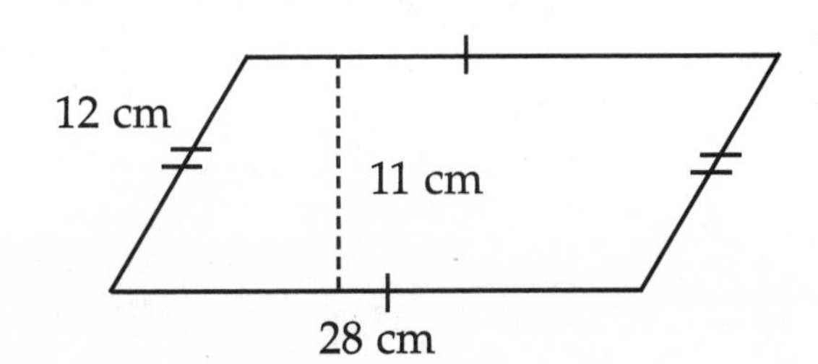

8. Find the third side; round answer to tenths.

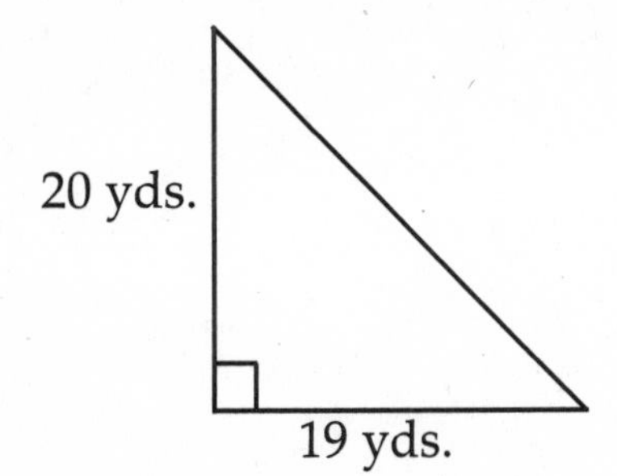

9. Find the area of the triangle in Problem 8.

10. a. Find the perimeter of the trapezoid shown.

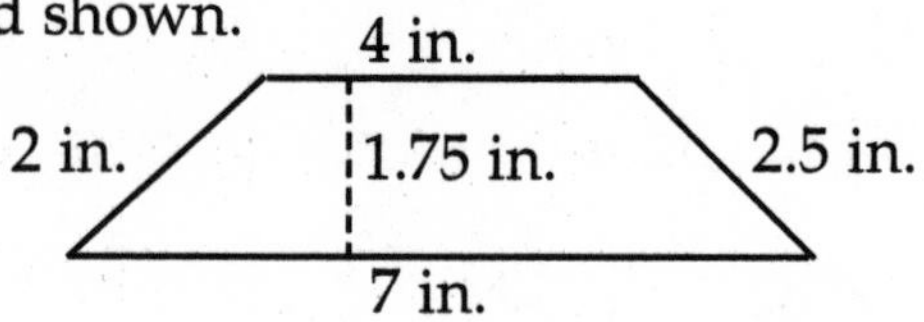

b. Find the area.

11. Change 1,125 square feet to square yards using a conversion fraction.

12. Find the volume of the rectangular solid shown.

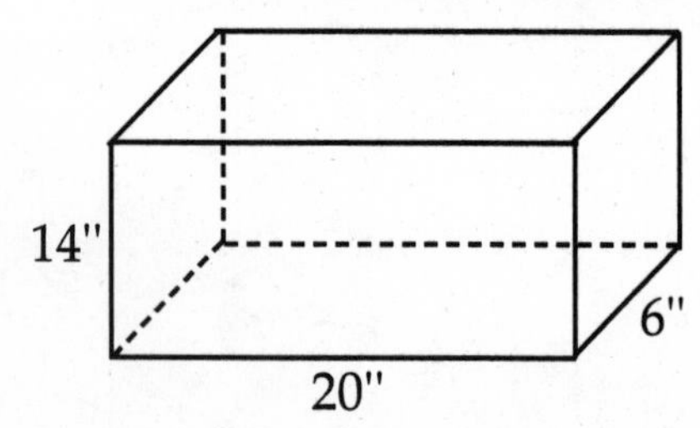

13. Find the total surface area of the rectangular solid in Problem 12.

14. Find the capacity or volume of the attic shown.

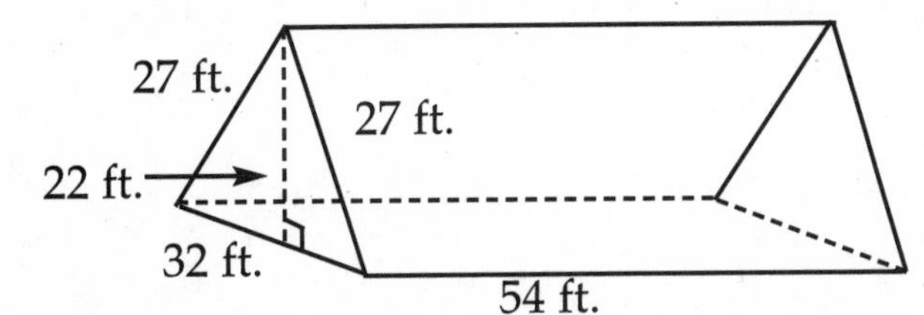

15. Find the surface area of the roof in Problem 14. Do not include the ends.

16. a. Name of figure: ______________________

b. Find the volume. ______________________

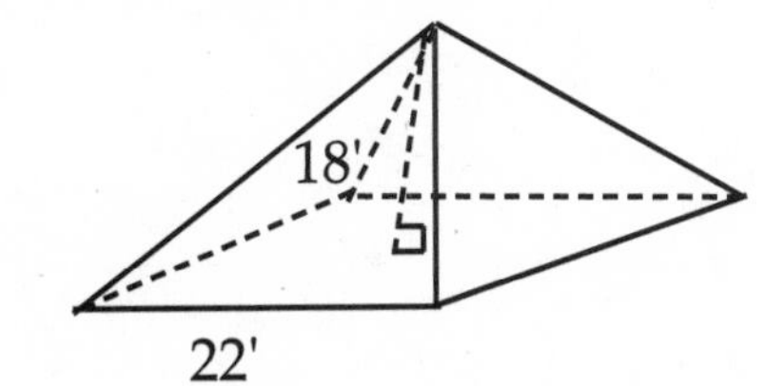

17. The diameter of the cylinder shown is 8 inches; the height is 16 inches. Find the volume.

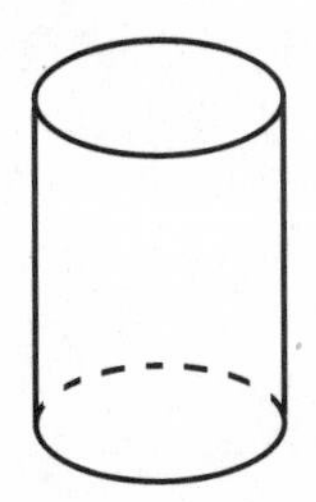

18. Find the total area of the figure in Problem 17.

19. The spherical tank shown has a 36-ft. diameter. Find the volume.

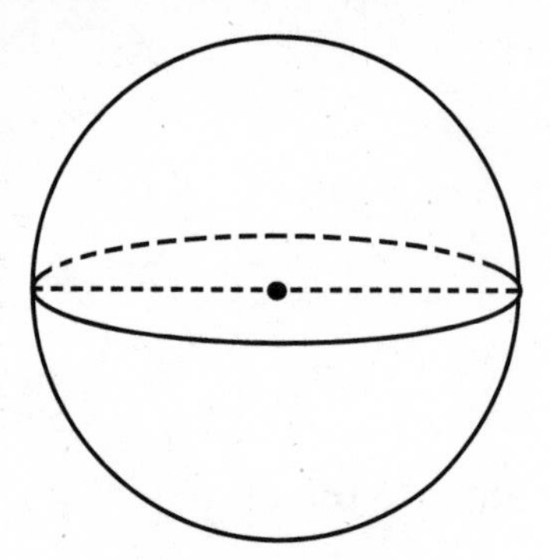

20. Find the surface area of the sphere in Problem 19.

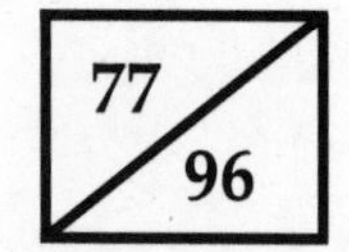

Date ____________________________

Score ____________________________

Name ____________________

Match the following items (each answer, 2 points).

1. ________ means
2. ________ coefficient
3. ________ like terms
4. ________ commutative property for multiplication
5. ________ algebra
6. ________ negative numbers
7. ________ $10c = 17$
8. ________ solution set
9. ________ a formula
10. ________ associative property for addition
11. ________ division property
12. ________ subtraction property

a. "the product of ten and a number is seventeen"
b. $a + 6 = 9; a = 3$
c. $V = Bh$
d. first and fourth terms of a proportion
e. $5y = 15; y = 3$
f. less than zero
g. uses unknowns
h. the 7 in the term $7xyz$
i. $a \cdot b = b \cdot a$
j. second and third terms of a proportion
k. $\{5\}$ of $2x = 10$
l. $3w - 2w + 8w$
m. $x + (y + z) = (x + y) + z$

Write the correct letter on the line (each answer, 2 points).

13. The expression $(a \cdot b) \cdot c = a \cdot (b \cdot c)$ is an example of ________ .

    a. the associative property for addition
    b. the commutative property for addition
    c. the commutative property for multiplication
    d. the associative property for multiplication

14. The coefficient of the term $104abc$ is ________ .

    a. 10
    b. 104
    c. abc
    d. bc

15. Using the distributive property with the equation $ab(x + y)$ results in ________ .

    a. $abx + y$
    b. $x + aby$
    c. $abx + aby$
    d. $x + y$

16. In the equation $A = bh$, if $b = 6$ and $h = 7$, then $A =$ ________ .

a. 6
b. 7
c. 67
d. 42

17. An example of unlike terms is ________ .

a. $a + b + c$
b. $2t + 3t - t$
c. $4w + 4w + w$
d. $8xyz - 7xyz + xyz$

18. The solution set of $a + 2a = 6$ is ________ .

a. {2}
b. {9}
c. {13}
d. {6}

19 Which of the following expressions commutative for addition?

a. $a - b = b - a$
b. $c + d = d + c$
c. $b - a = a - b$
d. $e \div f = f \div e$

20. The phrase "three more than twice a number" is ________ .

a. $3h + 2$
b. $2h + 3$
c. $3h - 2$
d. $2h - 3$

21. The solution set of "two more than three times a number is twenty" is ________ .

a. {6}
b. {7}
c. {8}
d. {5}

22. The fourth term of the proportion $e{:}f = g{:}h$ is________ .

a. e
b. f
c. g
d. h

Work the following problems (each answer, 3 points).

23. The missing terms in $6(a + 6b - 12) =$ ____ + ____ – ____ are ____________________ .

24. Show that $12 \cdot 6 = 6 \cdot 12$. ________________________________

25. Combine $14q - 6q + 8q + q - 10q$. ________________________________

26. Simplify $6ab - 2y - 3y + 2ab + 6y$. ________________________________

27. Solve and check $4x - 5 = 19$.

28. Solve and check $5a + 2 = 2a + 11$. ____________________

29. Solve and check $\frac{1}{5}b + 2 = 8$. ____________________

30. Solve the formula $A = \frac{1}{2}bh$ for A if $b = 6''$ and $h = 8''$. ____________________

31. Change $9d + 6 = 3d - d$ to words. ____

____

32. What number must be added to 21 to equal 87? ____________________

33. Find two consecutive integers whose sum is 25. ____________________

34. An engine makes 400 revolutions per minute. How many revolutions will it make in 20 minutes?

____

35. Jane's age is $\frac{1}{7}$ the age of her grandmother. If Jane is 11, how old is her grandmother?

____

66/83

Date _______________

Score _______________

Name ____________________

Match these items (each answer, 2 points).

1. ______ $A = \pi r^2$
2. ______ product
3. ______ composite number
4. ______ improper fraction
5. ______ mode
6. ______ commutative
7. ______ integer
8. ______ ratio
9. ______ numeral
10. ______ function
11. ______ percent
12. ______ variable
13. ______ quotient
14. ______ MCM
15. ______ π
16. ______ mixed number
17. ______ 100_2
18. ______ proportion
19. ______ means
20. ______ extremes

a. occurs the most
b. the answer in division
c. an unknown
d. ordered pair
e. the numerator is less than the denominator
f. Roman numeral
g. the first and fourth terms in a proportion
h. out of 100
i. $a + b = b + a$
j. 3.14
k. the second and third terms in a proportion
l. the numerator is more than the denominator
m. a fraction
n. the answer in multiplication
o. 4
p. a nonprime number
q. positive and negative whole numbers and zero
r. area of a circle
s. $a:b = c:d$
t. symbol for a number
u. a number and a fraction

Complete these items (each answer, 3 points).

21. The number 4,168 written in expanded form is ____________________.

22. Multiply $3\frac{7}{8} \cdot 1\frac{2}{3}$. ____________________

23. Divide $4\frac{1}{4}$ by $2\frac{2}{3}$. ____________________

24. Write the numerals from one to eight in base 2.

____

25. Solve and check $4d - 1 = 9 - d$.

26. Solve and check $2y - y + 6y + 4 = 12 - y$.

27. If one candy bar costs 22¢, how much do 12 dozen candy bars cost?

____

28. The LCM of 4 and 28 is ______________.

29. Multiply

$$\begin{array}{r} 52.18 \\ \underline{54.9} \end{array}$$

30. Divide $1.71\overline{)9{,}063}$

31. Find the prime factors of 54. ______________________

32. Change $\frac{1}{2}$, 0.04, 1, 0.0008, and 2.6 to percents. ______________________

33. Divide $4\frac{1}{8}$ by 7. ______________________

34. Find the area of a circle with radius 2.5 cm. ($A = \pi r^2$; $\pi = 3.14$)

____

35. Use the distributive property to find the product of $(4x + 3y)\ (5a - 4d)$.

____

36. Arrange the fractions $\frac{2}{3}$, $\frac{4}{5}$, 1, 8, $2\frac{3}{4}$, $\frac{14}{7}$, $\frac{0}{2}$, and $\frac{7}{8}$ in order from largest to smallest. ______________________

Write the correct letter on the line (each answer, 2 points).

37. The decimal 0.075 written as a percent is_______ .

a.	7.5%	c.	0.75%
b.	75%	d.	750%

38. The LCD for $\frac{5}{12}$ and $\frac{4}{45}$ is _______ .

a.	180	c.	45
b.	90	d.	12

39. In the problem, 24% of $B = 48$, $B =$ _______ .

a.	20	c.	2
b.	200	d.	10

40. The number, $4 \cdot 10^7$, written in numerals is _______ .

a.	47	c.	4,00,000
b.	40,000,000	d.	4,000

41. The value of 2 in the numeral 2,116,891 is _______ .

a.	2,000	c.	2,000,000
b.	20,000	d.	200,000

42. The mode of this list of numbers, 5, 7, 4, 7, 5, 6, 8, 5, and 6, is _______.

a.	4	c.	6
b.	5	d.	7

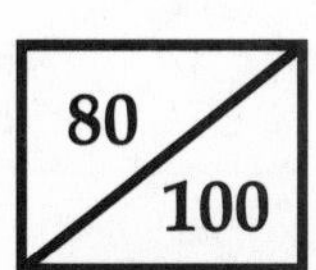

Date ______________________

Score ______________________

1. i
2. d
3. j
4. c
5. e
6. g
7. k
8. a
9. h
10. f
11. six thousand, three hundred eighty-four
12. 12,944
13. 2,800
14. 9
15. 9
16. 1,518
17. 20,808
18.
```
     2,365
   x   476
     14190
    16555
    9460
 1,125,740
```
19.
```
       286
  31)8,866
     62
     266
     248
      186
      186
        0
```
20.
```
       184 14/25
  25)4,614
     25
     211
     200
      114
      100
       14
```
21.
```
   6,865
 x    94
   27460
  61785
 645,310
```
22. 4,897
23. 21,810
24. a
25. a
26. a
27. c
28. c
29. b
30.
```
   52,738
 - 47,916
    4,822 miles
```
31.
```
  $ 71
    157
     21
     15
  $ 264
```
32.
```
   17
  x 3
  $51
```
33. 746 = 700 + 40 + 6

1. o
2. m
3. n
4. a
5. l
6. b
7. d
8. j
9. g
10. c
11. h
12. i
13. Egyptian
14. commutative
15. The addition table for Mod 4 is shown.

+	0	1	2	3
0	0	1	2	3
1	1	2	3	0
2	2	3	0	1
3	3	0	(1)	2

$3 + 2 = 1$

16. {2, 4}
17. Yes; since the last three digits, 296, are divisible by 8, 8,473,296 is divisible by 8.
18. 16
19. $6 = 2 \times 3$
    $9 = 3 \times 3 = 3^2$
    $\text{LCM} = 2 \times 3 \times 3 = 18$
20. The Fibonacci sequence is 1, 1, 2, 3, 5, 8, 13, 21, . . .
    The sixth number is 13.
21. 2; twin primes are primes differing by 2.
22. $15 = 1111_2$

8's	4's	2's	1's
1	1	1	1

23. $25 = 11001_2$

16's	8's	4's	2's	1's
1	1	0	0	1

24. $30 = 11110_2$

16's	8's	4's	2's	1's
1	1	0	0	1

25. $35 = 100011_2$

32's	16's	8's	4's	2's	1's
1	0	0	0	1	1

26. 7,000,000,077
27. 77,007,070
28. 700,700
29. 70,707
30. {5, 10, 15, 20, 25, 30, 40, 50}
31. {10, 20}
32. {5, 15, 25}
33. { } or ϕ

1. k
2. j
3. g
4. b
5. i
6. a
7. d
8. e
9. c
10. h
11. $\frac{7}{9}$
12. $9 \times 5 = 45$
$\frac{8 \times 5}{9 \times 5} = \frac{40}{45}$
13. $\frac{65 \div 13}{13 \div 13} = \frac{5}{1} = 5$
14. Multiply the whole number by the denominator of the fraction and add the numerator to it.
$7\frac{5}{6} = \frac{(7 \times 6) + 5}{6}$
$= \frac{42 + 5}{6}$
$= \frac{47}{6}$
15. $411 \div 100 = 100\overline{)411}$ $= 4\frac{11}{100}$
$\quad 400$
$\quad 11$
16. $\frac{4}{5} = 4 \div 5 = 5\overline{)4.0} = 0.8$
$\quad 40$
$\quad 0$
17. Drop the percent (%) sign and move the decimal point two places to the left
7% = 0.07
18. Change 30% to a fraction. Drop the percent (%) sign and write 100 in the denominator.
$30\% = \frac{30}{100}$
Reduce the fraction
$\frac{30 \div 10}{100 \div 10} = \frac{3}{10}$
19. 55 to 120 = 55:120 or $\frac{55}{120}$
$55 = \textcircled{5} \times 11$
$120 = 2 \times 2 \times 2 \times 3 \times 5$
$= 2^3 \times 3 \times \textcircled{5}$
GCF = 5
$\frac{55 \div 5}{120 \div 5} = \frac{11}{24}$ or 11:24
20. $10{,}000 = 10 \times 10 \times 10 \times 10$
$= 10^4$
21. 1 m = 100 cm
12(1) m = 12(100) cm
12 m = 1,200 cm
22. Move the decimal point two places to the right and add the percent (%) sign.
0.0023 = 0.23%
23. Drop the percent (%) sign and move the decimal point two places to the left.
369% = 3.69
24. $\frac{3}{14} + \frac{4}{14} = \frac{7}{14}$
$\frac{7 \div 7}{14 \div 7} = \frac{1}{2}$
25. 1 inch = 150 miles
11(1) inches = 11(150) miles
11 inches = 1,650 miles

26. $\frac{1}{4} = 1 \div 4 =$

$$\begin{array}{r} 0.25 \\ 4\overline{)1.00} \\ \underline{8} \\ 20 \\ \underline{20} \\ 0 \end{array}$$

27. $54 = (2) \times (3 \times 3) \times 3 = 2 \times 3^3$

$90 = (2) \times (3 \times 3) \times 5$

$= 2 \times 3^2 \times 5$

$\text{GCF} = 2 \times 3 \times 3 = 2 \times 3^2$

$= 2 \times 9 = 18$

$\frac{54 \div 18}{90 \div 18} = \frac{3}{5}$

28. $\frac{21}{5} = 21 \div 5 =$

$$\begin{array}{r} 4\frac{1}{5} \\ 5\overline{)21} \\ \underline{20} \\ 1 \end{array}$$

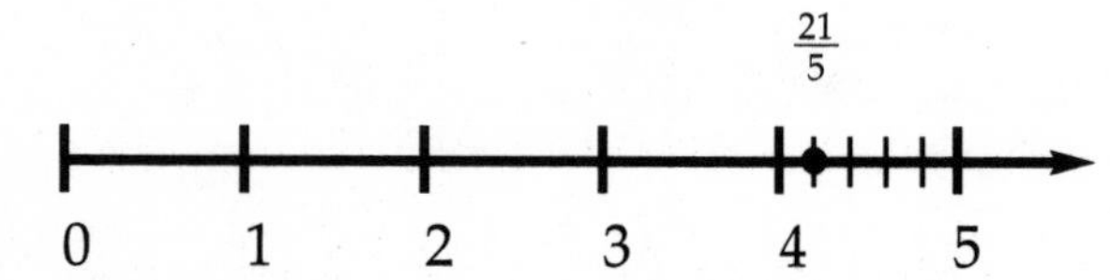

29. Examples:

$\frac{7}{4} = 7 \div 4 =$

$$\begin{array}{r} 1.75 \\ 4\overline{)7.00} \\ \underline{4} \\ 30 \\ \underline{28} \\ 20 \\ \underline{20} \\ 0 \end{array}$$

$\frac{7}{4} = 7 \div 4 =$

$$\begin{array}{r} 1\frac{3}{4} \\ 4\overline{)7} \\ \underline{4} \\ 3 \end{array}$$

$\frac{7}{4} = 1.75 = 175\%$

$\frac{7 \times 2}{4 \times 2} = \frac{14}{8}$

$\frac{7 \times 3}{4 \times 3} = \frac{21}{12}$

$\frac{7 \times 4}{4 \times 4} = \frac{28}{16}$

$\frac{7 \times 5}{4 \times 5} = \frac{35}{20}$

29. cont.

$\frac{7 \times 6}{4 \times 6} = \frac{42}{24}$

$\frac{7 \times 7}{4 \times 7} = \frac{49}{28}$

$\frac{7 \times 8}{4 \times 8} = \frac{56}{32}$

30. a

31. c

32. a

$\frac{3}{8} = 3 \div 8 =$

$$\begin{array}{r} 0.375 \\ 8\overline{)3.000} \\ \underline{24} \\ 60 \\ \underline{56} \\ 40 \\ \underline{40} \\ 0 \end{array}$$

33. b

9 out of 100 = $\frac{9}{100}$

$\frac{9}{100} = 9 \div 100 = 0.09$

34. c

To change 0.007 to a percent, move the decimal point two places to the right and add a percent (%) sign.

0.007 = 0.7%, which is less than one.

35. c

36. a

1. $$\begin{array}{r} 7\frac{3}{10} \\ +\ 11\frac{9}{10} \\ \hline 18\frac{12}{10} \end{array} = 19\frac{2}{10} = 19\frac{1}{5}$$

2. Bob: $$\begin{array}{r} \$108.48 \\ +\ 104.29 \\ \hline \$212.77 \end{array}$$

Mary: $$\begin{array}{r} \$100.48 \\ +\ 109.42 \\ \hline \$209.90 \end{array}$$

$$\begin{array}{r} \$212.77 \\ -\ 209.90 \\ \hline \$\ \ 2.87 \end{array}$$

Bob has $2.87 more than Mary.

3. $$\begin{array}{r} 1\frac{1}{4} \\ +1\frac{1}{4} \\ \hline 2\frac{2}{4} \end{array} = 2\frac{1}{2} \text{ cups}$$

4. $$\begin{array}{r} \$20.00 \\ -\ 16.25 \\ \hline \$\ 3.75 \end{array}$$

5. $$\begin{array}{rcl} 3\frac{1}{4} & = & \overset{2}{\cancel{3}}\,\frac{\overset{5}{\cancel{1}}}{4} \\ -\ 2\frac{1}{2} & = & 2\frac{2}{4} \\ \hline & & \frac{3}{4} \text{ of an hour} \end{array}$$

6. $$\begin{array}{r} \frac{1}{8} \\ +\ \frac{1}{8} \\ \hline \frac{2}{8} \end{array} = \frac{1}{4}$$

7. $$\begin{array}{rcl} 6\frac{3}{4} & = & 6\frac{21}{28} \\ +\ 14\frac{4}{7} & = & 14\frac{16}{28} \\ \hline & & 20\frac{37}{28} = 21\frac{9}{28} \end{array}$$

8. $$\begin{array}{rcl} 16\frac{3}{4} & = & 16\frac{15}{20} \\ +\ 27\frac{7}{10} & = & 27\frac{14}{20} \\ \hline & & 43\frac{29}{20} = 44\frac{9}{20} \end{array}$$

9. $$\begin{array}{r} \frac{5}{6} \\ -\ \frac{1}{6} \\ \hline \frac{4}{6} \end{array} = \frac{2}{3}$$

10. $$\begin{array}{r} \overset{6}{\cancel{7}}\,\frac{\overset{11}{\cancel{3}}}{8} \\ -\ 2\frac{5}{8} \\ \hline 4\frac{6}{8} \end{array} = 4\frac{3}{4}$$

11. $$\begin{array}{rcl} 17\frac{3}{10} & = & \overset{16}{\cancel{17}}\,\frac{\overset{39}{\cancel{9}}}{30} \\ -\ 11\frac{5}{6} & = & 11\frac{25}{30} \\ \hline & & 5\frac{14}{30} = 5\frac{7}{15} \end{array}$$

12. $$\begin{array}{r} 17.56 \\ 21.7 \\ 5.23 \\ \hline 44.49 \end{array}$$

13. $$\begin{array}{r} 77.234 \\ 25.4 \\ 4{,}564.43 \\ \hline 4{,}667.064 \end{array}$$

14. $$\begin{array}{r} 625.34 \\ 66.66 \\ 243.517 \\ \hline 935.517 \end{array}$$

15. $$\begin{array}{rcl} 17\frac{5}{12} & = & 17\frac{25}{60} \\ 25\frac{4}{5} & = & 25\frac{48}{60} \\ 67\frac{3}{4} & = & 67\frac{45}{60} \\ \hline & & 109\frac{118}{60} = 110\frac{58}{60} \\ & & = 110\frac{29}{30} \end{array}$$

16.
$$\begin{array}{r} 452\frac{2}{3} = 452\frac{28}{42} \\ +\ 25\frac{6}{7} = \ \ 25\frac{36}{42} \\ \hline 477\frac{64}{42} \\ -\ 467\frac{13}{14} = 467\frac{39}{42} \\ \hline 10\frac{25}{42} \end{array}$$

17.
$$\begin{array}{r} 4{,}573.65 \\ +\ \ 473.7 \\ \hline 5{,}047.350 \\ -\ 4{,}925.752 \\ \hline 121.598 \end{array}$$

18.
$$\begin{array}{r} 64.75 \\ +\ 25.7 \\ \hline 90.45 \\ -\ 65.9 \\ \hline 24.55 \end{array}$$

19.
$$\begin{array}{r} 0.0070 \\ -\ 0.0035 \\ \hline 0.0035 \end{array}$$

20.
$$\begin{array}{r} 17.3520 \\ -\ 1.2376 \\ \hline 16.1144 \end{array}$$

1. reciprocal

2. improper fractions

3. $\frac{3}{\cancel{8}_4} \times \frac{\cancel{6}^3}{7} = \frac{9}{28}$

4. $\frac{\cancel{2}^1}{3} \times \frac{5}{\cancel{6}_3} = \frac{5}{9}$

5. $\frac{5}{8} \times \frac{1}{6} = \frac{5}{48}$

6. $\frac{4}{9} \times \frac{1}{5} = \frac{4}{45}$

7. $3\frac{1}{3} \div 2\frac{1}{3} =$

$\frac{10}{3} \div \frac{7}{3} =$

$\frac{10}{\cancel{3}_1} \times \frac{\cancel{3}^1}{7} = \frac{10}{7} = 1\frac{3}{7}$

8. $4\frac{1}{6} \div 10 =$

$\frac{25}{6} \div \frac{10}{1} =$

$\frac{\cancel{25}^5}{6} \times \frac{1}{\cancel{10}_2} = \frac{5}{12}$

9. $3\frac{2}{3} \div 1\frac{1}{3} =$

$\frac{11}{3} \div \frac{4}{3} =$

$\frac{11}{\cancel{3}_1} \times \frac{\cancel{3}^1}{4} = \frac{11}{4} = 2\frac{3}{4}$

10. $5\frac{1}{4} \div 2\frac{3}{4} =$

$\frac{26}{5} \div \frac{11}{4} =$

$\frac{26}{5} \times \frac{4}{11} = \frac{104}{55} = 1\frac{49}{55}$

11.
```
   519.7
 x 0.031
    5197
  15591
 16.1107
```

12.
```
   0.30
 x 0.03
 0.0090 = 0.009
```

13.
```
  402.4
 x 0.24
  16096
  8048
 96.576
```

14.
```
  0.051
 x   15
   0255
  0051
 00.765 = 0.765
```

15.
```
             1.66
 0.065)0.107.90
          65
          429
          390
           390
           390
             0
```

16.
```
          5,502
 15.1)83,080.2
        755
         758
         755
          302
          302
            0
```

17.

$$\begin{array}{r} 24.55 \\ 45.3\overline{)1112.115} \\ \underline{906} \\ 2061 \\ \underline{1812} \\ 2491 \\ \underline{2265} \\ 2265 \\ \underline{2265} \\ 0 \end{array}$$

18. Amount of change:

$$\begin{array}{r} 30 \\ \underline{-\ 15} \\ 15 \end{array}$$

Rate: $\frac{15}{30} = \frac{1}{2} = 0.5$

$= 50\%$ decrease

19. Amount of change:

$$\begin{array}{r} 17.00 \\ \underline{-\ 13.09} \\ 3.91 \end{array}$$

Rate: $\frac{3.91}{17} =$

$$\begin{array}{r} 0.23 \\ 17\overline{)3.91} \\ \underline{34} \\ 51 \\ \underline{51} \\ 0 \end{array}$$

$0.23 = 23\%$ decrease

20. $R \times B = N$
$R = 71\% = 0.71$
$B = 3.2$
$0.71 \times 3.2 = N$

$$\begin{array}{r} 0.71 \\ \underline{\times\ \ 3.2} \\ 142 \\ \underline{213} \\ 2.272 = N \end{array}$$

21. $R \times B = N$
$B = 12.5$
$N = 6.2$
$R \times 12.5 = 6.2$
$R = \frac{6.2}{12.5}$

$$\begin{array}{r} 0.496 \\ 12.5\overline{)62.000} \\ \underline{500} \\ 1200 \\ \underline{1125} \\ 750 \\ \underline{750} \\ 0 \end{array}$$

$R = 0.496 = 49.6\%$

22. $R \times B = N$
$R = 0.05\% = 0.0005$
$B = 62$
$0.0005 \times 62 = N$

$$\begin{array}{r} 0.0005 \\ \underline{\times\ \ \ \ \ 62} \\ 00010 \\ \underline{00030} \\ 00.0310 \end{array}$$

$N = 0.031$

23. $R \times B = N$
$R = 32\% = 0.32$
$N = 18.304$
$0.32 \times B = 18.304$
$B = \frac{18.304}{0.32}$

$$\begin{array}{r} 57.2 \\ 0.32\overline{)18.304} \\ \underline{160} \\ 230 \\ \underline{224} \\ 64 \\ \underline{64} \\ 0 \end{array}$$

$B = 57.2$

24. $R \times B = N$
$R = 31.1\% = 0.311$
$N = 0.18971$
$0.311 \times B = 0.18971$
$B = \frac{0.18971}{0.311}$

$$\begin{array}{r} 0.61 \\ 0.311\overline{)0.18971} \\ \underline{1866} \\ 311 \\ \underline{311} \\ 0 \end{array}$$

$B = 0.61$

25. $R \times B = N$
$B = 32$
$N = 0.06$
$R \times 32 = 0.06$
$R = \frac{0.06}{32}$

$$\begin{array}{r} 0.001875 \\ 32\overline{)0.060000} \\ \underline{32} \\ 280 \\ \underline{256} \\ 240 \\ \underline{224} \\ 160 \\ \underline{160} \\ 0 \end{array}$$

$R = 0.001875 = 0.1875\%$

1. $f(n) = 2(4) - 3$
   $= 8 - 3$
   $= 5$

2. $f(n) = 2(5) - 3$
   $= 10 - 3$
   $= 7$

3. $f(n) = 2(6) - 3$
   $= 12 - 3$
   $= 9$

4. $f(n) = 2(7) - 3$
   $= 14 - 3$
   $= 11$

5. $f(n) = 2(8) - 3$
   $= 16 - 3$
   $= 13$

6. through 10.

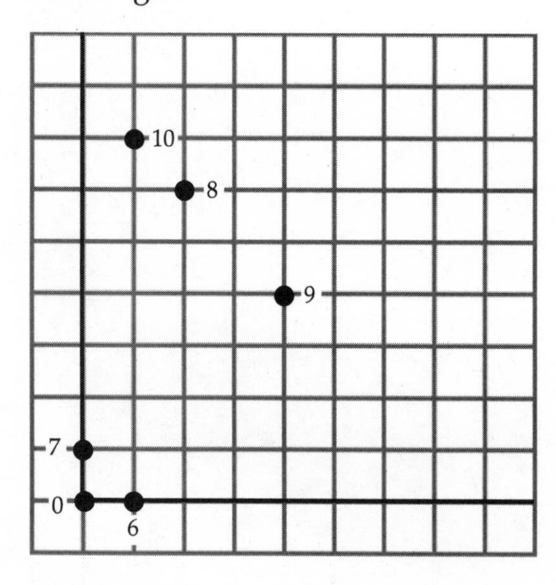

For Problems 11 through 13, the numbers for $f(n)$ are consecutive.

11. 3

12. 4

13. 5

14. $f(n) = \frac{n}{2} - 1$

15.

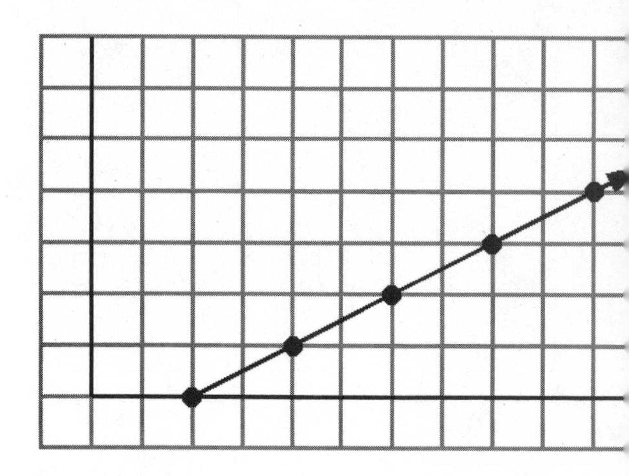

16.

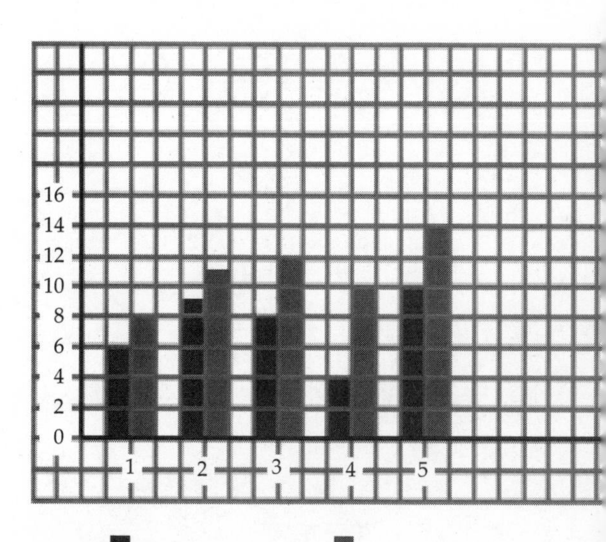

17.

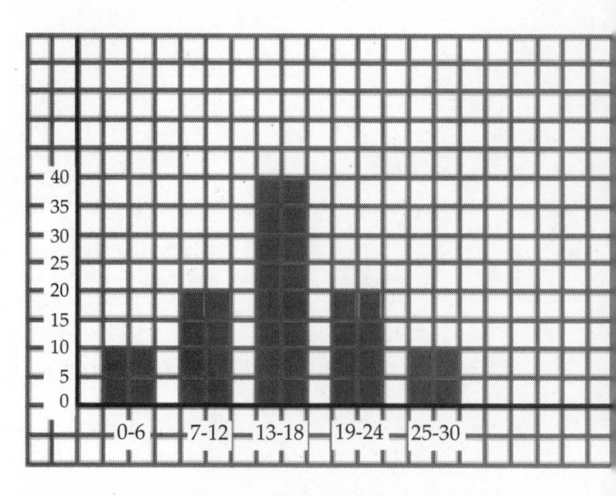

18. Frequency = 6
 $I \times F = 5 \times 6 = 30$

19. Frequency = 10
 $I \times F = 15 \times 10 = 150$

20. Frequency = 18
 $I \times F = 25 \times 18 = 450$

21. Frequency = 10
 $I \times F = 35 \times 10 = 350$

22. Frequency = 6
 $I \times F = 45 \times 6 = 270$

23. mean = $\frac{\text{sum of elements}}{\text{number of elements}}$
 $= \frac{1{,}250}{50}$
 $= 25$

24. median = 20–29 interval

25. mode = 20-29 interval

26.

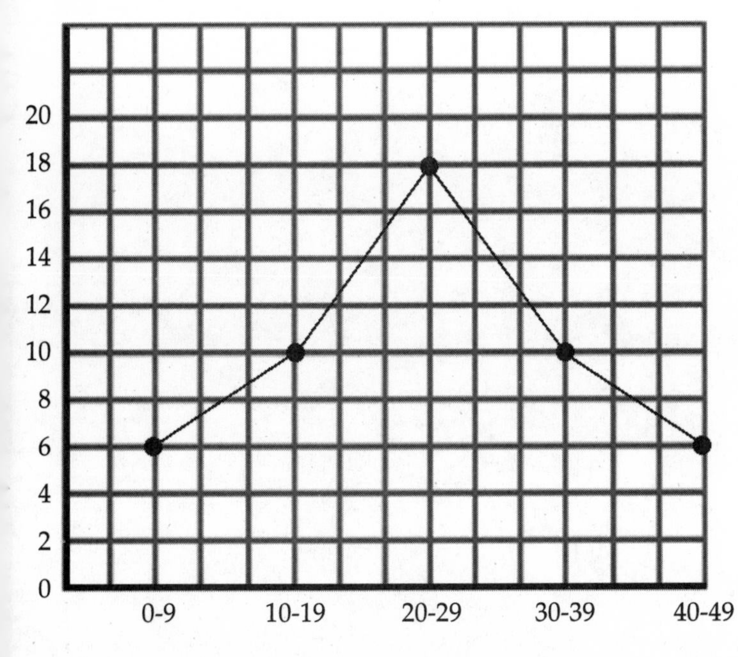

27. $P(\text{one red}) = \frac{2}{8} = \frac{1}{4}$

28. $P(\text{one white}) = \frac{3}{8}$

29. $P(\text{one blue}) = \frac{3}{8}$

30. $P(\text{number } 6) = \frac{1}{8}$

31. $P(\text{white and red}) = \frac{3}{8} \bullet \frac{2}{7}$
 $= \frac{6}{56} = \frac{3}{28}$

32. $P(\text{white or blue}) = \frac{3}{8} + \frac{3}{8}$
 $= \frac{6}{8} = \frac{3}{4}$

33. The number of balls numbered with a number greater than five is 3.
 $P(1 \text{ ball} > 5) = \frac{3}{8}$

34. The number of balls numbered with a number less than four is 3.
 $P(1 \text{ ball} < 4) = \frac{3}{8}$

35. The prime numbered balls are 2, 3, 5, and 7 (four balls).
 $P(\text{prime number}) = \frac{4}{8}$

36. $P(\text{odd or even}) = 1$

37. $P(\text{red and white and blue})$
 $= \frac{\cancel{2}^{1}}{\cancel{8}_{4}} \bullet \frac{3}{7} \bullet \frac{\cancel{3}^{1}}{\cancel{6}_{2}} = \frac{3}{56}$

38. $P(4 \text{ and } 5 \text{ and } 6)$
 $= \frac{1}{8} \bullet \frac{1}{7} \bullet \frac{1}{6} = \frac{1}{336}$

Math 807 Alternate Test Key

1. -14

2. 7

3. $3, 1, -1, -3, -5, \ldots$

4 $-3, -2, -1, 0$

5. $=$

6. $<$

7.

-4 -3 -2 -1 0 1 2 3 4

8.

-4 -3 -2 -1 0 1 2 3 4

9. $(4, -2)$

10. $(-3, 0)$

11. $-10 + 7 = -3$

12.
$$\begin{aligned} &14 + (-5) \\ &= 14 - 5 \\ &= 9 \end{aligned}$$

13.
$$\begin{aligned} &0 - (-11) \\ &= 0 + 11 \\ &= 11 \end{aligned}$$

14. $6 - 9 = -3$

15. $-4 \bullet (-2) = 8$

16. $3 \bullet (-7) = -21$

17. $8^2 = 8 \times 8 = 64$

18.
$$\begin{aligned} (-8)^{-2} &= \frac{1}{(-8)^2} \\ &= \frac{1}{(-8) \bullet (-8)} \\ &= \frac{1}{64} \end{aligned}$$

19. $-3 \div 0$ is undefined (cannot divide by zero)

20. $-27 \div 3 = -9$

21.
$$\begin{aligned} &a + bc \\ &= 4 + (-1) \bullet 2 \\ &= 4 - 2 \\ &= 2 \end{aligned}$$

22.
$$\begin{aligned} &ab - c \\ &= 4 \bullet (-1) - 2 \\ &= -4 - 2 \\ &= -6 \end{aligned}$$

23.
$$\begin{aligned} 5h^3 &= 5(3)^3 \\ &= 5 \bullet 3 \bullet 3 \bullet 3 \\ &= 135 \end{aligned}$$

24.
$$\begin{aligned} f^2 + h^g &= (-2)^2 + 3^0 \\ &= -2 \bullet (-2) + 1 \\ &= 4 + 1 \\ &= 5 \end{aligned}$$

25.
$$\begin{aligned} r + s &= 4 \\ 2 + s &= 4 \\ s &= 4 - 2 \\ s &= 2 \end{aligned}$$
$(2, 2)$

$$\begin{aligned} -3 + s &= 4 \\ s &= 4 - (-3) \\ s &= 4 + 3 \\ s &= 7 \end{aligned}$$
$(-3, 7)$

$$\begin{aligned} r + 0 &= 4 \\ r &= 4 \end{aligned}$$
$(4, 0)$

26. $y = -2x$
$y = -2 \bullet (-1)$
$y = 2$
(-1, 2)

$y = -2 \bullet 0$
$y = 0$
(0, 0)

$y = -2 \bullet 7$
$y = -14$
(7, -14)

27. $n = m + 3$ for $-5 < m < 1$
for $m = -5$: $n = -5 + 3$
$n = -2$
(-5, -2)

for $m = -4$: $n = -4 + 3$
$n = -1$
(-4, -1)

for $m = -3$: $n = -3 + 3$
$n = 0$
(-3, 0)

for $m = -2$: $n = -2 + 3$
$n = 1$
(-2, 1)

for $m = -1$: $n = -1 + 3$
$n = 2$
(-1, 2)

for $m = 0$: $n = 0 + 3$
$n = 3$
(0, 3)

for $m = 1$: $n = 1 + 3$
$n = 4$
(1, 4)

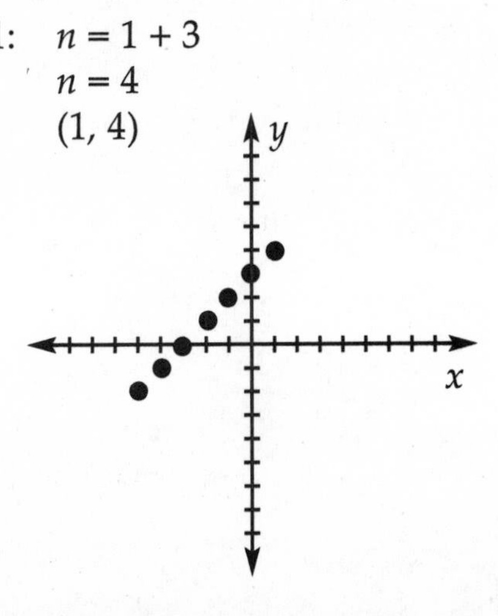

28. $b = a^2 - 3$ for $-3 \le a \le 2$
for $a = -3$: $b = (-3)^2 - 3$
$b = 9 - 3$
$b = 6$
(-3, 6)

for $a = -2$: $b = (-2)^2 - 3$
$b = 4 - 3$
$b = 1$
(-2, 1)

for $a = -1$: $b = (-1)^2 - 3$
$b = 1 - 3$
$b = -2$
(-1, -2)

for $a = 0$: $b = 0^2 - 3$
$b = 0 - 3$
$b = -3$
(0, -3)

for $a = 1$: $b = 1^2 - 3$
$b = 1 - 3$
$b = -2$
(1, -2)

for $a = 2$: $b = 2^2 - 3$
$b = 4 - 3$
$b = 1$
(2, 1)

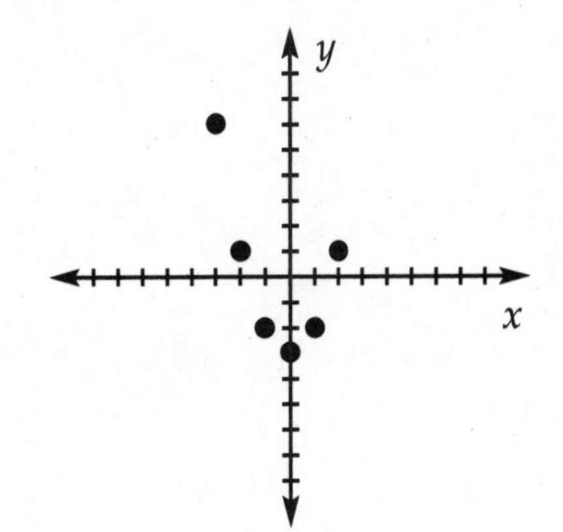

1. Arithmetic method:

```
   2    3.   3
 √5    42.   89

   4
43|1   42
  |1   29
  ------------
       |13   89
463    |13   89
       ----------
              0
```

Divide-and-average method:
$23^2 = 529$; $24^2 = 576$
The square root of 542.89 is between 23 and 24; therefore, try halfway in-between and estimate at 23.5.
Divide 23.5 into 542.89.

```
       23.1
 23.5)542.89
      470
       728
       705
        239
        235
          4
```

Average 23.5 and 23.1
$\frac{23.5 + 23.1}{2} = \frac{46.6}{2} = 23.3$

Check the answer by multiplying
23.3 x 23.3

```
   23.3
 x 23.3
    699
   699
  466
 542.89
```

The answer, therefore, is 23.3.
$A = l \bullet w$
$A = 11 \bullet 5$
$A = 165$ ft.^{2}

2. 11 ft. • 15 ft. = 165 ft.^{2}

3. a. P of
room = $(2 \bullet l) + (2 \bullet w)$
$P = (2 \bullet 15) + (2 \bullet 11)$
$P = 30 + 22$
$p = 52$ ft.

door = 30 in.
door = $\frac{30}{12} = 2.5$ ft.

length of baseboard
= perimeter of room – door
length of baseboard
= 52 – 2.5
= 49.5 ft.

b. cost = 1.20 49.5
cost = $59.40

4. $C = \pi \bullet d$
$C = 3.14 \bullet 24$
$C = 75.36$ ft.
or
$C = 2\,\pi\, r$
$r = \frac{24}{2} = 12$ ft.

$C = 2 \bullet 3.14 \bullet 12$
$C = 3.14 \bullet 24$
$C = 75.36$ ft.

5. $A = \pi \bullet r^2$
$r = \frac{24}{2} = 12$ ft.
$A = 3.14 \bullet (12)^2$
$A = 3.14 \bullet 144$
$A = 452.16$ ft.^{2}

6. $A = \frac{1}{2} \bullet b \bullet h$
$A = \frac{1}{2} \bullet 38 \bullet 16$
$A = 19 \bullet 16$
$A = 304\ m^2$

7. a. parallelogram
   b. $A = b \bullet h$
   $A = 28 \bullet 11$
   $A = 308 \text{ cm}^2$

8. $a^2 + b^2 = c^2$
   $a = 19$, $b = 20$, c is the third side
   $19^2 + 20^2 = c^2$
   $361 + 400 = c^2$
   $761 = c^2$
   $c = \sqrt{761}$
   $c \doteq 27.58622$
   $c \doteq 27.6$ yds.

9. $A = \frac{1}{2} \bullet b \bullet \text{h}$
   $A = \frac{1}{2} \bullet 19 \bullet 20$
   $A = 19 \bullet 10$
   $A = 190 \text{ yd.}^2$

10. a. P = sum of sides
    $P = 4 + 2.5 + 7 + 2$
    $P = 15.5$ in.
    b. $A = \frac{1}{2} \bullet h \bullet (b1 + b2)$
    $A = \frac{1}{2} \bullet 1.75 \bullet (7 + 4)$
    $A = \frac{1}{2} \bullet 1.75 \bullet 11$
    $A = \frac{1}{2} \bullet 11 \bullet 1.75$
    $A = 5.5 \bullet 1.75$
    $A = 9.625 \text{ in.}^2$

11. $1{,}125 \text{ ft.}^2 \bullet \frac{1 \text{ yd.} \bullet 1 \text{ yd.}}{3 \text{ ft.} \bullet 3 \text{ ft.}}$
    $= 1{,}125 \cancel{\text{ft.}^2} \bullet \frac{1 \text{ yd.}^2}{9 \cancel{\text{ft.}^2}}$
    $= 1{,}125 \backslash 9 \text{ yd.}^2$
    $= 125 \text{ yd.}^2$

12. $V = l \bullet w \bullet h$
    $V = 20 \bullet 6 \bullet 14$
    $V = 120 \bullet 14$
    $V = 1{,}680 \text{ in.}^3$

13. top $= 20 \bullet 6 = 120 \text{ in.}^2$
    bottom $= 20 \bullet 6 = 120 \text{ in.}^2$
    left side $= 14 \bullet 6 = 84 \text{ in.}^2$
    right side $= 14 \bullet 6 = 84 \text{ in.}^2$
    front $= 14 \bullet 20 = 280 \text{ in.}^2$
    back $= 14 \bullet 20 = 280 \text{ in.}^2$
    total surface area $= 968 \text{ in.}^2$

14. $V = \frac{1}{2} \bullet b_\Delta \bullet h_\Delta \bullet h_p$
    $V = \frac{1}{2} \bullet 32 \bullet 22 \bullet 54$
    $V = 16 \bullet 22 \bullet 54$
    $V = 19{,}008 \text{ ft.}^3$

15. Roof is two rectangles 27′ x 54′ each.
    $A_{\text{rectangle}} = l \bullet w$
    $A = 27 \bullet 54$
    $A = 1{,}458 \text{ ft.}^2$
    $A_{\text{roof}} = 2 \bullet 1{,}458$
    $A_{\text{roof}} = 2{,}916 \text{ ft.}^2$

16. a. pyramid
    b. $V = \frac{1}{3} B \bullet h$
    $B = s^2$
    $B = 22^2$
    $B = 484$
    $V = \frac{1}{3} \bullet 484 \bullet 18$
    $V = 484 \bullet 6$
    $V = 2{,}904 \text{ ft.}^3$

17. $V = \pi \bullet r^2 \bullet h$
    $r = \frac{8}{2} = 4$ in
    $V = 3.14 \bullet 4^2 \bullet 16$
    $V = 3.14 \bullet 16 \bullet 16$
    $V = 3.14 \bullet 256$
    $V = 803.84 \text{ in.}^3$

18. $TA = 2 \bullet \pi \bullet r \bullet (r + h)$

$r = \frac{8}{2} = 4$ in.

$TA = 2 \bullet 3.14 \bullet 4 \bullet (4 + 16)$

$TA = 2 \bullet 3.14 \bullet 4 \bullet (20)$

$TA = 6.28 \bullet 80$

$TA = 502.4$ in.^{2}

19. $V = \frac{4}{3} \bullet \pi \bullet r^3$

$r = \frac{36}{2} = 18$ ft.

$V = \frac{4}{3} \bullet 3.14 \bullet 18^3$

$V = \frac{4}{3} \bullet 3.14 \bullet 5{,}832$

$V = \frac{4}{3} \bullet 18{,}312.48$

$V = \frac{73{,}249.92}{3}$

$V = 24{,}416.64$ ft.^{3}

20. $S = 4 \bullet \pi \bullet r^2$

$r = \frac{36}{2} = 18$ ft.

$S = 4 \bullet 3.14 \bullet 18^2$

$S = 4 \bullet 3.14 \bullet 324$

$S = 12.56 \bullet 324$

$S = 4{,}069.44$ ft.^{2}

1. j
2. h
3. l
4. i
5. g
6. f
7. a
8. k
9. c
10. m
11. e
12. b
13. d
14. b
15. c
16. d
17. a
18. a
19. b
20. b
21. a
22. d
23. $6a + 36b - 72$
24. $12 \bullet 6 = 72$; $6 \bullet 12 = 72$;
    $72 = 72$, so the statement is true
25. $7q$
26. $6ab - 2y - 3y + 2ab + 6y$
    $= 6ab + 2ab - 2y - 3y + 6y$
    $= 8ab + y$
27. $$\begin{array}{r} 4x - 5 = 19 \\ 5 = 5 \\ \hline 4x = 24 \end{array}$$
    Solution: $x = 6$
    Check: $4 \bullet 6 - 5 = 19$
    $24 - 5 = 19$
    $19 = 19$
28. $$\begin{array}{r} 5a + 2 = 2a + 11 \\ -2 = -2 \\ \hline 5a = 2a + 9 \\ -2a = -2a \\ \hline 3a = 9 \end{array}$$
    Solution $a = 3$
    Check $5 \bullet 3 + 2 = 2 \bullet 3 + 11$
    $15 + 2 = 6 + 11$
    $17 = 17$
29. $$\begin{array}{r} \frac{1}{5}b + 2 = 8 \\ -2 = -2 \\ \hline \frac{1}{5}b = 6 \end{array}$$
    $5 \bullet \frac{1}{5}b = 5 \bullet 6$
    Solution: $b = 30$
    Check: $\frac{1}{5} \bullet 30 + 2 = 8$
    $6 + 2 = 8$
    $8 = 8$
30. $A = \frac{1}{2} \bullet 6 \bullet 8$
    $= \frac{1}{2} \bullet \frac{6}{1} \bullet \frac{8}{1}$
    $= 24''$

31. Nine times a number plus six is the difference of three times a number minus the number.

32. Let n = the number

$n + 21 = 87$

$n = 87 - 21$

$n = 66$

33. Let n = one integer

$n + 1$ = next consecutive integer

$n + n + 1 = 25$

$2n + 1 = 25$

$2n = 25 - 1$

$2n = 24$

$n = \frac{24}{2}$

$n = 12$

$n + 1 = 13$

The integers are 12 and 13.

Check: $\begin{array}{r} 12 \\ +\ 13 \\ \hline 25 \end{array}$

34. Let r = revolutions in 20 minutes

$\frac{400}{1} = \frac{r}{20}$

$r \bullet 1 = 400 \bullet 20$

$r = 8{,}000$ revolutions

35. Let a = grandmother's age

$\frac{1}{7} \bullet a = 11$

$\frac{a}{7} = 11$

$a = 11 \bullet 7$

$a = 77$ years old

1. r
2. n
3. p
4. l
5. a
6. i
7. q
8. m
9. t
10. d
11. h
12. c
13. b
14. f
15. j
16. u
17. o
18. s
19. k
20. g
21. 4,000 + 100 + 60 + 8
22. $3\frac{7}{8} \bullet 1\frac{2}{3}$

    $= \frac{31}{8} \bullet \frac{5}{3}$

    $= \frac{155}{24} = 6\frac{11}{24}$

23. $4\frac{1}{4} \div 2\frac{2}{3}$

    $= \frac{17}{4} \div \frac{8}{3}$

    $= \frac{17}{4} \bullet \frac{3}{8}$

    $= \frac{51}{32} = 1\frac{19}{32}$

24. 1 =

1's
1

= 1_2

2 =

2's	1's
1	0

= 10_2

3 =

2's	1's
1	1

= 11_2

4 =

4's	2's	1's
1	0	0

= 100_2

5 =

4's	2's	1's
1	0	1

= 101_2

6 =

4's	2's	1's
1	1	0

= 110_2

7 =

4's	2's	1's
1	1	1

= 111_2

8 =

8's	4's	2's	1's
1	0	0	0

= 1000_2

25. $4d - 1 = 9 - d$

    $1 = 1$ (Addition)

    $4d = 10 - d$

    $d = d$ (Addition)

    $5d = 10$

    $\frac{5d}{5} = \frac{10}{5}$ (Division)

    $d = 2$

26. $2y - y + 6y + 4 = 12 - y$

(Combine terms) $7y + 4 = 12 - y$

(Subtraction) $-4 = -4$

$7y = 8 - y$

(Addition) $y = y$

$8y = 8$

(Division) $\frac{8y}{8} = \frac{8}{8}$

$y = 1$

27. 22¢ = 0.22

12 dozen = 12 • 12 = 144

0.22 • 144 = 31.68

The cost is $31.68

28. $4 = 2 \bullet 2$

$28 = 2 \bullet 2 \bullet 7$

$LCM = 2 \bullet 2 \bullet 7 = 28$

29.
```
   52.18
     549
   46962
  20872
 26090
 2864.682
```

30.
```
        5,300.
 1.71)9063.00.
      855
       513
       513
         0
```

31. $54 = 2 \bullet 3 \bullet 3 \bullet 3$

32. $\frac{1}{2} = 50\%$

0.04: Move the decimal point two places to the right and add the percent
0.04 = 4%

1: Move the decimal point two places to the right and add the percent sign.
1. = 100%

32 cont.

0.0008: Move the decimal point two places to the right and add the percent sign.
0.0008 = 0.08%

2.6: Move the decimal point two places to the right and add the percent sign.
2.6 = 260%

33. $4\frac{1}{8} \div 7$

$= \frac{33}{8} \div \frac{7}{1}$

$= \frac{33}{8} \bullet \frac{1}{7}$

$= \frac{33}{56}$

34. $A = \pi r^2; \pi = 3.14$

$A = 3.14 \bullet (2.5)^2$

$A = 314 \bullet 6.25$

$A = 19.625$ or 19.6 sq. cm

35. $(4x + 3y)\ (5a - 4d)$

$= 4x \bullet 5a + 4x\,(-4d) + 3y \bullet 5a + 3y \bullet (-4d)$

$= 20xa - 16xd + 15ya - 12yd$

36. Write the whole numbers in order from largest to smallest first. The whole numbers are 1, 8, $\frac{14}{7}$ (equals 2), and $\frac{0}{2}$ (equals 0).

$8, \frac{14}{7}, 1, \frac{0}{2}$

The fractions $\frac{2}{3}, \frac{4}{5}, 2\frac{3}{4}$, and $\frac{7}{8}$ are left. Since $2\frac{3}{4}$ is the largest fraction (it is a mixed number), write it where it belongs in the list.

$8, 2\frac{3}{4}, \frac{14}{7}, 1, \frac{0}{2}$

36. cont.
Now find the lowest common denominator (LCD) of the three fractions left.

$\frac{2}{3}$: $3 = 1 \bullet 3$

$\frac{4}{5}$: $5 = 1 \bullet 5$

$\frac{7}{8}$: $8 = 2 \bullet 2 \bullet 2$

$\text{LCD} = 3 \bullet 5 \bullet 2 \bullet 2 \bullet 2 = 120$

Raise each fraction to higher terms with a denominator of 120.

$3 \bullet 40 = 120$, so $\frac{2}{3} = \frac{2 \bullet 40}{3 \bullet 40} = \frac{80}{120}$

$5 \bullet 24 = 120$, so $\frac{4}{5} = \frac{4 \bullet 24}{5 \bullet 24} = \frac{96}{120}$

$8 \bullet 15 = 120$, so $\frac{7}{8} = \frac{7 \bullet 15}{8 \bullet 15} = \frac{105}{120}$

The fractions $\frac{2}{3}$, $\frac{4}{5}$, and $\frac{7}{8}$ arranged in order from largest to smallest are $\frac{7}{8}$ (or $\frac{105}{120}$), $\frac{4}{5}$ (or $\frac{96}{120}$), and $\frac{2}{3}$ (or $\frac{80}{120}$).
The complete list of fractions arranged in order from largest to smallest is 8, 2 $\frac{3}{4}$, $\frac{14}{7}$, 1, $\frac{7}{8}$, $\frac{4}{5}$, $\frac{2}{3}$, $\frac{0}{2}$.

37. To write 0.075 as a percent, move the decimal point two places to the right and add the percent sign.
$0.075 = 7.5\%$ (a)

38. $12 = 2 \bullet 2 \bullet 3$
$45 = 3 \bullet 3 \bullet 5$
$\text{LCD} = 2 \bullet 2 \bullet 3 \bullet 3 \bullet 5 = 180$ (a)

39.
$$\begin{aligned} \% \bullet B &= N \\ 24\% &= 0.24 \\ 0.24 \bullet B &= 48 \\ \frac{0.24 \bullet B}{0.24} &= \frac{48}{0.24} \\ B &= \frac{48}{0.24} \\ B &= 200 \text{ (b)} \end{aligned}$$

40. 10^7 = a 1 followed by 7 zeros, or 10,000,000. $4 \bullet 10{,}000{,}000 = 40{,}000{,}000$ (b)

41. c

42. The mode is the numeral that occurs the most frequently, which is 5 (occurs three times) (b).

LIFEPAC

ANSWER KEYS

NUMBER SYMBOLS
Digits
0
1
2
3
4
6
7
8
9

NUMBER-OPERATION SYMBOLS		
Symbol	Meaning	Example
$+$	addition	$1 + 2 = 3$
$-$	subtraction	$3 - 2 = 1$
x	multiplication	$2 \times 3 = 6$ 5 $2 \bullet 3 = 6$
( )		$(2)\ (3) = 6$
$\div$	division	$6 \div 3 = 2$

SET-OPERATIONS SYMBOLS		
$\cup$	union	$A \cup B = M$
$\cap$	intersection	$A \cap B = K$

PUNCTUATION SYMBOLS		
Symbol	Meaning	Example
.	decimal point	$\pi = 3.1416$
,	comma	$A = \{3, 4, 5\}$
( )	parenthesis	$2 + (3 + 1) = 6$
[ ]	brackets	$2 + [1 + (3 + 1)] - 7$
{ }	braces	$\{1, 2\} = \{2, 1\}$

NUMBER-RELATION SYMBOLS		
Symbol	Meaning	Example
$=$	is equal to	$2 + 3 = 5$
$\neq$	is not equal to	$8 + 1\ \ 7$
$<$	is less than	$3 < 4$
$\not<$	is not less than	$3\ \ 2$
$>$	is greater than	$9 > 7$
$\not>$	is not greater than	$7 \not> 9$
$\leq$	is less than or equal to	$4 \leq 4$
$\not\leq$	is not less than or equal to	$4 \not\leq 3$
$\geq$	is greater than or equal to	$6 \geq 5$
$\not\geq$	is not greater than or equal to	$5 \not\geq 6$

SET-RELATION SYMBOLS		
Symbol	Meaning	Example
$\in$	is an element of	$1 \in \{1, 2\}$
$\notin$	is not an element of	$3 \notin \varnothing$
$\subset$	is a subset of	$\varnothing$ CA
$\not\subset$	is not a subset of	$\{1, 2\} \not\subset \{2,3\}$
$\leftrightarrow$	is equivalent to	$\{1, 2\} \leftrightarrow \{a, b\}$
$\not\leftrightarrow$	is not equivalent to	$A \not\leftrightarrow 3.4$
$=$	is equal to	$A = A$
$\neq$	is not equal to	$\{1, 2\} \neq \varnothing$

I. SECTION ONE

1.1 thirty-six

1.2 fourteen

1.3 seventy-two

1.4 one hundred two

1.5 one hundred eleven

1.6 two hundred thirty-six

1.7 one thousand, four hundred

1.8 two thousand, six hundred ten

1.9 three thousand, seven hundred twenty-six

1.10 fourteen thousand, six hundred twenty-five

1.11 twenty-five thousand, one hundred eighteen

1.12 four hundred twenty-five thousand, six hundred sixty-five

1.13 six hundred fourteen thousand, eight hundred sixty-two

1.14 13

1.15 27

1.16 59

1.17 206

1.18 324

1.19 588

1.20 2,645

1.21 7,066

1.22 15,830

1.23 27,421

1.24 133,711

1.25 325,262

1.26 1,500,899

1.27 3,437,171

1.28 400 + 10 + 8

1.29 700 + 20 + 6

1.30 5,000 + 400 + 60 + 8

1.31 8,000 + 100 + 70 + 5

1.32 6,429

1.33 38,175

1.34 8,736

1.35 27,529

1.36 600 + 20 + 9

1.37 2,000 + 100 + 40 + 7

1.38 a. 800,000,000
b. eight hundred million

1.39 a. 5,000,000,000
b. five billion

1.40 6,000,000,000

1.41 4,000,000,000,000

1.42 5,000,000,000,000,000

1.43 30

1.44 90

1.45 200

1.46 900

1.47 3,000

1.48 2,000

1.49 30,000

1.50 80,000

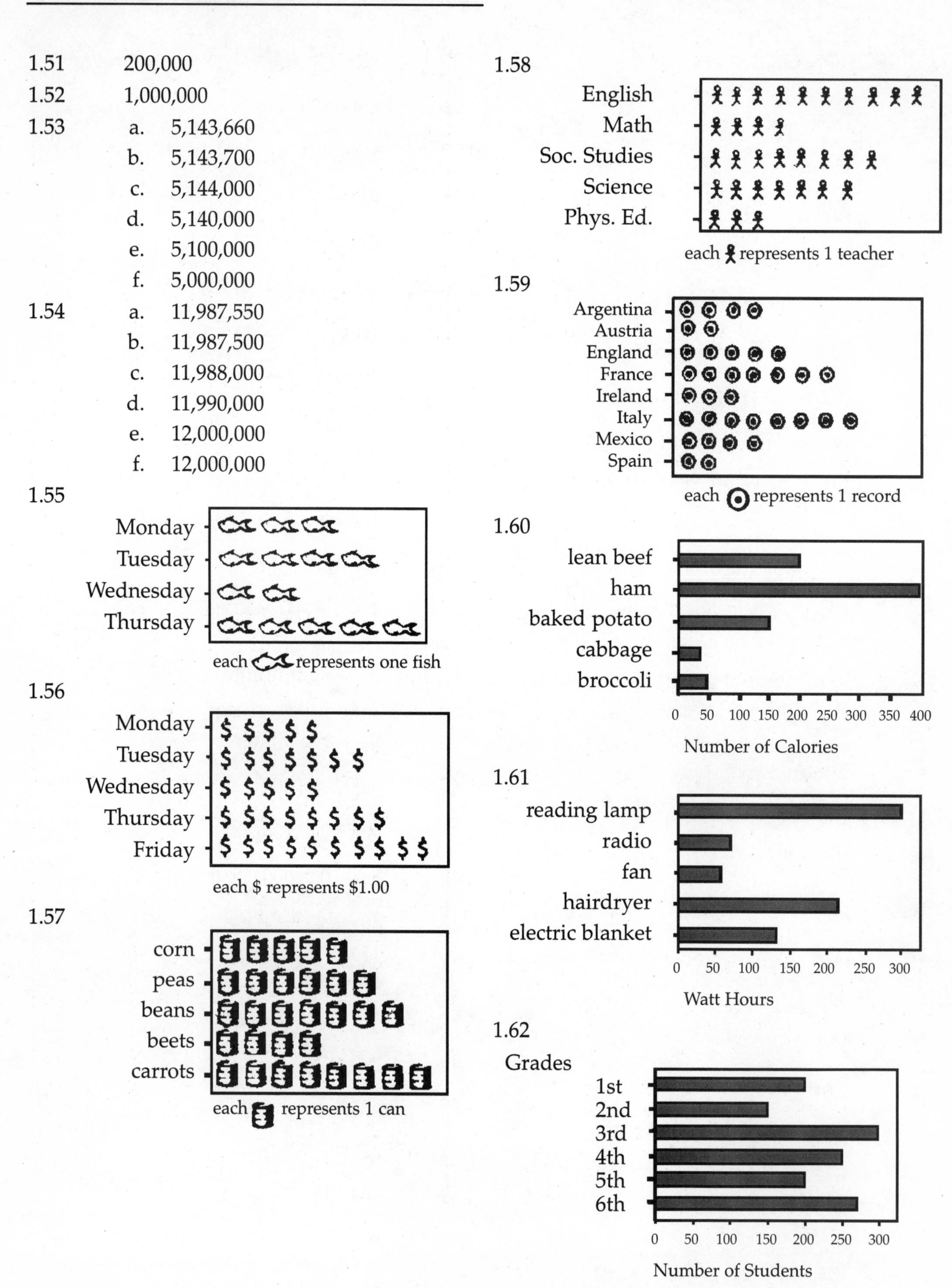
1.51 200,000
1.52 1,000,000
1.53 a. 5,143,660
b. 5,143,700
c. 5,144,000
d. 5,140,000
e. 5,100,000
f. 5,000,000
1.54 a. 11,987,550
b. 11,987,500
c. 11,988,000
d. 11,990,000
e. 12,000,000
f. 12,000,000
1.55
Monday
Tuesday
Wednesday
Thursday
each represents one fish
1.56
Monday
Tuesday
Wednesday
Thursday
Friday
each $ represents $1.00
1.57
corn
peas
beans
beets
carrots
each represents 1 can
1.58
English
Math
Soc. Studies
Science
Phys. Ed.
each represents 1 teacher
1.59
Argentina
Austria
England
France
Ireland
Italy
Mexico
Spain
each represents 1 record
1.60
lean beef
ham
baked potato
cabbage
broccoli
0 50 100 150 200 250 300 350 400
Number of Calories
1.61
reading lamp
radio
fan
hairdryer
electric blanket
0 50 100 150 200 250 300
Watt Hours
1.62
Grades
1st
2nd
3rd
4th
5th
6th
0 50 100 150 200 250 300
Number of Students

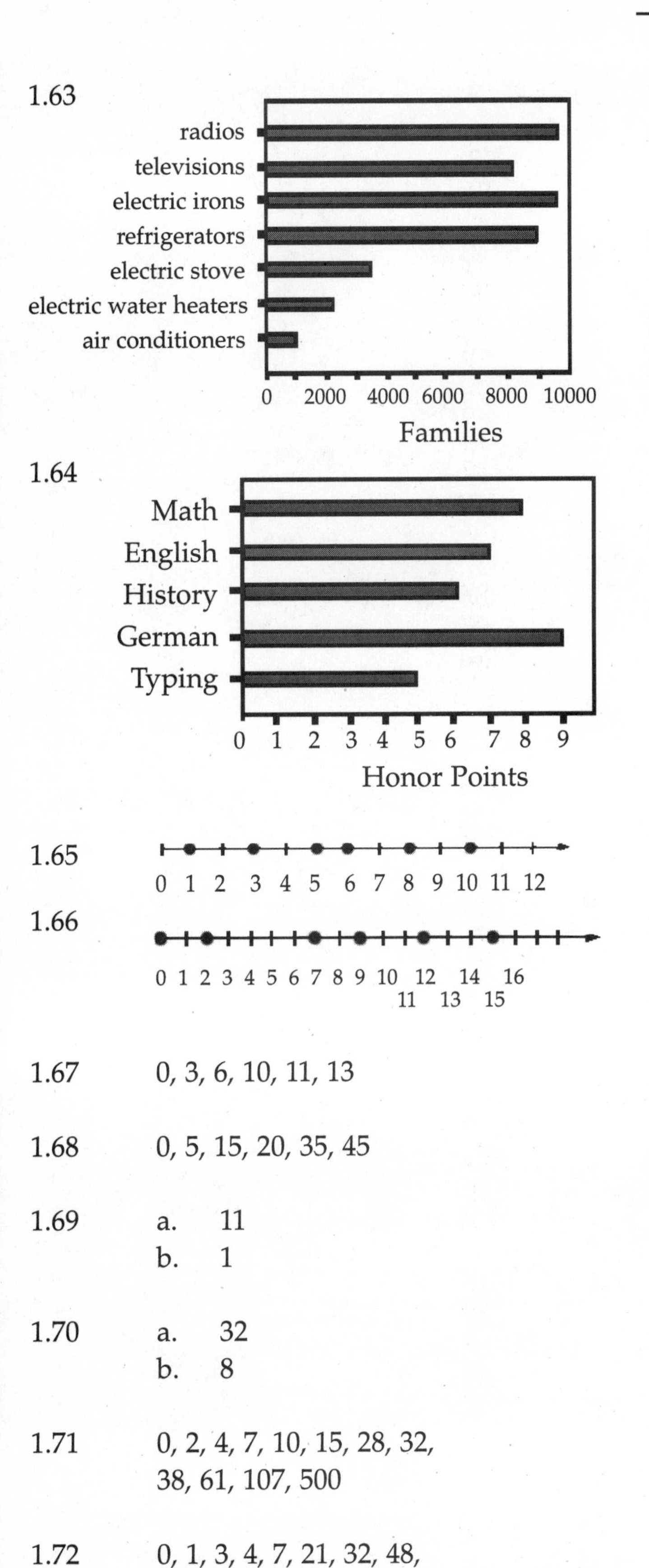

1.67 0, 3, 6, 10, 11, 13

1.68 0, 5, 15, 20, 35, 45

1.69 a. 11
b. 1

1.70 a. 32
b. 8

1.71 0, 2, 4, 7, 10, 15, 28, 32, 38, 61, 107, 500

1.72 0, 1, 3, 4, 7, 21, 32, 48, 140, 141, 316, 566

1.73 0 2 4 6 8 10 12 14 16 18 20

1.74 0 3 6 9 12 15 18 21 24 27

II. SECTION TWO

2.1 a. addend
b. addition sign
c. addend
d. equal sign
e. sum

2.2 a. addend
b. addition sign
c. addend
d. addition sign
e. addend
f. equal sign
g. sum

2.3 Example:
6 + 7 + 5 = 18

2.4 Example:
4 + 8 + 9 + 10 = 31

2.5 Example:
3 + 4 + 4 = 11

2.6 Example:
2 + 2 + 4 + 7 = 15

2.7

+	1	6	3	4	2	0	9	7	5	8
3	4	9	6	7	5	3	12	10	8	11
7	8	13	10	11	9	7	16	14	12	15
8	9	14	11	12	10	8	17	15	13	16
1	2	7	4	5	3	1	10	8	6	9
0	1	6	3	4	2	0	9	7	5	8
4	5	10	7	8	6	4	13	11	9	12
2	3	8	5	6	4	2	11	9	7	10
9	10	15	12	13	11	9	18	16	14	17
6	7	12	9	10	8	6	15	13	11	14
5	6	11	8	9	7	5	14	12	10	13

Problem	Answer
2.8	24 31 55
2.9	51 47 98
2.10	15 78 93
2.11	42 37 79
2.12	92 40 132
2.13	59 27 86
2.14	86 88 174
2.15	13 56 69
2.16	7 13 26 46
2.17	8 21 74 103
2.18	10 86 42 138
2.19	22 53 68 143
2.20	61 5 38 104
2.21	48 60 32 140
2.22	28 19 57 104
2.23	77 87 97 261
2.24	7 6 81 46 140
2.25	10 21 46 89 166
2.26	42 39 71 96 248
2.27	18 28 38 48 132
2.28	113
2.29	139
2.30	131
2.31	110
2.32	137

2.33	173
2.34	164
2.35	138
2.36	100
2.37	250
2.38	260
2.39	241
2.40	367
2.41	987
2.42	1,405
2.43	1,346
2.44	1,344
2.45	904
2.46	1,887
2.47	911
2.48	699
2.49	178
2.50	1,473
2.51	1,127
2.52	951
2.53	1,332
2.54	1,847
2.55	1,673
2.56	705
2.57	1,399
2.58	2,565
2.59	2,775
2.60	418 667 1,085
2.61	262 475 737
2.62	287 776 1,063
2.63	971 26 400 1,397
2.64	829 941 72 1,842
2.65	714 866 278 1,858
2.66	104 342 566 47 1,059
2.67	366 571 147 835 1,919
2.68	781 668 147 544 2,140
2.69	9,006
2.70	9,879
2.71	9,992
2.72	12,122

2.73 5,202

2.74 9,857

2.75 23,049

2.76 51,879

2.77 426,753

2.78 828,948

2.79 953,910

2.80 163,123

2.81 10,940,811

2.82 9,799,087

2.83 27,249,421

2.84 68,295

2.85 16,347

2.86 19,301

2.87 20,418

2.88 28,599

2.89 989

2.90 13,366

2.91 27,814,639

2.92 7,069

2.93 32,430,528

2.94 66,591,587

2.95
a. minuend
b. minus sign
c. subtrahend
d. equal sign
e. difference

2.96
a. minuend
b. minus sign
c. subtrahend
d. equal sign
e. difference

2.97 56 – 45 = 11

2.98 496 – 360 = 136

2.99

–	2	6	1	4	8	0	5	7	3	9
4	8	2	7	0	4	6	1	3	9	5
1	1	5	0	3	7	9	4	6	2	8
6	6	0	5	8	2	4	9	1	7	3
9	3	7	2	5	9	1	6	8	4	0
8	4	8	3	6	0	2	7	9	5	1
0	2	6	1	4	8	0	5	7	3	9
2	0	4	9	2	6	8	3	5	1	7
5	7	1	6	9	3	5	0	2	8	4
7	5	9	4	7	1	3	8	0	6	2
3	9	3	8	1	5	7	2	4	0	6
	12	16	11	14	18	10	15	17	13	19

2.100 21

2.101 22

2.102 212

2.103 111

2.104 1,111

2.105 1,112

2.106 21,211

2.107 21,221

2.108 111,111

2.109 222,222

2.110 1,111,222

2.111 211

2.112
$$\begin{array}{r} 827 \\ \underline{616} \\ 211 \end{array}$$

2.113
$$\begin{array}{r} 6,700,667 \\ \underline{600,546} \\ 6,100,121 \end{array}$$

2.114
$$\begin{array}{r} 4,862,789 \\ \underline{3,051,678} \\ 1,811,111 \end{array}$$

2.115
$$\begin{array}{r} 5,664 \\ \underline{553} \\ 5,111 \end{array}$$

2.116
$$\begin{array}{r} 381,486 \\ \underline{270,371} \\ 111,115 \end{array}$$

2.117
$$\begin{array}{r} 14,614,016 \\ \underline{3,503,005} \\ 11,111,011 \end{array}$$

2.118
$$\begin{array}{r} 990 \\ \underline{260} \\ 730 \end{array}$$

2.119
$$\begin{array}{r} 53,596 \\ \underline{42,412} \\ 11,184 \end{array}$$

2.120
$$\begin{array}{r} 1,841,465 \\ \underline{730,014} \\ 1,111,451 \end{array}$$

2.121
$$\begin{array}{r} 3,347 \\ \underline{2,247} \\ 1,100 \end{array}$$

2.122 6

2.123 6

2.124 17

2.125 27

2.126 76

2.127 39

2.128 88

2.129 279

2.130 188

2.131 908

2.132 1,859

2.133 4,909

2.134 4,188

2.135 7,445

2.136 173

2.137 2,819

2.138
$$\begin{array}{r} 14,614 \\ \underline{12,716} \\ 1,898 \end{array}$$

2.139
$$\begin{array}{r} 18,112 \\ \underline{14,783} \\ 3,329 \end{array}$$

2.140
$$\begin{array}{r} 27,814 \\ \underline{19,128} \\ 8,686 \end{array}$$

2.141
$$\begin{array}{r} 91,450 \\ \underline{85,642} \\ 5,808 \end{array}$$

2.142
$$\begin{array}{r} 28,411 \\ \underline{9,526} \\ 18,885 \end{array}$$

2.143
$$\begin{array}{r} 33,486 \\ \underline{3,399} \\ 30,087 \end{array}$$

2.144 51,101
3,846
47,255

2.145 70,000
9,999
60,001

2.146 143,142
132,263
10,879

2.147 314,281
126,882
187,399

2.148 866,100
47,286
818,814

2.149 541,862
352,973
188,889

2.150
a. multiplier
b. times sign
c. multiplicand
d. equal sign
e. product

2.151
a. multiplier
b. times sign
c. multiplicand
d. equal sign
e. product

2.152 5 x 9 = 45

2.153 4 x 5 = 20

2.154 4 + 4 + 4 + 4 + 4 + 4 = 24
or 6 + 6 + 6 + 6 = 24

2.155 5 + 5 + 5 +5 + 5 + 5 + 5 + 5 = 40
or 8 + 8 + 8 + 8 + 8 = 40

2.156 8 + 8 + 8 + 8 + 8 + 8 + 8 + 8 + 8 = 72
or 9 + 9 + 9 + 9 + 9 + 9 + 9 + 9 = 72

2.157 6 + 6 + 6 + 6 + 6 + 6 + 6 = 42
or 7 + 7 + 7 + 7 + 7 + 7 = 42

2.158 36

2.159 18

2.160

X	4	6	0	3	5	9	2	1	7	8
5	20	30	0	15	25	45	10	5	35	40
1	4	6	0	3	5	9	2	1	7	8
6	24	36	0	18	30	54	12	6	42	48
8	32	48	0	24	40	72	16	8	56	64
2	8	12	0	6	10	18	4	2	14	16
0	0	0	0	0	0	0	0	0	0	0
4	16	24	0	12	20	36	8	4	28	32
7	28	42	0	21	35	63	14	7	49	56
9	36	54	0	27	45	81	18	9	63	72
3	12	18	0	9	15	27	6	3	21	24

2.161 96

2.162 77

2.163 117

2.164 147

2.165 270

2.166 76

2.167 196

2.168 498

2.169 231

2.170 448

2.171 84

2.172 0

2.173
86
10
00
86
860

2.174
11
11
11
11
121

2.175
12
12
24
12
144

2.176
14
12
28
14
168

2.177
21
15
105
21
315

2.178
34
18
272
34
612

2.179
26
24
104
52
624

2.180
14
31
14
42
434

2.181
60
42
120
240
2,520

2.182
47
56
282
235
2,632

2.183
92
81
92
736
7,452

2.184
47
47
329
188
2,209

2.185
76
76
456
532
5,776

2.186
65
56
390
325
3,640

2.187
93
39
837
279
3,627

2.188
98
97
686
882
9,506

2.189
76
28
608
152
2,128

2.190
88
66
528
528
5,808

2.191	852
2.192	1,728
2.193	3,115
2.194	6,651
2.195	1,768
2.196	3,580
2.197	11,896
2.198	10,712
2.199	26,568
2.200	291,375
2.201	563,013

2.202
136
10
000
136
1,360

2.203
247
13
741
247
3,211

2.204
385
26
2310
770
10,010

2.205
815
37
5705
2445
30,155

2.206
476
48
3808
1904
22,848

2.207
1,817
50
0000
9085
90,850

2.208
2,184
70
0000
15288
152,880

2.209
3,817
27
26719
7634
103,059

2.210
5,866
48
46928
23464
281,568

2.211
7,125
91
7125
64125
648,375

2.212
5,777
85
28885
46216
491,045

2.213
237
114
948
237
237
27,018

2.214
414
206
2484
000
828
85,284

2.215
```
    591
    321
   ----
    591
   1182
  1773
  -------
  189,711
```

2.216
```
    800
    500
   ----
    000
   000
  4000
  -------
  400,000
```

2.217
```
    148
    800
   ----
    000
   000
  1184
  -------
  118,400
```

2.218
```
    725
    639
   ----
   6525
  2175
 4350
 -------
 463,275
```

2.219
```
    776
    839
   ----
   6984
  2328
 6208
 -------
 651,064
```

2.220
```
    778
    489
   ----
   7002
  6224
 3112
 -------
 380,442
```

2.221
```
    999
    100
   ----
    000
   000
  999
  ------
  99,900
```

2.222
```
   1,486
     106
   -----
    8916
   0000
  1486
  -------
  157,516
```

2.223
```
   2,007
     286
   -----
   12042
  16056
  4014
  -------
  574,002
```

2.224
```
     4,116
       313
     -----
     12348
     4116
   12348
   ---------
   1,288,308
```

2.225
```
     5,257
       527
     -----
     36799
    10514
   26285
   ---------
   2,770,439
```

2.226
```
     7,412
       689
     -----
     66708
    59296
   44472
   ---------
   5,106,868
```

2.227
```
     5,000
       800
     -----
      0000
     0000
   40000
   ---------
   4,000,000
```

2.228
```
     1,146
     2,810
     -----
      0000
     1146
    9168
   2292
   ---------
   3,220,260
```

2.229
2,014
× 3,126
12084
4028
2014
6042
6,295,764

2.230
3,428
× 5,867
23996
20568
27424
17140
20,112,076

2.231
5,000
× 7,000
0000
0000
0000
35000
35,000,000

2.232
6,141
× 2,842
12282
24564
49128
12282
17,452,722

2.233
14,660
× 3,147
102620
58640
14660
43980
46,135,020

2.234
17,817
× 4,814
71268
17817
142536
71268
85,771,038

2.235
a. dividend
b. division sign
c. divisor
d. equal sign
e. quotient

2.236
a. dividend
b. division sign
c. divisor
d. equal sign
e. quotient

2.237 $51 \div 5 = 10\frac{1}{5}$ or $5\overline{)51}$ with quotient $10\frac{1}{5}$

2.238 $75 \div 8 = 9\frac{3}{8}$ or $8\overline{)75}$ with quotient $9\frac{3}{8}$

2.239 5

2.240 8

2.241 9

2.242 6

2.243 6

2.244 7

2.245 8

2.246 7

2.247 7

2.248 6

2.249 $5\overline{)47}$ = 9, $9 = 9\frac{2}{5}$
45
2

2.250 $8\overline{)74}$ = 9, $9 = 9\frac{2}{8} = 9\frac{1}{4}$
72
2

2.251
$$\begin{array}{r} 5 = 5\tfrac{2}{9} \\ 9\overline{)47} \\ \underline{45} \\ 2 \end{array}$$

2.252
$$\begin{array}{r} 4 = 4\tfrac{2}{4} = 4\tfrac{1}{2} \\ 4\overline{)18} \\ \underline{16} \\ 2 \end{array}$$

2.253
$$\begin{array}{r} 4 = 4\tfrac{5}{6} \\ 6\overline{)29} \\ \underline{24} \\ 5 \end{array}$$

2.254
$$\begin{array}{r} 20 \\ 5\overline{)100} \\ \underline{10} \\ 00 \\ \underline{00} \\ 0 \end{array}$$

2.255
$$\begin{array}{r} 21 \\ 4\overline{)84} \\ \underline{8} \\ 04 \\ \underline{4} \\ 0 \end{array}$$

2.256
$$\begin{array}{r} 24 \\ 6\overline{)144} \\ \underline{12} \\ 24 \\ \underline{24} \\ 0 \end{array}$$

2.257
$$\begin{array}{r} 16 \\ 8\overline{)128} \\ \underline{8} \\ 48 \\ \underline{48} \\ 0 \end{array}$$

2.258
$$\begin{array}{r} 14 \\ 7\overline{)98} \\ \underline{7} \\ 28 \\ \underline{28} \\ 0 \end{array}$$

2.259
$$\begin{array}{r} 34 \\ 3\overline{)102} \\ \underline{9} \\ 12 \\ \underline{12} \\ 0 \end{array}$$

2.260
$$\begin{array}{r} 23 = 23\tfrac{2}{4} = 23\tfrac{1}{2} \\ 4\overline{)94} \\ \underline{8} \\ 14 \\ \underline{12} \\ 2 \end{array}$$

2.261
$$\begin{array}{r} 15 = 15\tfrac{1}{7} \\ 7\overline{)106} \\ \underline{7} \\ 36 \\ \underline{35} \\ 1 \end{array}$$

2.262
$$\begin{array}{r} 34 = 34\tfrac{8}{9} \\ 9\overline{)314} \\ \underline{27} \\ 44 \\ \underline{36} \\ 8 \end{array}$$

2.263
$$\begin{array}{r} 89 = 89\tfrac{1}{5} \\ 5\overline{)446} \\ \underline{40} \\ 46 \\ \underline{45} \\ 1 \end{array}$$

2.264
$$\begin{array}{r} 757 \\ 2\overline{)1{,}514} \\ \underline{14} \\ 11 \\ \underline{10} \\ 14 \\ \underline{14} \\ 0 \end{array}$$

2.265
$$\begin{array}{r} 566 \\ 4\overline{)2{,}264} \\ \underline{20} \\ 26 \\ \underline{24} \\ 24 \\ \underline{24} \\ 0 \end{array}$$

2.266

$690 = 690\frac{3}{6} = 690\frac{1}{2}$

6)4,143
36
54
54
03
0
3

2.267

$3{,}880 = 3{,}880\frac{2}{3}$

3)11,642
9
26
24
24
24
02
0
2

2.268

$4{,}496 = 4{,}496\frac{5}{7}$

7)31,477
28
34
28
67
63
47
42
5

2.269

30
10)300
30
00
00
0

2.270

24
12)288
24
48
48
0

2.271

11
11)121
11
11
11
0

2.272

13
13)169
13
39
39
0

2.273

20
20)400
40
00
00
0

2.274

20
31)620
62
00
00
0

2.275

2
28)56
56
0

2.276

21
25)525
50
25
25
0

2.277

15
15)225
15
75
75
0

2.278

23
17)391
34
51
51
0

2.279

14
23)322
23
92
92
0

2.280

12
56)672
56
112
112
0

2.281

58
14)812
70
112
112
0

2.282

33
22)726
66
66
66
0

2.283

15
45)675
45
225
225
0

2.284

10 = 10 $\frac{16}{24}$ = 10 $\frac{2}{3}$
24)256
24
16
00
16

2.285

21 = 21 $\frac{9}{17}$
17)366
34
26
17
9

2.286

14 = 14 $\frac{2}{41}$
41)576
41
166
164
2

2.287

18 = 18 $\frac{10}{33}$
33)604
33
274
264
10

2.288

10 = 10 $\frac{5}{50}$ = 10 $\frac{1}{10}$
50)505
50
05
0
5

2.289

89
14)1,246
112
126
126
0

2.290

75
27)2,025
189
135
135
0

2.291

59
41)2,419
205
369
369
0

2.292

23
56)1,288
112
168
168
0

2.293

74 = 74 $\frac{1}{29}$
29)2,147
203
117
116
1

2.294

63 = 63 $\frac{2}{35}$
35)2,207
210
107
105
2

2.295

563 = 563 $\frac{1}{12}$
12)6,757
60
75
72
37
36
1

2.296
756
24)18,144
168
134
120
144
144
0

2.297
476
49)23,324
196
372
343
294
294
0

2.298
$1,237 = 1,237\frac{2}{56} = 1,237\frac{1}{28}$
56)69,274
56
132
112
207
168
394
392
2

2.299
23
122)2,806
244
366
366
0

2.300
45
221)9,945
884
1105
1105
0

2.301
19
412)7,828
412
3708
3708
0

2.302
42
238)9,996
952
476
476
0

2.303
$53 = 53\frac{2}{135}$
135)7,157
675
407
405
2

2.304
$34 = 34\frac{3}{288} = 34\frac{1}{96}$
288)9,795
864
1155
1152
3

2.305
$49 = 49\frac{4}{186} = 49\frac{2}{93}$
186)9,118
744
1678
1674
4

2.306
75
206)15,450
1442
1030
1030
0

2.307
344
162)55,728
486
712
648
648
648
0

2.308
34
875)29,750
2625
3500
3500
0

2.309
```
        47
811 )38,117
     3244
      5677
      5677
         0
```

2.310
```
        36
405 )14,580
     1215
      2430
      2430
         0
```

2.311
```
       475
112 )53,200
     448
      840
      784
       560
       560
         0
```

2.312
```
       422
133 )56,126
     532
      292
      266
       266
       266
         0
```

2.313
```
       228
244 )55,632
     488
      683
      488
      1952
      1952
         0
```

2.314
```
        53 = 53 3/558 = 53 1/186
558 )29,577
     2790
      1677
      1674
         3
```

2.315
```
        87 = 87 4/452 = 87 1/113
452 )39,328
     3616
      3168
      3164
         4
```

2.316
```
        75 = 75 6/336 = 75 1/56
336 )25,206
     2352
      1686
      1680
         6
```

2.317
```
        50 = 50 249/752
752 )37,849
     3760
       249
       000
       249
```

2.318
```
       1,500
105 )157,500
     105
      525
      525
        00
        00
         00
         00
          0
```

2.319
```
       1,616
204 )329,664
     204
     1256
     1224
       326
       204
      1224
      1224
         0
```

2.320
```
        689
413 )284,557
     2478
      3675
      3304
       3717
       3717
          0
```

2.321
```
        415
378 )156,870
     1512
       567
       378
      1890
      1890
         0
```

2.322

```
          907
873 )791,811
     7857
      611
      000
     6111
     6111
        0
```

2.323

```
          871
782 )681,122
     6256
      5552
      5474
        782
        782
          0
```

2.324

```
          316
568 )179,488
     1704
      908
      568
     3408
     3408
        0
```

2.325

```
              248 = 248 2,464/4,521
4,521 )1,123,672
        9042
        21947
        18084
         38632
         36168
          2464
```

$248 = 248\frac{2,464}{4,521}$

2.326

```
        4,156
238 )989,128
     952
      371
      238
     1332
     1190
      1428
      1428
         0
```

2.327

```
                751 = 751 4,725/4,792
4,792 )3,603,517
        33544
         24911
         23960
          9517
          4792
          4725
```

$751 = 751\frac{4,725}{4,792}$

2.328

```
                224 = 224 4,064/7,522 = 224 2,032/3,761
7,522 )1,688,992
        15044
         18459
         15044
          34152
          30088
           4064
```

$224 = 224\frac{4,064}{7,522} = 224\frac{2,032}{3,761}$

III. SECTION THREE

3.1

```
  39
+ 27
  66 books in the Bible
```

3.2

```
14
18
27
36
95 acres
```

3.3

```
$ 27
  30
  35
  40
  31
$163 total
```

3.4

```
  912
  905
  910
  895
  962
  969
5,553 years
```

3.5

```
$1,000
   625
   200
   326
$2,151 total monthly income
```

3.6
```
  400
  500
  269
  347
-----
1,516 total enrollment
```

3.7
```
 31
 28
 31
 30
 31
 30
----
181 days
```

3.8
```
 14
 17
 20
 18
 15
 23
 22
----
129 total points
```

3.9
```
 43,730
 22,200
 40,500
 76,500
 64,300
 60,500
 85,200
 45,600
 64,400
 53,400
 45,400
--------
601,730 families in Israel
```

3.10
```
$18,500
    670
    300
-------
$19,470 is next year's salary
```

3.11
```
$160
  75
----
$ 85  left
```

3.12
```
O.T.:  39
N.T. : 27
       --
       12
```
The Old Testament has 12 more books than the New Testament.

3.13
```
$ 200      $ 425
+  28      - 228
-----      -----
$ 228      $ 197 to borrow
```

3.14
```
1,400
  814
-----
  586 miles
```

3.15
```
93,000,000
   240,000
----------
92,760,000 miles
```

3.16
```
60,141
51,467
------
 8,674 miles
```

3.17
```
$169,414,816,115
  141,668,127,237
-----------------
$  27,746,688,878 more
```

3.18
```
816
747
---
 69 more girls than boys
```

3.19
```
  4        6
  1        7
  3        2
  5        8
  9        1
 --       --
 22 got off   24 got on

 24       17
-22      + 2
---      ---
  2       19 passengers
```

3.20
```
 5        6
 2        4
 4        7
 4        5
 3        5
 5        8
 4        5
 5        6
 3        4
--       --
35 par   50 was Jack's score

     50
    -35
    ---
     15 strokes above par
```

3.21
```
  11
  25
 ---
  55
  22
----
$275
```

3.22
```
  40
   3
----
$120
```

3.23
```
 50      25     10
x 3     x 6    x 8
----    ----   ----
150 ¢   150 ¢  80 ¢

 12      14
x 5     x 1
----    ----
60 ¢    14 ¢

150
150
 80
 60
 14
---
454 pennies
```

3.24
```
 $3      $15
x 6     + 18
---     ----
$18      $33   was the cost for
               the tape recorder
```

3.25
```
   $ 826
     471
   -----
     826
   5782
  3304
--------
$389,046
```

3.26
```
   287      209
   x 8     x 10
------    -----
$2,296      000
           209
         ------
         $2,090

   273      164
   x 4      x 6
------    -----
$1,092     $984

$2,296
 2,090
 1,092
   984
------
$6,462  total
```

3.27
```
Sue:     18
        x 5
        ---
         90  points

Bill:    17
        x 5
        ---
         85  points

Judy:    14
        x 5
        ---
         70  points

Jake:    20
        x 5
        ---
        100  points

Ben:      5
        x 5
        ---
         25  points

total points:   90
                85
                70
               100
                25
               ---
               370  points
```

3.28
```
$     86,000,000
              15
----------------
     430000000
     86000000
----------------
$1,290,000,000
```

3.29
```
a.   $ 640
         6
    ------
    $3,840  per month

b.  $3,840
         9
   -------
   $34,560  per school year
```

3.30
```
first place:      10
                 x 2
                 ---
                  20 points

second place:  5 points

third place:       4
                 x 2
                 ---
                   8 points

fourth place:      3
                 x 3
                 ---
                   9 points
```

3.30 cont.

total points:
20
5
8
9
42 points

3.31
23 miles per gallon
56)1,288
112
168
168
0

3.32
96
x 2
192 miles both ways

12 gallons
16)192
16
32
32
0

3.33
$32 each
3)96
9
06
6
0

3.34
53 bushels
22)1,166
110
66
66
0

3.35
4 acres
43,000)172,000
172000
0

3.36
24
x 6
144 cookies needed

4 pounds
36)144
144
0

3.37
350 miles per hour
7)2,450
2 1
35
35
00
0
0

3.38
$17 \frac{1}{1{,}147¢}$ = about 17¢
1,147)19,500
1147
8030
8029
1

3.39
58 feet
12)696
60
96
96
0

3.40
45 bushels
56)2,520
224
280
280
0

3.41
12
x 6
72 inches

3.42
36
x 7
252 inches

3.43
100
x 27
700
200
2,700 cm

3.44
54 dm
100)5,400
500
400
400
0

3.45

```
              52 miles
5,280)274,560
       26400
        10560
        10560
            0
```

3.46

```
  1,760
 x   12
   3520
  1760
 21,120 yards
```

3.47

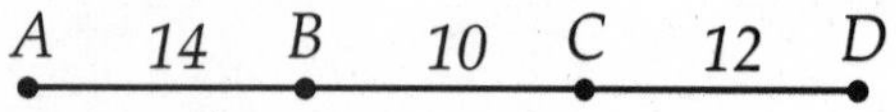

A 14 B 10 C 12 D

AD = *AB* + *BC* + *CD*
AD = 14 + 10 + 12
AD = 36 inches

```
    3  AD = 3 feet
12 )36
    36
     0
```

3.48

E 270 F 174 G 256 H

EH = *EF* + *FG* + *GH*
EH = 270 + 174 + 256
EH = 700 mm

100 mm = 1 dm

```
      7  EH = 7 dm
100 )700
     700
       0
```

3.49

perimeter = *AB* + *BC* + *CA*
perimeter = 14 + 17 + 27

a. perimeter = 58 yards

b 1 yard = 3 feet

```
  58
 x 3
 174 feet
```

3.50

perimeter = *DE* + *EF* + *FG* + *GD*
perimeter = 14 + 14 +14 + 14
or 4 x 14

a. perimeter = 56 cm

b. 1 cm = 10 mm

```
  56
x 10
  00
 56
 560 mm
```

3.51

```
   56 meters
3 )168
   15
    18
    18
     0
```

3.52

```
   68 inches
4 )272
   24
    32
    32
     0
```

3.53

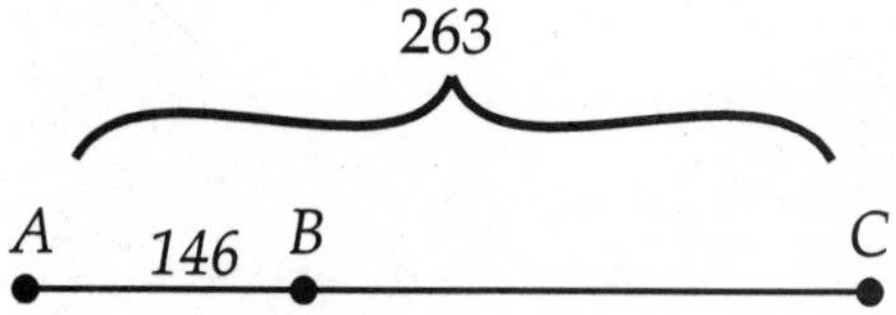

A 146 B C (AC = 263)

BC = *AC* – *AB*
BC = 263 – 146
BC = 117 mm

3.54

53 = 19 + 23 + third side
53 = 42 + third side
third side = 11 feet

3.55

```
   106
 x 100
   000
  000
 106
 10,600 cm
```

100 cm = 1m

```
         106 meters
100 )10,6000
     100
       60
       00
       600
       600
         0
```

3.56

```
    5 feet = length of
9 )45        one revolution
   45
    0
```

1 foot = 12 inches

```
 12
x 5
 60 inches
```

3.57

```
 55°
 76°
 49°
180° are in a triangle
```

3.58 area = $l \times w$
area = 75 x 47
area = 3,525 sq. cm

3.59 area = $\frac{b \times a}{2}$
area = $\frac{27 \times 18}{2}$
area = $\frac{486}{2}$
area = 243 sq. in.

3.60 a. 5
b.
```
   108°
5 )540
   5
   04
    0
    40
    40
     0
```

3.61 total area = area triangle *ABE* + area triangle *BED* + area triangle *BCD*
total area = 54 + 97 + 66
total area = 217 sq. in.

3.62 a.
```
   180° are in a semicircle
2 )360
   2
   16
   16
    00
    00
     0
```
b. area of circle = 2 x area of semi-circle
area of circle = 2 x 156
area of circle = 312 sq. cm

3.63 a. 12
b.
```
  12
 x 9
108 inches
```

3.64 a. 6
b. area of each face = 3 x 3
area of each face = 9 sq. yds.
area of all faces of the cube = 9 x 6 = 54 sq. yds.

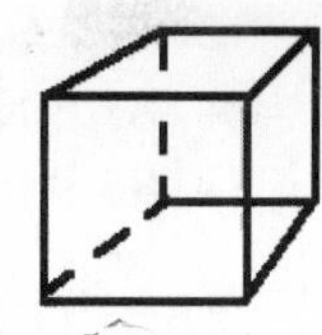

3.65 a. area of base = 42 x 42
area of base = 1,764 sq. m
b. 4

3.66 volume = $l \times w \times h$
volume = 15 x 12 x 10
volume = 1,800 cubic feet
Note: Solutions to Problems 3.67 through 3.86 are provided for students who work the problems but do not have access to calculators.

3.67
```
 8,146
 4,116
    26
47,814
   615
------
60,717
```

3.68
```
  2,481
 37,112
  6,199
      5
     27
814,145
-------
859,969
```

3.69
```
    1,146
    4,576
    8,771
6,776,789
       87
---------
6,791,369
```

3.70
55,146
81,804
94,988
27,177
259,115

3.71
44,116
– 35,277
8,889

3.72
666,114
– 477,228
188,886

3.73
47,776,100
– 9,149,376
38,626,724

3.74
10,000,000
– 1,123,411
8,876,589

3.75
8,816
14
35264
8816
123,424

3.76
1,141
27
7987
2282
30,807

3.77
8
7
56

214
56
1284
1070
11,984

3.78
14
21
14
28
294

294
74
1176
2058
21,756

21,756
85
108780
174048
1,849,260

3.79
563
89)50,107
445
560
534
267
267
0

3.80
4,226
458)1,935,508
1832
1035
916
1190
916
2748
2748
0

3.81
86
1,126)96,836
9008
6756
6756
0

3.82
155
15,556)2,411,180
15556
85558
77780
77780
77780
0

3.83
671
55
3355
3355
36,905

36,905
8
295,240

26,840
11)295,240
22
75
66
92
88
44
44
00
00
0

3.84

```
   4,761       62
 + 8,664     + 13
  13,425       75

       179
75 )13,425
     75
     592
     525
       675
       675
         0
```

3.85

```
  15       225
  15        15
  75      1125
 15       225
 225      3,375

 3,375        50,625
    15            15
 16875        253125
 3375         50625
 50,625       759,375
```

3.86

```
   86          1,194
  427        −    83
  681          1,111
 1,194

           1
1,111 )1,111
        1111
           0
```

3.87

11	4	9
6	8	10
7	12	5

sum = 24

3.88

17	10	15
12	14	16
13	18	11

sum = 42

3.89

12	5	10
7	9	11
8	13	6

sum = 27

3.90

28	21	26
23	25	27
24	29	22

sum = 75

3.91 Example:

15	8	13
10	12	14
11	16	9

sum = 36

3.92

	4	6	
1	1/2	1/8	3
4	0/4	0/6	1
3	0/8	1/2	2
	5	2	

The product is 14,352.

3.93

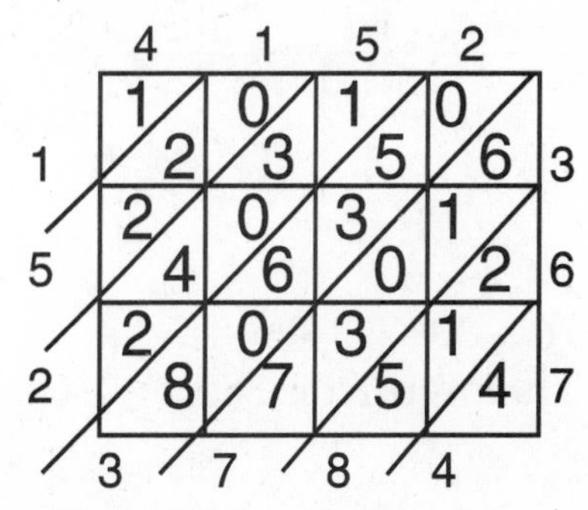

The product is 1,523,784.

3.94

5 4 6 7

5 | 4/5 | 3/6 | 5/4 | 6/3 | 9

3 | 4/0 | 3/2 | 4/8 | 5/6 | 8

6 | 1/0 | 0/8 | 1/2 | 1/4 | 2

8 5 9 4

The product is 5,368,594.

3.95

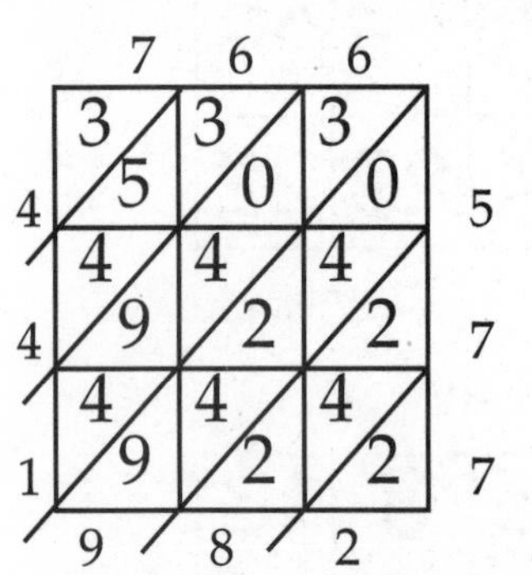

The product is 441,982.

3.96 Example:

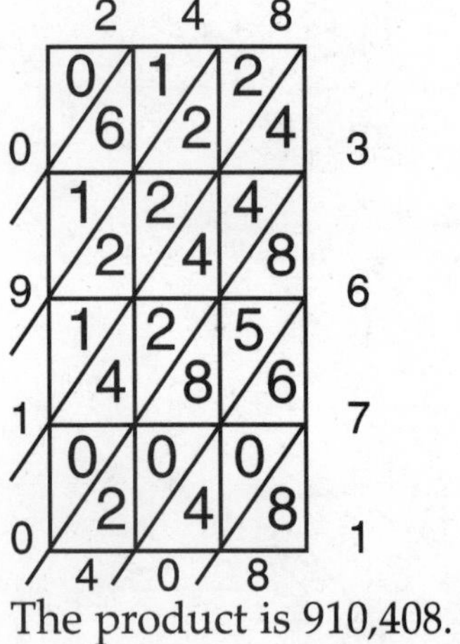

The product is 910,408.

3.97

Column 1	Column 2
1	76
2	152
4	304

1 + 4 = 5
76 + 304 = 380
The product is 380.

3.98

Column 1	Column 2
1	85
2	170
4	340
8	680

1 + 4 + 8 = 13
85 + 340 + 680 = 1,105
The product is 1,105.

3.99

Column 1	Column 2
1	151
2	302
4	604
8	1,208
16	2,416
32	4,832

1 + 8 + 32 = 41
151 + 1,208 + 4,832 = 6,191
The product is 6,191.

3.100

Column 1	Column 2
1	247
2	494
4	988
8	1,976
16	3,952
32	7,904

32 + 2 = 34
7,904 + 494 = 8, 398
The product is 8,398.

I. SECTION ONE

1.1 Across from left to right:

$4 = 4$

$4 \times 0 = 0$

$8 - 4 = 4$

$5 - 1 = 4$

$\frac{1{,}842 - 1{,}834}{2} = \frac{8}{2} = 4$

$4 \times 1 = 4$

$3 \times 1\frac{1}{3} = 3 \times \frac{4}{3} = 4$

$2 + 2 = 4$

$3 + 1 = 4$

$1 + 1 + 1 + 1 = 4$

$12 \times \frac{1}{3} = 4$

| | | | = 4 tally marks

$72 \div 18 = 4$

$6 - 2 = 4$

IV = 4 in Roman numerals

$628.424 \div 157.106 = 4$

4 x 0 does not represent four.

1.2 Examples:
19 x 2, 40 – 2, 36 + 2, and
14 + 24

1.3 2

1.4 $(3 \underline{+} 5) \underline{\times} 2 = 16$

1.5 $(3 \underline{+} 5) \underline{\div} 2 = 4$

1.6 $(3 \underline{\times} 5) \underline{+} 2 = 17$

1.7 $(3 \underline{\div} 5) \underline{\times} 2 = 1.2$

1.8 false

1.9 true

1.10 true

1.11 true

1.12 true

1.13 false

1.14 true

1.15 10

1.16 16

1.17 26

1.18 5

1.19 8

1.20 2

1.21 70

1.22 a. $9 \times 2 = 18$

$3 \times (7 - 1) = 3 \times 6 = 18$

$\frac{4}{3} + \frac{50}{3} + \frac{54}{3} = 18$

$29 - 11 = 18$

$\frac{60}{5} + \frac{30}{5} = \frac{90}{5} = 18$

$0 + 18 = 18$

$(3 + 5) + 10 = 8 + 10 = 18$

$35 - (10 + 7) = 35 - 17$
$= 18$

$9.72 + 8.28 = 18.00$

$(4 \times 2) + (5 \times 2) = 8 + 10$
$= 18$

$15 + 3 = 18$

$(90 \times \frac{4}{9}) \times (0.3 \times 1.5)$
$= 40 \times 0.45 = 18$

b. $\frac{9 - 6}{10 - 5} = \frac{3}{5}$

$12 \times 0.05 = 0.6 = \frac{6}{10} = \frac{3}{5}$

$\frac{7}{5} - \frac{4}{5} = \frac{3}{5}$

$(\frac{1}{50} + \frac{3}{25}) + (\frac{2}{5} + \frac{3}{50})$

$= (\frac{1}{50} + \frac{6}{50}) + (\frac{20}{50} + \frac{3}{50})$

$= \frac{7}{50} + \frac{23}{50} = \frac{30}{50} = \frac{3}{5}$

$\frac{6}{10} = \frac{3}{5}$

$\frac{6 + 0}{10 + 0} = \frac{6}{10} = \frac{3}{5}$

1.22 cont.

$$\frac{3 \times 978}{5 \times 978} = \frac{3}{5}$$

$$\frac{(9+3)-(2 \times 4.5)}{(18-3) \div (7-4)} = 1\frac{2-9}{15 \div 3} = \frac{3}{5}$$

c. $\frac{3-3}{842} = \frac{0}{842} = 0$

$$\frac{7+3}{15-5} - \frac{52}{26 \times 2} =$$

$$\frac{10}{10} - \frac{52}{52} = 1 - 1 = 0$$

$$\frac{0}{18} = 0$$

$8 - 8 = 0$

$(74 - 70) - 4 = 4 - 4 = 0$

$$\frac{2}{3} - \frac{58}{87} = \frac{2}{3} - \frac{2 \times 29}{3 \times 29} =$$

$$\frac{2}{3} - \frac{2}{3} = 0$$

1.23 a. 3
b. 10
c. 17

1.24 Either order:
a. *A*
b. *D*

Either order:
c. *B*
d. *C*

1.25 𓏺𓏺𓏺𓏺𓏺

1.26 𓎆𓎆𓎆𓎆

1.27 𓎆

1.28 𓎆𓏺𓏺

1.29 𓎆𓎆𓎆𓎆𓎆𓎆𓎆𓎆𓎆𓏺𓏺𓏺𓏺𓏺𓏺𓏺𓏺𓏺

1.30 𓂭𓎆𓏺

1.31 𓆐𓂭𓂭𓆼𓆼𓆼𓆼𓍢𓏺

1.32 𓆏𓆏𓆏𓆏𓆏𓆏𓆼𓆼𓆼𓆼𓎆

1.33 𓆼𓆼𓍢𓍢𓍢𓍢𓍢𓍢𓍢𓍢𓎆𓎆𓎆𓎆𓎆𓎆𓏺𓏺𓏺𓏺

1.34 𓂭𓂭𓂭𓍢𓍢𓍢𓍢𓎆𓎆𓏺

1.35 𓆼𓍢𓍢𓍢𓍢𓍢𓍢𓍢𓍢𓍢𓎆𓎆𓎆𓎆𓎆𓎆𓎆𓏺𓏺𓏺

1.36 ∠̇′ β′

1.37 ξ′ ε′

1.38 ρ′ ξ′ γ′

1.39 ͵α ρ′∠′ α′

1.40 ͵∠ ͵α ξ′ ε′

1.41 LV

1.42 LXXXII

1.43 VL

1.44 CCXXXI

1.45 6

1.46 15

1.47 52

1.48 200

1.49 152

1.50 4

1.51 9

1.52 14

1.53 59

1.54 1,104

1.55 659

1.56 1,614

1.57 𓎆𓏺𓏺𓏺𓏺𓏺

1.58 𓏺𓏺𓏺𓏺

1.59 𓎆𓎆𓏺

1.60 𓎆𓎆𓎆𓎆𓏺𓏺

1.61 𓎆𓎆𓎆𓏺𓏺

1.62 Example:
Roman numerals generally take less space and the figures are easier to draw.

1.63 Four hundred thirty-two

1.64 Six thousand, four hundred eighty-three

1.65 Two hundred seven

1.66 Four hundred sixty-seven thousand, one hundred forty-nine

1.67 Fifty thousand, four hundred fifteen

1.68 Five hundred four thousand

1.69 One million, six hundred three thousand, seven hundred ninety-three

1.70 Four million, eight hundred fifty thousand, seven

1.71 Twenty-nine million, six hundred two thousand, one hundred

1.72 Seven hundred ninety-six million, one hundred fifty-three thousand, two hundred five

1.73 Three million, one thousand, twenty

1.74 Four billion, three hundred eighty-five million, four hundred ninety-six thousand, one hundred thirty-two

1.75 348

1.76 2,059

1.77 302

1.78 98,240

1.79 6,739

1.80 400,803

1.81 7,640,078

1.82 496,714,500

1.83 5,800,741,017

1.84 1,000,017,000,030,000

1.85 $(4 \times 10^2) + (3 \times 10^1) + 2$

1.86 $(6 \times 10^3) + (4 \times 10^2)$
$+ (8 \times 10^1) + 3$

1.87 $(2 \times 10^2) + 7$

1.88 $(4 \times 10^5) + (6 \times 10^4)$
$+ (7 \times 10^3) + (1 \times 10^2)$
$+ (4 \times 10^1) + 9$

1.89 $(5 \times 10^4) + (4 \times 10^2)$
$+ (1 \times 10^1) + 5$

1.90 $(5 \times 10^5) + (4 \times 10^3)$

1.91 $(1 \times 10^6) + (6 \times 10^5)$
$+ (3 \times 10^3) + (7 \times 10^2)$
$+ (9 \times 10^1) + 3$

1.92 $(4 \times 10^6) + (8 \times 10^5)$
$+ (5 \times 10^4) + 7$

1.93 $(2 \times 10^7) + (9 \times 10^6)$
$+ (6 \times 10^5) + (2 \times 10^3)$
$+ (1 \times 10^2)$

1.94 $(7 \times 10^8) + (9 \times 10^7)$
$+ (6 \times 10^6) + (1 \times 10^5)$
$+ (5 \times 10^4) + (3 \times 10^3)$
$+ (2 \times 10^2) + 5$

1.95 $(3 \times 10^6) + (1 \times 10^3)$
$+ (2 \times 10^1)$

1.96 $(4 \times 10^9) + (3 \times 10^8)$
$+ (8 \times 10^7) + (5 \times 10^6)$
$+ (4 \times 10^5) + (9 \times 10^4)$

+ $(6 \times 10^3) + (1 \times 10^2)$
+ $(3 \times 10^1) + 2$

1.97 100,000 or hundred thousands

1.98 762

1.99 975,431

1.100 9,630

1.101 88,653,100

1.102 2^5

1.103 5^4

1.104 25^6

1.105 9^3

1.106 a. 16
b. 3

1.107 a. 3
b. 24

1.108 a. 321
b. 5

1.109 true

1.110 false

1.111 1011_2

1.112 1100_2

1.113 1101_2

1.114 1110_2

1.115 1111_2

1.116 10000_2

1.117 10001_2

1.118 10010_2

1.119 10011_2

1.120 10100_2

1.121 10101_2

1.122 10110_2

1.123 10111_2

1.124 11000_2

1.125 11001_2

1.126 11010_2

1.127 11011_2

1.128 11100_2

1.129 11101_2

1.130 11110_2

1.131 110010_2

1.132 10000000_2

1.133 101000000_2

1.134 55

1.135 65

1.136 1

1.137 0

1.138 3

1.139 a. 0
b. 1
c. 10

1.140 a. 0
b. 0

II. SECTION TWO

2.1 {q, r, s, t, u, v, w, x, y, z}

2.2 {January, June, July}

2.3 Example:
{Mr. Coe, Mrs. Doe, Mrs. Foe}

2.4 {Matthew, Mark, Luke, John}

2.5 {The vowels in our alphabet}

2.6 {The digits used in base 10}

2.7 {Months beginning with letter M}

2.8 {Symbols for operations with the real numbers}

2.9 {Symbols used in Roman system for numbers}

2.10 Examples:
{1st base, 2nd base, 3rd base}

2.11 Example:
{The bases in a ball game}

2.12 Examples:
{2, 4, 6, . . .},
{1, 3, 5, 7, . . .},
{1, 2, 3, 4, . . .}

2.13 $9 \in T$

2.14 $11 \notin E$

2.15 $r \notin Z$

2.16 $n \in A$

2.17 infinite

2.18 finite

2.19 finite

2.20 finite

2.21 infinite

2.22 finite

2.23 8 + 1 = 2 + 7
9 = 9

2.24 10 + 3 = 8 + 5
13 = 13

2.25 15 + 2 = 11 + 6
17 = 17

2.26 41 + 9 = 35 + 15
50 = 50

2.27 50 + 2 = 49 + 3
52 = 52

2.28 14 + 9 = 7 + 16
23 = 23

2.29

```
   77        38
 x 38      x 77
  616       266
 231       266
2,926     2,926
```

2.30

```
   87        78
 x 78      x 87
  696       546
 609       624
6,786     6,786
```

2.31

```
  167        32
 x 32     x 167
  334       224
 501       192
5,344      32
          5,344
```

2.32

```
  409        17
 x 17     x 409
 2863       153
 409       680
6,953     6,953
```

2.33

```
  276        24
 x 24     x 276
 1104       144
 552        168
6,624       48
          6,624
```

2.34 35 x 5 = 7 x 25
175 = 175

2.35 18 x 2 = 9 x 4
36 = 36

2.36 14 x 0 = 7 x 0
0 = 0

2.37 18 x 3 = 6 x 9
54 = 54

2.38 not commutative

2.39 commutative

2.40 not commutative

2.41 not commutative

2.42 commutative

2.43 not commutative

2.44 not commutative

2.45 B

2.46 C

2.47 D

2.48 A

2.49 A

2.50 D

2.51 B

2.52 A

2.53
a. $(13 - 5) - 2$
$8 - 2 = 6$
b. $13 - (5 - 2)$
$13 - 3 = 10$
c. no

2.54
a. $(48 \div 12) \div 2$
$4 \div 2 = 2$
b. $48 \div (12 \div 2)$
$48 \div 6 = 8$
c. no

2.55 {10, 15, 20, 25}

2.56 {0, 10, 100, 1,000}

2.57 {1, 2, 3, 101, 102, 103, 104}

2.58 {6, 7, 8, 9, 10}

2.59 {20}

2.60 { }

2.61 {100, 1,000}

2.62 { }

2.63
a. $S \cup T$ = {Tom, Dick, Harry, Joe, George}
b. $S \cap T$ = {Dick, Joe}

2.64
a. $W \cup X$ = {M, T, W, Th, F, Sat., Sun.}
b. $W \cap X$ = {Friday}

2.65
a. $G \cup H$ = {ping, pong, bong,bow, yes, pang, wow, no}
b. $G \cap H$ = {bong}

2.66
a. $I \cup J$ = {red, white, blue}
b. $I \cap J$ = {red, white, blue}

2.67
$R \cup S$ = {1, 2, 3, 4, 5, 6}
$S \cup R$ = {1, 2, 3, 4, 5, 6}

2.68
$R \cap S$ = {3, 4}
$S \cap T$ = {6}
$R \cap (S \cap T)$ = { }
$(R \cap S) \cap T$ = { }

2.69
S = {3, 4, 5, 6}
$\varnothing$ = { }
$S \cap \varnothing = \varnothing$

2.70
$R \cap S$ = {1, 2, 3, 4, 5, 6}
$(R \cup S) \cup T$= {1, 2, 3, 4, 5, 6, 7, 8, . . . 20}
$S \cup T$ = {3, 4, 5, 6, 7, 8,. . .20}
$R \cup (S \cup T)$ = {1, 2, 3, 4,. . .20}

2.71

2.72 From the Venn diagram in Problem 2.71, 27 students are members of the gymnastic team and/or the chess team.

$$\begin{array}{r} 800 \\ -\ 27 \\ \hline 773 \end{array}$$

773 students are not members of either team.

2.73 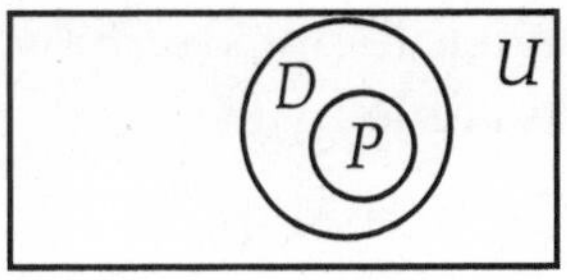

2.74 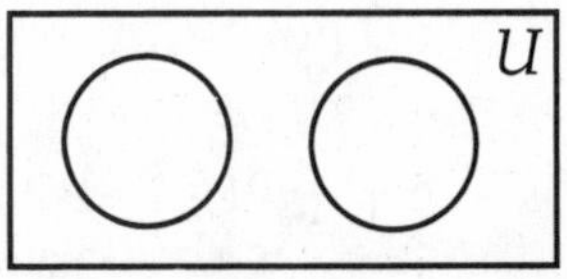

2.75 {2, 3, 4, 5, 6, 7, 8, 9}

2.76 {4, 7}

2.77 {2, 5, 6, 8}

2.78 {3, 9, 4, 7, 6}

2.79 {2, 5, 3, 9, 6, 8}

2.80 {2, 5, 8}

III. SECTION THREE

3.1 2, 3, 5, 7, 11, 13, 17, 19, 23, 29, 31, 37, 41, 43, 47, 53, 59, 61, 67, 71, 73, 79, 83, 89, 97, 101, 103, 107, 109, 113, 127, 131, 137, 139, 149, 151, 157, 163, 167, 173, 179, 181, 191, 193, 197, 199

3.2 90, 91, 92, 93, 94

3.3 2

3.4 2, 3, 5, and 7

3.5 $2^2 \times 3^2$ (or $2 \times 2 \times 3 \times 3$)

3.6 Example:
When you reach the number whose square is the same or greater than the number you are using as a limit, you will have crossed out all composite numbers.

3.7 3, 5; 5, 7; 11, 13; 17, 19;
29, 31; 41, 43; 59, 61;
71, 73; 101, 103; 107, 109;
137, 139; 149, 151;
179, 181; 191, 193; 197, 199

3.8 Example:
Since 2 is the only even prime, and since when two numbers differ by 3 one of them must be even, 2 and 5 are the only primes differing by 3.

3.9 2, 3, 4, 6, 8

(2:) The units' digit, 4, is even, so 24 is divisible by 2.

(3:) The sum of the digits = 2 + 4 = 6. 6 is divisible by 3.

(4:) 24 ÷ 4 = 6, so 24 is divisible by 4.

5: The units' digit is neither 0 nor 5, so 5 is not a divisor.

(6:) Factors of 6 are 2 and 3. Since 24 is divisible by 2 and 3, it is also divisible by 6.

(8:) Factors of 8 are 2 and 4. Since 24 is divisible by 2 and 4, it is also divisible by 8.

9: The sum of the digits is 6. 6 is not divisible by 9, so 24 is not divisible by 9.

10: Factors of 10 are 2 and 5. Since 24 is divisible by 2 but not by 5, it is not divisible by 10.

3.10 2, 4, 8

(2:) The units' digit, 6, is even, so 16 is divisible by 2.

3: The sum of the digits = 1 + 6 = 7. 7 is not divisible by 3, so 16 is not divisible by 3.

(4:) 16 ÷ 4 = 4, so 16 is divisible by 4.

5: The units' digit is neither 0 nor 5, so 5 is not a divisor.

3.10 cont.

6: Factors of 6 are 2 and 3. Since 16 is divisible by 2 but not by 3, it is not divisible by 6.

(8:) Factors of 8 are 2 and 4. Since 16 is divisible by 2 and 4, it is also divisible by 8.

9: The sum of the digits is 7. 7 is not divisible by 9, so 16 is not divisible by 9.

10: Factors of 10 are 2 and 5. Since 16 is divisible by 2 but not by 5, it is not divisible by 10.

3.11 2, 5, 10

(2:) The units' digit is 0, so 50 is divisible by 2.

3: The sum of the digits = 5 + 0 = 5. 5 is not divisible by 3, so 50 is not divisible by 3.

4: $50 \div 4 = 12\frac{1}{2}$, so 50 is not divisible by 4.

(5:) The units' digit is 0, so 5 is a factor.

6: Factors of 6 are 2 and 3. Since 50 is divisible by 2 but not by 3, it is not divisible by 6.

8: Factors of 8 are 2 and 4. Since 50 is divisible by 2 but not by 4, it is not divisible by 8.

9: The sum of the digits is 5. 5 is not divisible by 9, so 50 is not divisible by 9.

(10:) Factors of 10 are 2 and 5. Since 50 is divisible by 2 and 5, it is also divisible by 10.

3.12 3

2: The units' digit, 7, is not even, so 87 is not divisible by 2.

(3:) The sum of the digits = 8 + 7 = 15. 15 is divisible by 3, so 87 is divisible by 3.

4: $87 \div 4 = 21\frac{3}{4}$, so 87 is not divisible by 4.

5: The units' digit is neither 0 nor 5, so 5 is not a divisor.

6 Factors of 6 are 2 and 3. Since 87 is divisible by 3 but not by 2, it is not divisible by 6.

8: Factors of 8 are 2 and 4. Since 87 is not divisible by 2 or 4, it is not divisible by 8.

9: The sum of the digits is 15. 15 is not divisible by 9, so 87 is not divisible by 9.

10: Factors of 10 are 2 and 5. Since 87 is not divisible by 2 or 5, it is not divisible by 10.

3.13 2, 3, 6

(2:) The units' digit, 4, is even, so 174 is divisible by 2.

(3:) The sum of the digits = 1 + 7 + 4 = 12. 12 is divisible by 3, so 174 is divisible by 3.

4: The last two digits, 74, divided by 4 = $18\frac{1}{2}$, so 174 is not divisible by 4.

5: The units' digit is neither 0 nor 5, so 5 is not a divisor.

(6:) Factors of 6 are 2 and 3. Since 174 is divisible by 2 and 3, it is also divisible by 6.

8: Factors of 8 are 2 and 4. Since 174 is divisible by 2 but not by 4, it is not divisible by 8.

9: The sum of the digits is 12. 12 is not divisible

3.13 cont.

by 9, so 174 is not divisible by 9.

10: Factors of 10 are 2 and 5. Since 174 is divisible by 2 but not by 5, it is not divisible by 10.

3.14 2, 3, 4, 6, 8

(2:) The units' digit, 2, iseven, so 312 is divisible by 2.

(3:) The sum of the digits $= 3 + 1 + 2 = 6$. 6 is divisible by 3, so 312 is divisible by 3.

(4:) The last two digits, 12, divided by $4 = 3$, so 312 is divisible by 4.

5: The units' digit is neither 0 nor 5, so 5 is not a divisor.

(6:) Factors of 6 are 2 and 3. Since 312 is divisible by 2 and 3, it is also divisible by 6.

(8:) Factors of 8 are 2 and 4. Since 312 is divisible by 2 and 4, it is also divisible by 8.

9: The sum of the digits is 6. 6 is not divisible by 9, so 312 is not divisible by 9.

10: Factors of 10 are 2 and 5. Since 312 is divisible by 2 but not by 5, it is not divisible by 10.

3.15 2, 5, 10

(2:) The units' digit is 0, so 250 is divisible by 2.

3: The sum of the digits $= 2 + 5 + 0 = 7$. 7 is not divisible by 3, so 250 is not divisible by 3.

4: The last two digits, 50, divided by $4 = 12\frac{1}{2}$, so 250 is not divisible by 4.

(5:) The units' digit is 0, so 5 is a divisor.

6: Factors of 6 are 2 and 3. Since 250 is divisible by 2 but not by 3, it is not divisible by 6.

8: Factors of 8 are 2 and 4. Since 250 is divisible by 2 but not by 4, it is not divisible by 8

9: The sum of the digits is 7. 7 is not divisible by 9, so 250 is not divisible by 9.

(10:) Factors of 10 are 2 and 5. Since 250 is divisible by 2 and 5, it is also divisible by 10.

3.16 3

2: The units' digit, 7, is not even, so 597 is not divisible by 2.

(3:) The sum of the digits $= 5 + 9 + 7 = 21$. 21 is divisible by 3, so 597 is divisible by 3.

4: The last two digits, 97, divided by $4 = 24\frac{1}{4}$, so 597 is not divisible by 4.

5: The units' digit is neither 0 nor 5, so 5 is not a divisor.

6: Factors of 6 are 2 and 3. Since 597 is divisible by 3 but not by 2, it is not divisible by 6.

8: Factors of 8 are 2 and 4. Since 597 is not divisible by 2 or by 4, it is not divisible by 8.

9: The sum of the digits is 21. 21 is not divisible by 9, so 597 is not divisible by 9.

10: Factors of 10 are 2 and 5. Since 597 is not divisible by 2 or 5, it is not divisible by 10.

3.17 2

(2:) The units' digit, 2, is even, so 4,382 is divisible by 2.

3: The sum of the digits $= 4 + 3 + 8 + 2 = 17$. 17 is not divisible by 3, so 4,382 is not divisible by 3.

4: The last two digits, 82, divided by $4 = 20\frac{1}{2}$, so 4,382 is not divisible by 4.

5: The units' digit is neither 0 nor 5, so 5 is not a divisor.

6: Factors of 6 are 2 and 3. Since 4,382 is divisible by 2 but not by 3, it is not divisible by 6.

8: Factors of 8 are 2 and 4. Since 4,382 is divisible by 2 but not by 4, it is not divisible by 8.

9: The sum of the digits is 17. 17 is not divisible by 9, so 4,382 is not divisible by 9.

10: Factors of 10 are 2 and 5. Since 4,382 is divisible by 2 but not by 5, it is not divisible by 10.

3.18 3

2: The units' digit, 3, is not even, so 8,643 is not divisible by 2.

(3:) The sum of the digits $= 8 + 6 + 4 + 3 = 21$. 21 is divisible by 3, so 8,643 is divisible by 3.

4: The last two digits, 43, divided by $4 = 10\frac{3}{4}$, so 8,643 is not divisible by 4.

5: The units' digit is neither 0 nor 5, so 5 is not a divisor.

6: Factors of 6 are 2 and 3. Since 8,643 is divisible by 3 but not by 2, it is not divisible by 6.

8: Factors of 8 are 2 and 4. Since 8,643 is not divisible by 2 or 4, it is not divisible by 8.

9: The sum of the digits is 21. 21 is not divisible by 9, so 8,643 is not divisible by 9.

10: Factors of 10 are 2 and 5. Since 8,643 is not divisible by 2 or 5, it is not divisible by 10.

3.19 2, 3, 4, 6, 8

(2:) The units' digit, 8, is even, so 10,488 is divisible by 2.

(3:) The sum of the digits $= 1 + 0 + 4 + 8 + 8 = 21$. 21 is divisible by 3, so 10,488 is divisible by 3.

(4:) The last two digits, 88, divided by $4 = 22$, so 10,488 is divisible by 4.

5: The units' digit is neither 0 nor 5, so 5 is not a divisor.

(6:) Factors of 6 are 2 and 3. Since 10,488 is divisible by 2 and 3, it is also divisible by 6.

(8:) The last three digits, 488, divided by $8 = 61$, so 10,488 is divisible by 8.

9: The sum of the digits is 21. 21 is not divisible by 9, so 10,488 is not divisible by 9.

3.19 cont.

10: Factors of 10 are 2 and 5. Since 10,488 is divisible by 2 but not by 5, it is not divisible by 10.

3.20 2, 3, 4, 6

(2:) The units' digit, 2, is even, so 23,412 is divisible by 2.

(3:) The sum of the digits = 2 + 3 + 4 + 1 + 2 = 12. 12 is divisible by 3, so 23,412 is divisible by 3.

(4:) The last two digits, 12, divided by 4 = 3, so 23,412 is divisible by 4.

5: The units' digit is neither 0 nor 5, so 5 is not a divisor.

(6:) Factors of 6 are 2 and 3. Since 23,412 is divisible by 2 and 3, it is also divisible by 6.

8: The last three digits, 412, divided by 8 = $51\frac{1}{2}$, so 23,412 is not divisible by 8.

9: The sum of the digits is 12. 12 is not divisible by 9, so 23,412 is not divisible by 9.

10: Factors of 10 are 2 and 5. Since 23,412 is divisible by 2 but not 5, it is not divisible by 10.

3.21 2, 3, 4, 5, 6, 8, 9, 10

(2:) The units' digit, 0, so 7,296,480 is divisible by 2.

(3:) The sum of the digits = 7 + 2 + 9 + 6 + 4 + 8 + 0 = 36. 36 is divisible by 3, so 7,296,480 is divisible by 3.

(4:) The last two digits, 80, divided by 4 = 20, so 7,296,480 is divisible by 4.

(5:) The units' digit is 0, so 5 is a divisor.

(6:) Factors of 6 are 2 and 3. Since 7,296,480 is divisible by 2 and 3, it is also divisible by 6.

(8:) The last three digits, 480, divided by 8 = 60, so 7,296,480 is divisible by 8.

(9:) The sum of the digits is 36. 36 is divisible by 9, so 7,296,480 is divisible by 9.

(10:) Factors of 10 are 2 and 5. Since 7,296,480 is divisible by 2 and 5, it is also divisible by 10.

3.22 2, 4, 5, 8, 10

(2:) The units' digit, 0, so 5128,600 is divisible by 2.

3: The sum of the digits = 1 + 2 + 8 +6 +0 + 0 = 17. 17 is not divisible by 3, so 128,600 is not divisible by 3.

(4:) The last three digits, 600, divided by 4 = 150, so 128,600 is divisible by 4.

(5:) The units' digit is 0, so 5 is a divisor.

6: Factors of 6 are 2 and 3. Since 128,600 is divisible by 2 but not by 3, it is not divisible by 6.

(8:) The last three digits, 600, divided by 8 = 75, so 128,600 is divisible by 8.

9: The sum of the digits is 17. 17 is not divisible by 9, so 128,600 is not divisible by 9.

3.22 cont.

(10:) Factors of 10 are 2 and 5. Since 128,600 is divisible by 2 and 5, it is also divisible by 10.

3.23 Factors of 15 are 3 and 5.
24 is not divisible by 15.
16 is not divisible by 15.
50 is divisible by 5 but not by 3; not divisible by 15.
87 is divisible by 3 but not by 5; not divisible by 15.
174 is divisible by 3 but not by 5; not divisible by 15.
312 is divisible by 3 but not by 5; not divisible by 15.
250 is divisible by 5 but not by 3; not divisible by 15.
597 is divisible by 3 but not by 5; not divisible by 15.
4,382 is not divisible by 3 or 5; not divisible by 15.
8,643 is divisible by 3 but not by 5; not divisible by 15.
10,488 is divisible by 3 but not by 5; not divisible by 15.
23,412 is divisible by 3 but not by 5; not divisible by 15.
7,296,480 is divisible by 3 and 5; not divisible by 15.
128,600 is divisible by 5 but not by 3; not divisible by 15.
The only number given divisible by 15 is (7,296,480.)

3.24 Factors of 18 are 2 and 9.
24 is not divisible by 18.
16 is not divisible by 18.
50 is divisible by 2 but not by 9; not divisible by 18.
87 is not divisible by 2 or 9; not divisible by 18.
174 is divisible by 2 but not by 9; not divisible by 18.
312 is divisible by 2 but not by 9; not divisible by 18.
250 is divisible by 2 but not by 9; not divisible by 18.

3.24 cont.

597 is not divisible by 2 or 9; not divisible by 18.
4,382 is divisible by 2 but not by 9; not divisible by 18.
8,643 is not divisible by 2 or 9; not divisible by 18.
10,488 is divisible by 2 but not by 9; not divisible by 18.
23,412 is divisible by 2 but not by 9; not divisible by 18.
7,296,480 is divisible by 2 and 9; divisible by 18.
128,600 is divisible by 2 but not by 9; not divisible by 18.
The only number given divisible by 18 is (7,296,480.)

3.25 Factors of 45 are 5 and 9.

24 is not divisible by 45.

16 is not divisible by 45.

50 is not divisible by 45.

87 is not divisible by 5 or 9; not divisible by 45.
174 is not divisible by 5 or 9; not divisible by 45.
312 is not divisible by 5 or 9; not divisible by 45.
250 is divisible by 5 but not by 9; not divisible by 45.

597 is not divisible by 5 or 9; not divisible by 45.

4,382 is not divisible by 5 or 9; not divisible by 45.

8,643 is not divisible by 5 or 9; not divisible by 45.

10,488 is not divisible by 5 or 9; not divisible by 45.

3.25 cont.

23,412 is not divisible by 5 or 9; not divisible by 45.
7,296,480 is divisible by 5 and 9; divisible by 45.
128,600 is divisible by 5 but not by 9; not divisible by 45.

The only number given divisible by 45 is 7,296,480.

3.26 Factors of 48 are 6 and 8.
24 is not divisible by 48.
16 is not divisible by 48.
50 is not divisible by 48.
87 is not divisible by 6 or 8; not divisible by 48.
174 is divisible by 6 but not by 8; not divisible by 48.
312 is divisible by 6 and 8; but not by 48.
250 is not divisible by 6 or 8; not divisible by 48.
597 is not divisible by 6 or 8; not divisible by 48.
4,382 is not divisible by 6 or 8; not divisible by 48.
8,643 is not divisible by 6 or 8; not divisible by 48.
10,488 is divisible by 6 and 8; but not by 48.
23,412 is divisible by 6 but not by 8; not divisible by 48.

7,296,480 is divisible by 6 and 8; divisible by 48.
128,600 is divisible by 8 but not by 6; not divisible by 48.

The only number given divisible by 48 is 7,296,480.

3.27 2 x 13

3.28 3 x 13

3.29 2 x 3 x 13

3.30 3 x 17

3.31 3 x 19

3.32 3 x 31

3.33 2^3 x 3 x 5

3.34 2 x 191

3.35 2 x 3 x 7 x 11

3.36 1 x 1 = 1

3.37 2 x 2 = 4

3.38 3 x 3 = 9

3.39 4 x 4 = 16

3.40 5 x 5 = 25

3.41 6 x 6 = 36

3.42 7 x 7 = 49

3.43 8 x 8 = 64

3.44 9 x 9 = 81

3.45 10 x 10 = 100

3.46
$$\begin{array}{r} 11 \\ \underline{11} \\ 11 \\ \underline{11} \\ 121 \end{array}$$

3.47
$$\begin{array}{r} 12 \\ \underline{12} \\ 24 \\ \underline{12} \\ 144 \end{array}$$

3.48
$$\begin{array}{r} 13 \\ \underline{13} \\ 39 \\ \underline{13} \\ 169 \end{array}$$

3.49
$$\begin{array}{r} 14 \\ \underline{14} \\ 56 \\ \underline{14} \\ 196 \end{array}$$

3.50 $\begin{array}{r} 15 \\ \underline{15} \\ 75 \\ \underline{15} \\ 225 \end{array}$

3.51 $\begin{array}{r} 16 \\ \underline{16} \\ 96 \\ \underline{16} \\ 256 \end{array}$

3.52 $\begin{array}{r} 17 \\ \underline{17} \\ 119 \\ \underline{17} \\ 289 \end{array}$

3.53 $\begin{array}{r} 18 \\ \underline{18} \\ 144 \\ \underline{18} \\ 324 \end{array}$

3.54 $\begin{array}{r} 19 \\ \underline{19} \\ 171 \\ \underline{19} \\ 361 \end{array}$

3.55 $\begin{array}{r} 20 \\ \underline{20} \\ 00 \\ \underline{40} \\ 400 \end{array}$

3.56 $\begin{array}{r} 21 \\ \underline{21} \\ 21 \\ \underline{42} \\ 441 \end{array}$

3.57 $\begin{array}{r} 22 \\ \underline{22} \\ 44 \\ \underline{44} \\ 484 \end{array}$

3.58 $\begin{array}{r} 23 \\ \underline{23} \\ 69 \\ \underline{46} \\ 529 \end{array}$

3.59 $\begin{array}{r} 24 \\ \underline{24} \\ 96 \\ \underline{48} \\ 576 \end{array}$

3.60 $\begin{array}{r} 25 \\ \underline{25} \\ 125 \\ \underline{50} \\ 625 \end{array}$

3.61 $\sqrt{625} = 25$

3.62 $\sqrt{81} = 9$

3.63 $\sqrt{144} = 12$

3.64 $\sqrt{289} = 17$

3.65 2 x 2 x 2 x 2 x 2 = 32

3.66 2 x 2 x 3 x 3 x 3
= 4 x 27
= 108

3.67 2 x 3 x 5 x 5
= 6 x 25
= 150

3.68 5 x 5 x 13
= 25 x 13
= 325

3.69 11 x 11 x 3 x 3
= 121 x 9
= 1,089

3.70 17 x 17 x 19
= 289 x 19
= 5,491

3.71 2 x 2 x 2 x 2 x 2 x 3 x 3
= 32 x 9
= 288

3.72 2 x 2 x 7 x 7 x 7
= 4 x 343
= 1,372

3.73 $2 \times 2 \times 2 \times 4 \times 4 \times 4$
$\times 4 \times 4 \times 6 \times 6 \times 6$
$= 8 \times 1{,}024 \times 216$
$= 1{,}769{,}472$

3.74 $36 = 2 \times 2 \times 3 \times 3 = 2^2 \times 3^2$

3.75 $45 = 3 \times 3 \times 5 = 3^2 \times 5$

3.76 $36 = 2^2 \times \textcircled{3^2}$
$45 = \textcircled{3^2} \times 5$
GCF $= 3^2 = 9$

3.77 $20 = 2 \times 2 \times 5 = \textcircled{2^2} \times \textcircled{5}$
$60 = 2 \times 2 \times 3 \times 5 = \textcircled{2^2} \times 3 \times \textcircled{5}$
GCF $= 2^2 \times 5 = 20$

3.78 $14 = 2 \times 7$
$15 = 3 \times 5$
GCF $= 1$

3.79 $275 = 5 \times 5 \times 11 = \textcircled{5^2} \times \textcircled{11}$
$1{,}100 = 5 \times 5 \times 2 \times 2 \times 11$
$= \textcircled{5^2} \times 2^2 \times \textcircled{11}$
GCF $= 5^2 \times 11 = 275$

3.80 $546 = 2 \times \textcircled{3} \times 7 \times \textcircled{13}$
$1{,}521 = 3 \times \textcircled{3} \times 13 \times \textcircled{13}$
$= 3^2 \times 13^2$
GCF $= 3 \times 13 = 39$

3.81 $12 = 2 \times 2 \times 3 = 2^2 \times 3$
$15 = 3 \times 5$
no

3.82 $19 = 1 \times 19$
$57 = 3 \times 19$
no

3.83 $98 = 2 \times 7 \times 7 = 2 \times 7^2$
$343 = 7 \times 7 \times 7 = 7^3$
no

3.84 $77 = 7 \times 11$
$132 = 2 \times 2 \times 3 \times 11$
no

3.85 $100 = 2 \times 2 \times 5 \times 5 = 2^2 \times 5^2$
$117 = 3 \times 3 \times 13 = 3^2 \times 13$
yes

3.86 $99 = 3 \times 3 \times 11 = 3^2 \times 11$
$104 = 2 \times 2 \times 2 \times 13$
$= 2^3 \times 13$
yes

3.87 $66 = 2 \times 3 \times 11$
$77 = 7 \times 11$
no

3.88 $125 = 5 \times 5 \times 5 = 5^3$
$136 = 2 \times 2 \times 2 \times 17$
$= 2^3 \times 17$
yes

3.89 $86 = 2 \times 43$
$103 = 1 \times 103$
yes

3.90 $85 = 5 \times 17$
$102 = 2 \times 3 \times 17$
no

3.91 $65 = 5 \times 13$
$273 = 3 \times 7 \times 13$
no

3.92 $79 = 1 \times 79$
$97 = 1 \times 97$
yes

3.93 $8 = 2^3$
$20 = 2^2 \times 5$
$30 = 2 \times 3 \times 5$
GCF $= 2$

3.94 $34 = 2 \times 17$
$51 = 3 \times 17$
$68 = 2^2 \times 17$
GCF $= 17$

3.95 $56 = 2^3 \times 7$
$66 = 2 \times 3 \times 11$
$76 = 2^2 \times 19$
GCF $= 2$

3.96 $24 = 2^3 \times 3$
$48 = 2^4 \times 3$
$96 = 2^5 \times 3$
$100 = 2^2 \times 5^2$
GCF $= 2^2 = 4$

3.97
$52 = 2^2 \times 13$
$78 = 2 \times 3 \times 13$
$104 = 2^3 \times 13$
$130 = 2 \times 5 \times 13$
$GCF = 2 \times 13 = 26$

3.98
$36 = 2^2 \times 3^2$
$54 = 2 \times 3^3$
$90 = 2 \times 3^2 \times 5$
$180 = 2^2 \times 3^2 \times 5$
$GCF = 2 \times 3^2 = 2 \times 9 = 18$

3.99
$11^2 = 121$
1, 11, and 121 are the only factors of 121. Since the prime factors have no factors, there can be no additional factors.

3.100
Example:
$2 \times 3 \times 5 = 30$
The factors of 30 are 1, 2, 3, 5, 6, 10, 15, and 30.
30 has 8 divisors.

3.101
$630 \overline{)875}$ = 1 R245
875 − 630 = 245

$245 \overline{)630}$ = 2 R140
630 − 490 = 140

$140 \overline{)245}$ = 1 R105
245 − 140 = 105

$105 \overline{)140}$ = 1 R35
140 − 105 = 35

$(35) \overline{)105}$ = 3 R0
105 − 105 = 0

The GCF is 35.
Check: $630 = 2 \times 5 \times 7 \times 3^2$
$875 = 5^3 \times 7$
$GCF = 5 \times 7 = 35$

3.102
$864 \overline{)936}$ = 1 R72
936 − 864 = 72

$(72) \overline{)864}$ = 12 R0
86 − 72 = 14; 144 − 144 = 0

The GCF is 72.
Check: $864 = 2^5 \times 3^3$
$936 = 2^3 \times 3^2 \times 13$
$GCF = 2^3 \times 3^2$
$= 8 \times 9$
$= 72$

3.103
$8{,}464 \overline{)11{,}500}$ = 1 R3,036
11,500 − 8464 = 3036

$3{,}036 \overline{)8{,}464}$ = 2 R2,392
8,464 − 6072 = 2392

$2{,}392 \overline{)3{,}036}$ = 1 R644
3,036 − 2392 = 644

$644 \overline{)2{,}392}$ = 3 R460
2,392 − 1932 = 460

$460 \overline{)644}$ = 1 R184
644 − 460 = 184

$184 \overline{)460}$ = 2 R92
460 − 368 = 92

$(92) \overline{)184}$ = 2 R0
184 − 184 = 0

The GCF = 92.

3.103 cont.

Check: $8,464 = 2^4 \times 23^2$

$11,500 = 2^2 \times 5^3 \times 23$

$GCF = 2^2 \times 23$

$= 4 \times 23 = 92$

3.104

$38 = 2 \times 19$

$57 = 3 \times 19$

$GCF = 19$

$\frac{38 \div 19}{57 \div 19} = \frac{2}{3}$

3.105

$68 = 2^2 \times 17$

$85 = 5 \times 17$

$GCF = 17$

$\frac{68 \div 17}{85 \div 17} = \frac{4}{5}$

3.106

$58 = 2 \times 29$

$87 = 3 \times 29$

$GCF = 29$

$\frac{58 \div 29}{87 \div 29} = \frac{2}{3}$

3.107

$57 = 3 \times 19$

$76 = 2^2 \times 19$

$GCF = 19$

$\frac{57 \div 19}{76 \div 19} = \frac{3}{4}$

3.108

$69 = 3 \times 23$

$92 = 2^2 \times 23$

$GCF = 23$

$\frac{69 \div 23}{92 \div 23} = \frac{3}{4}$

3.109

$52 = 2^2 \times 13$

$91 = 7 \times 13$

$GCF = 13$

$\frac{52 \div 13}{91 \div 13} = \frac{4}{7}$

3.110

$62 = 2 \times 31$

$93 = 3 \times 31$

$GCF = 31$

$\frac{62 \div 31}{93 \div 31} = \frac{2}{3}$

3.111

$85 = 5 \times 17$

$102 = 2 \times 3 \times 17$

$GCF = 17$

$\frac{85 \div 17}{102 \div 17} = \frac{5}{6}$

3.112

$60 = 2^2 \times 3 \times 5$

$126 = 2 \times 3^2 \times 7$

$LCM = 2^2 \times 3 \times 5 \times 3 \times 7$

$= 1,260$

3.113

$70 = 2 \times 5 \times 7$

$175 = 7 \times 5^2$

$LCM = 2 \times 5^2 \times 7 = 350$

3.114

$39 = 3 \times 13$

$65 = 5 \times 13$

$LCM = 3 \times 5 \times 13 = 195$

3.115

$12 = 2^2 \times 3$

$15 = 3 \times 5$

$18 = 2 \times 3^2$

$LCM = 2^2 \times 3^2 \times 5 = 180$

3.116

$18 = 3^2 \times 2$

$24 = 3 \times 2^3$

$63 = 3^2 \times 7$

$LCM = 3^2 \times 2^3 \times 7 = 504$

3.117

$75 = 3 \times 5^2$

$500 = 2^2 \times 5^3$

$90 = 2 \times 3^2 \times 5$

$LCM = 2^2 \times 3^2 \times 5^3 = 4,500$

3.118

multiples of 2: 2, 4, 6, 8, 10

multiples of 5: 5, 10

10 is the LCM

3.119

multiples of 3: 3, 6, 9, 12, 15, 18, 21

multiples of 7: 7, 14, 21

21 is the LCM

3.120 multiples of 17: 17, 34, 51, 68, 85, 102, 119, 136, 153, 170, 187, 204, 221, 238, 255, 272, 289, 306, 323
multiples of 19: 19, 38, 57, 76, 95, 114, 133, 152, 171, 190, 209, 228, 247, 266, 285, 304, 323
323 is the LCM

3.121 multiples of 2: 2, 4, 6, 8, 10, 12, 14, 16, 18, 20, 22, 24, 26, 28, 30, 32, 34, 36, 38, 40, 42, 44, 46, 48, 50, 52, 54, 56, 58, 60, 62, 64, 66, 68, 70, 72, 74, 76, 78, 80, 82, 84, 86, 88, 90, 92, 94, 96, 98, 100, 102, 104, 106, 108, 110, 112, 114, 116, 118, 120, 122, 124, 126, 128, 130, 132, 134, 136, 138, 140, 142, 144, 146, 148, 150, 152, 154, 156, 158, 160, 162, 164, 166, 168, 170, 172, 174, 176, 178, 180, 182, 184, 186, 188, 190, 192, 194, 196, 198, 200, 202, 204, 206, 208, 210
multiples of 3: 3, 6, 9, 12, 15, 18, 21, 24, 27, 30, 33, 36, 39, 42, 45, 48, 51, 54, 57, 60, 63, 66, 69, 72, 75, 78, 81, 84, 87, 90, 93, 96, 99, 102, 105, 108, 111, 114, 117, 120, 123, 126, 129, 132, 135, 138, 141, 144, 147, 150, 153, 156, 159, 162, 165, 168, 171, 174, 177, 180, 183, 186, 189, 192, 195, 198, 201, 204, 207, 210
multiples of 5: 5, 10, 15, 20, 25, 30, 35, 40, 45, 50, 55, 60, 65, 70, 75, 80, 85, 90, 95, 100, 105, 110, 115, 120, 125, 130, 135, 140, 145, 150, 155, 160, 165, 170, 175, 180, 185, 190, 195, 200, 205, 210

3.121 cont.
multiples of 7: 7, 14, 21, 28, 35, 42, 49, 56, 63, 70, 77, 84, 91, 98, 105, 112, 119, 126, 133, 140, 147, 154, 161, 168, 175, 182, 189, 196, 203, 210
210 is the LCM

3.122 multiples of 14: 14, 28
multiples of 28: 28
28 is the LCM

3.123 multiples of 4: 4, 8, 12
multiples of 6: 6, 12
multiples of 12: 12
12 is the LCM

3.124 multiples of 6: 6, 12, 18, 24, 30, 36, 42, 48, 54, 60, 66, 72, 78, 84, 90, 96, 102, 108, 114, 120
multiples of 8: 8, 16, 24, 32, 40, 48, 56, 64, 72, 80, 88, 96, 104, 112, 120
multiples of 10: 10, 20, 30, 40, 50, 60, 70, 80, 90, 100, 110, 120
120 is the LCM

3.125 You must multiply the prime numbers together.

3.126
a. no
b. Since the set of natural numbers is an infinite set, a greater number will always exist that is a multiple of a given natural number.

3.127 When the two numbers are the same.

3.128
a. 2 = 1 x 2
5 = 1 x 5
10 is the LCM

Add 10 minutes to 1:00. The buzzers will sound together again at 1:10.

3.128 cont.

b. the buzzers will sound every 10 minutes at 1:10, 1:20,1:30, 1:40, and 1:50. They will sound together 5 times.

3.129
$5 = 1 \times 5$
$12 = 2^2 \times 3$
$LCD = 5 \times 2^2 \times 3 = 60$

3.130
$8 = 2^3$
$10 = 2 \times 5$
$12 = 2^2 \times 3$
$LCD = 2^3 \times 5 \times 3 = 120$

3.131
$20 = 2^2 \times 5$
$12 = 2^2 \times 3$
$LCD = 2^2 \times 5 \times 3 = 60$

3.132
$12 = 2^2 \times 3$
$10 = 2 \times 5$
$LCD = 2^2 \times 3 \times 5 = 60$

3.133
$12 = 2^2 \times 3$
$32 = 2^5$
$LCD = 2^2 \times 3 \times 2^3 = 96$

3.134
$24 = 2^3 \times 3$
$16 = 2^4$
$12 = 2^2 \times 3$
$LCD = 2^3 \times 3 \times 2 = 48$

3.135
$6 = 2 \times 3$
$8 = 2^3$
$LCD = 2 \times 3 \times 2^2 = 24$

3.136 1 + 2 + 4 + 8 + 16 + 32 + 64 + 127 + 254 + 508 + 1,016 + 2,032 + 4,064 = 8,128

3.137 Count the number of letters in each word.

3.138 The Fibonacci sequence is 1, 1, 2, 3, 5, 8, 13, 21. . . and is found by adding the last two numbers to determine the next. Not counting number 1 twice, the 8th term is 13 + 21 = 34.
9th term: 21 + 34 = 55
10th term: 34 + 55 = 89
11th term: 55 + 89 = 144

3.139 12th term: 89 + 144 = 233
13th term: 144 + 233 = 377
We could have 377 rabbits.

3.140

1
1 1
1 2 1
1 3 3 1
1 4 6 4 1
1 5 10 10 5 1
1 6 15 20 15 6 1
1 7 21 35 35 21 7 1

3.141

2 3 4 5 6 7 8
12,345,679
× 9
111,111,111

3.142

1 2 2 3 4 4 5
0 1 1 1 2 2 2
12,345,679
× 63
37 037 037
74074074
777,777,777

3.143

1 1 2 2 3 3 4
0 1 1 2 2 3 3
12,345,679
× 54
49382716
61728395
666,666,666

3.144 If the multiplier is a multiple of 9, multiplying 12,345,679 will give a number whose digits are all the same.

3.145

$$\begin{array}{r} 11 \\ \underline{11} \\ 11 \\ \underline{11} \\ 121 \end{array}$$

3.146

$$\begin{array}{r} 111 \\ \underline{111} \\ 111 \\ 111 \\ \underline{111} \\ 12{,}321 \end{array}$$

3.147

$$\begin{array}{r} 1{,}111 \\ \underline{1{,}111} \\ 1111 \\ 1111 \\ 1111 \\ \underline{1111} \\ 1{,}234{,}321 \end{array}$$

3.148 Products are symmetrical and count.

3.149

(3 peach trees, 1 apple trees) (2 apples trees, 1 plum trees)

(3, 1) (2, 1)

The orchard has 3 peach trees for each apple tree, or 6 peach trees for 2 apple trees. Since the orchard has 2 apple trees for each plum tree, it has 6 peach trees for each plum tree.

(peach trees, plum trees)

(6, 1)

3.150 (2, 5)

3.151 (0, 1), (0, 2), (0, 3), (1, 1), (1, 2), (1, 3)

3.152 (1, 1), (1 2), (1, 3), (1, 4), (2, 1), (2, 2), (2, 3), (2, 4), (3, 1), (3, 2), (3, 3), (3, 4), (4, 1), (4, 2), (4, 3), (4, 4), (5, 1), (5, 2), (5, 3), (5, 4)

3.153

$(1, 2 \times 1 + 3) = (1, 2 + 3) = (1, 5)$

$(2, 2 \times 2 + 3) = (2, 4 + 3) = (2, 7)$

$(3, 2 \times 3 + 3) = (3, 6 + 3) = (3, 9)$

I. SECTION ONE

1.1 $\frac{2}{3}$; 2 tells how many parts and 3 tells the size of the parts.

1.2 $\frac{5}{6}$; 5 tells how many parts and 6 tells the size of the parts.

1.3 $\frac{5}{8}$

1.4 $\frac{5}{12}$

1.5 $\frac{5}{8}$ is larger than $\frac{5}{12}$ because eighths are larger than twelfths.

1.6 $\frac{5}{12}$ is smaller than $\frac{5}{8}$ because twelfths are smaller than eighths.

1.7 Example:

$\frac{1}{2} = \frac{1 \times 2}{2 \times 2} = \frac{2}{4}$

$\frac{1}{2} = \frac{1 \times 3}{2 \times 3} = \frac{3}{6}$

$\frac{1}{2} = \frac{1 \times 4}{2 \times 4} = \frac{4}{8}$

$\frac{1}{2} = \frac{1 \times 5}{2 \times 5} = \frac{5}{10}$

$\frac{1}{2} = \frac{1 \times 6}{2 \times 6} = \frac{6}{12}$

1.8 Example:

$\frac{3}{4} = \frac{3 \times 2}{4 \times 2} = \frac{6}{8}$

$\frac{3}{4} = \frac{3 \times 3}{4 \times 3} = \frac{9}{12}$

$\frac{3}{4} = \frac{3 \times 4}{4 \times 4} = \frac{12}{16}$

$\frac{3}{4} = \frac{3 \times 5}{4 \times 5} = \frac{15}{20}$

$\frac{3}{4} = \frac{3 \times 6}{4 \times 6} = \frac{18}{24}$

1.9 $\frac{12}{16}$

1.10 $\frac{25}{35}$

1.11 $\frac{2}{3}$

$\frac{4}{6} = \frac{4 \div 2}{6 \div 2} = \frac{2}{3}$

$\frac{8}{12} = \frac{8 \div 4}{12 \div 4} = \frac{2}{3}$

$\frac{6}{9} = \frac{6 \div 3}{9 \div 3} = \frac{2}{3}$

$\frac{12}{18} = \frac{12 \div 6}{18 \div 6} = \frac{2}{3}$

$\frac{14}{21} = \frac{14 \div 7}{21 \div 7} = \frac{2}{3}$

1.12 $\frac{3}{4}$

$\frac{6}{8} = \frac{6 \div 2}{8 \div 2} = \frac{3}{4}$

$\frac{9}{12} = \frac{9 \div 3}{12 \div 3} = \frac{3}{4}$

$\frac{12}{16} = \frac{12 \div 4}{16 \div 4} = \frac{3}{4}$

$\frac{15}{20} = \frac{15 \div 5}{20 \div 5} = \frac{3}{4}$

$\frac{21}{28} = \frac{21 \div 7}{28 \div 7} = \frac{3}{4}$

1.13 $\frac{6}{8} = \frac{6 \div 2}{8 \div 2} = \frac{3}{4}$

1.14 $\frac{12}{15} = \frac{12 \div 3}{15 \div 3} = \frac{4}{5}$

1.15 $\frac{20}{35} = \frac{20 \div 5}{35 \div 5} = \frac{4}{7}$

1.16 $\frac{18}{24} = \frac{18 \div 6}{24 \div 6} = \frac{3}{4}$

1.17 $\frac{28}{42} = \frac{28 \div 14}{42 \div 14} = \frac{2}{3}$

1.18 $\frac{42}{70} = \frac{42 \div 14}{70 \div 14} = \frac{3}{5}$

1.19 $\frac{51}{85} = \frac{51 \div 17}{85 \div 17} = \frac{3}{5}$

1.20 $\frac{35}{91} = \frac{35 \div 7}{91 \div 7} = \frac{5}{13}$

1.21 $\frac{10}{100} = \frac{10 \div 10}{100 \div 10} = \frac{1}{10}$

1.22 $\frac{18}{162} = \frac{18 \div 18}{162 \div 18} = \frac{1}{9}$

1.23 Example:

$\frac{2}{3} = \frac{2 \times 2}{3 \times 2} = \frac{4}{6}$

$\frac{2}{3} = \frac{2 \times 3}{3 \times 3} = \frac{6}{9}$

$\frac{2}{3} = \frac{2 \times 4}{3 \times 4} = \frac{8}{12}$

$\frac{2}{3} = \frac{2 \times 8}{3 \times 8} = \frac{16}{24}$

$\frac{2}{3} = \frac{2 \times 16}{3 \times 16} = \frac{32}{48}$

$\frac{2}{3} = \frac{2 \times 32}{3 \times 32} = \frac{64}{96}$

1.24 Example:

$\frac{5}{7} = \frac{5 \times 2}{7 \times 2} = \frac{10}{14}$

$\frac{5}{7} = \frac{5 \times 3}{7 \times 3} = \frac{15}{21}$

$\frac{5}{7} = \frac{5 \times 4}{7 \times 4} = \frac{20}{28}$

$\frac{5}{7} = \frac{5 \times 8}{7 \times 8} = \frac{40}{56}$

$\frac{5}{7} = \frac{5 \times 16}{7 \times 16} = \frac{80}{112}$

$\frac{5}{7} = \frac{5 \times 32}{7 \times 32} = \frac{160}{224}$

1.25
a. $12 \div 2 = 6$
$\frac{1}{2} = \frac{1 \times 6}{2 \times 6} = \frac{6}{12}$

b. $12 \div 4 = 3$
$\frac{2}{3} = \frac{2 \times 4}{3 \times 4} = \frac{8}{12}$

c. $12 \div 4 = 3$
$\frac{3}{4} = \frac{3 \times 3}{4 \times 3} = \frac{9}{12}$

d. $12 \div 6 = 2$
$\frac{5}{6} = \frac{5 \times 2}{6 \times 2} = \frac{10}{12}$

e. $12 \div 4 = 3$
$\frac{1}{4} = \frac{1 \times 3}{4 \times 3} = \frac{3}{12}$

1.26
a. $48 \div 6 = 8$
$\frac{1}{6} = \frac{1 \times 8}{6 \times 8} = \frac{8}{48}$

b. $48 \div 3 = 16$
$\frac{1}{3} = \frac{1 \times 16}{3 \times 16} = \frac{16}{48}$

c. $48 \div 16 = 3$
$\frac{3}{16} = \frac{3 \times 3}{16 \times 3} = \frac{9}{48}$

d. $48 \div 24 = 2$
$\frac{17}{24} = \frac{17 \times 2}{24 \times 2} = \frac{34}{48}$

e. $48 \div 3 = 16$
$\frac{2}{3} = \frac{2 \times 16}{3 \times 16} = \frac{32}{48}$

1.27 $15 \div 3 = 5$
$\frac{3}{5} = \frac{3 \times 5}{5 \times 5} = \frac{15}{25}$

1.28 $28 \div 7 = 4$
$\frac{7}{13} = \frac{7 \times 4}{13 \times 4} = \frac{28}{52}$

1.29
a. $100 \div 10 = 10$
$\frac{3}{10} = \frac{3 \times 10}{10 \times 10} = \frac{30}{100}$

b. $1{,}000 \div 10 = 100$
$\frac{3}{10} = \frac{3 \times 100}{10 \times 100} = \frac{300}{1{,}000}$

c. $10{,}000 \div 10 = 1{,}000$
$\frac{3}{10} = \frac{3 \times 1{,}000}{10 \times 1{,}000} = \frac{3{,}000}{10{,}000}$

1.30
a. $50 \div 10 = 5$
$\frac{7}{10} = \frac{7 \times 5}{10 \times 5} = \frac{35}{50}$

b. $100 \div 10 = 10$
$\frac{7}{10} = \frac{7 \times 10}{10 \times 10} = \frac{70}{100}$

c. $1{,}000 \div 10 = 100$
$\frac{7}{10} = \frac{7 \times 100}{10 \times 100} = \frac{700}{1{,}000}$

1.31 $\frac{7}{3} = 7 \div 3 = 2\frac{1}{3}$

1.32 $\frac{17}{5} = 17 \div 5 = 3\frac{2}{5}$

1.33 dividend

1.34 divisor

1.35 $\frac{28}{5}$

1.36 $\frac{56}{13}$

1.37 $\frac{5}{1}$

1.38 $\frac{41}{1}$

1.39 $\frac{26}{13} = 26 \div 13 = 2$

1.40 $\frac{150}{25} = 150 \div 25 = 6$

1.41 $\frac{8}{2} = 4$

$\frac{14}{7} = 2$

$\frac{5}{1} = 5$

$\frac{8}{4} = 2$

$\frac{42}{21} = 2$

$\frac{32}{8} = 4$

1.42 $\frac{14}{7} = 2$

$\frac{19}{19} = 1$

$\frac{27}{3} = 9$

$\frac{100}{25} = 4$

$\frac{85}{17} = 5$

$\frac{2}{1} = 2$

1.43 $\frac{9}{4} = 9 \div 4 = 2\frac{1}{4}$

1.44 $\frac{22}{5} = 22 \div 5 = 4\frac{2}{5}$

1.45 $\frac{72}{4} = 72 \div 4 = 18\frac{0}{4}$

1.46 $\frac{112}{6} = 112 \div 6 = 18\frac{4}{6} = 18\frac{2}{3}$

1.47 a. $\frac{1,461}{23}$

b. $63 = 63\frac{12}{23}$

$$\begin{array}{r} 63 \\ 23\overline{)1,461} \\ \underline{138} \\ 81 \\ \underline{69} \\ 12 \end{array}$$

1.48 a. $\frac{416}{12} = \frac{416 \div 4}{12 \div 4} = \frac{104}{3}$

b. $34 = 34\frac{8}{12} = 34\frac{2}{3}$

$$\begin{array}{r} 34 \\ 12\overline{)416} \\ \underline{36} \\ 56 \\ \underline{48} \\ 8 \end{array}$$

1.49 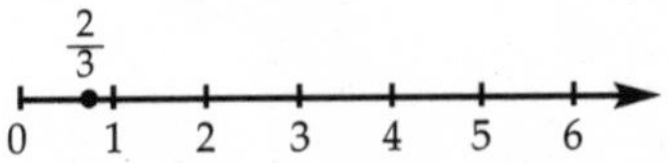

1.50 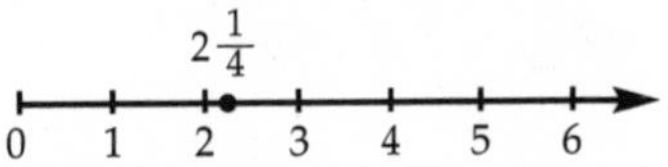

1.51 an unlimited number

1.52 an unlimited number

1.53 a. $1\frac{1}{2}$

b. $2\frac{3}{4}$

c. $4\frac{1}{3}$

d. $5\frac{3}{5}$

1.54 a. $1\frac{1}{3}$

b. $2\frac{1}{2}$

c. $4\frac{5}{6}$

d. $6\frac{3}{4}$

1.55 $1\frac{2}{3} = \frac{1 \times 3 + 2}{3} = \frac{3 + 2}{3} = \frac{5}{3}$

1.56 $3\frac{2}{5} = \frac{3 \times 5 + 2}{5} = \frac{15 + 2}{5} = \frac{17}{5}$

1.57 $6\frac{3}{4} = \frac{6 \times 4 + 3}{4} = \frac{24 + 3}{4} = \frac{27}{4}$

1.58 $12\frac{5}{7} = \frac{12 \times 7 + 5}{7} = \frac{84 + 5}{7} = \frac{89}{7}$

1.59 a. $2\frac{1}{2} = \frac{2 \times 2 + 1}{2} = \frac{4 + 1}{2} = \frac{5}{2}$

b. $4\frac{1}{3} = \frac{4 \times 3 + 1}{3} = \frac{12 + 1}{3} = \frac{13}{3}$

c. $6\frac{3}{4} = \frac{6 \times 4 + 3}{4} = \frac{24 + 3}{4} = \frac{27}{4}$

1.60 a. $2\frac{1}{3} = \frac{2 \times 3 + 1}{3} = \frac{6 + 1}{3} = \frac{7}{3}$

b. $3\frac{1}{2} = \frac{3 \times 2 + 1}{2} = \frac{6 + 1}{2} = \frac{7}{2}$

c. $5\frac{1}{4} = \frac{5 \times 4 + 1}{4} = \frac{20 + 1}{4} = \frac{21}{4}$

1.61 Fourths are greater than fifths, so $2\frac{3}{4} > 2\frac{3}{5}$.

1.62 Elevenths are greater than thirteenths, so $3\frac{2}{11} > 3\frac{2}{13}$.

1.63

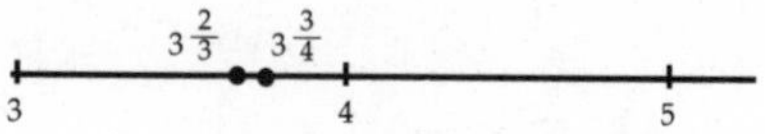

1.64

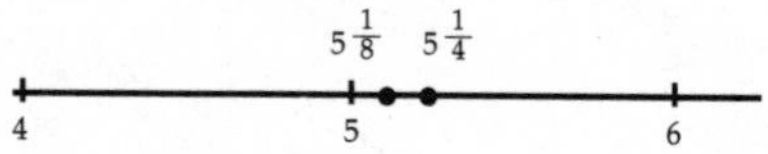

1.65 $\frac{1}{12}, \frac{1}{8}, \frac{1}{4}, \frac{1}{2}, 1\frac{1}{16}, 1\frac{1}{8}, 1\frac{1}{2}, 2\frac{1}{2},$

$3\frac{1}{4}, 4\frac{1}{3}$

1.66 $\frac{5}{8}, \frac{2}{3}, 1\frac{5}{8}, 2\frac{1}{4}, 2\frac{1}{2}, 3\frac{1}{2}, 4\frac{2}{5},$

$5\frac{2}{3}, 8\frac{1}{2}, 16\frac{1}{8}$

1.67 $\frac{4}{3} = 1\frac{1}{3}$

$\frac{7}{2} = 3\frac{1}{2}$

$\frac{32}{3} = 10\frac{2}{3}$

$15\frac{1}{8}, \frac{32}{3}, 9\frac{1}{2}, 3\frac{3}{4}, \frac{7}{2}, 3, 1\frac{1}{2},$

$\frac{4}{3}, \frac{1}{4}, 0$

1.68 $\frac{7}{2} = 3\frac{1}{2}$

$\frac{11}{4} = 2\frac{3}{4}$

$13\frac{7}{8}, 8, 6\frac{2}{3}, 5\frac{1}{4}, \frac{7}{2}, \frac{11}{4}, 2\frac{1}{16},$

$1\frac{1}{4}, \frac{1}{3}, 0$

1.69

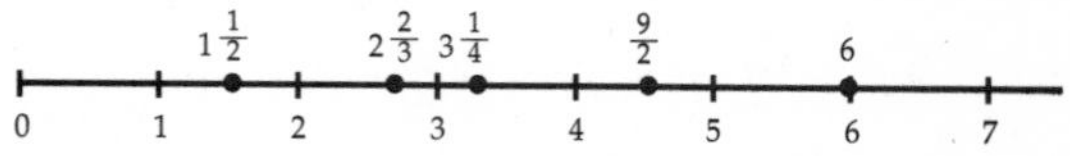

1.70

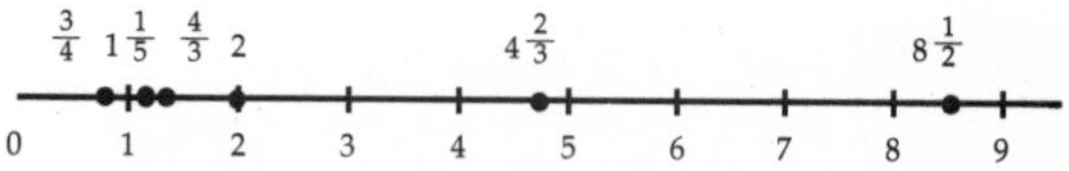

II. SECTION TWO

2.1 ten and six hundredths

2.2 twenty-seven and forty-eight hundredths

2.3 46.67

2.4 110.0100

2.5 a. 12.21

b. twelve and twenty-one hundredths

2.6 a. 206.213

b. two hundred six and two hundred thirteen thousandths

2.7 a. $\frac{237}{100} = 2\frac{37}{100} = 2.37$

b. two and thirty-seven hundredths

2.8 a. $\frac{1{,}467}{1{,}000} = 1\frac{467}{1{,}000} = 1.467$

b. one and four hundred sixty-seven thousandths

2.9 a. 3 is 3 ones
b. 2 is 2 tenths
c. 7 is 7 hundredths

2.10 a. 1 is 1 ten
b. 8 is 8 ones
c. 5 is 5 tenths
d. 7 is 7 hundredths
e. 4 is 4 thousandths

2.11 $30 + 4 + \frac{6}{10} + \frac{5}{100} =$
$34 + 0.6 + 0.05 = 34.65$

2.12 $500 + 60 + 7 + \frac{1}{10} + \frac{3}{100} + \frac{2}{1,000} =$
$567 + 0.1 + 0.03 + 0.002 = 567.132$

2.13 a. 2 is 20,000
b. 2 is 20
c. 2 is $\frac{2}{10}$ or 0.2
d. 2 is $\frac{2}{10,000}$ or 0.0002

2.14 a. 1 is 1 million
b. 1 is 1 thousand
c. 1 is 1 hundredth
d. 1 is 1 ten thousandth

2.15 0.13

2.16 18.2

2.17 3.062, 2.01, 1.111, 1.1, 1.01, 0.81, 0.03, 0.0009

2.18 0.0001, 0.009, 0.14, 1.1, 1.4, 2.014, 8.11, 11.11

2.19 2.6 is larger
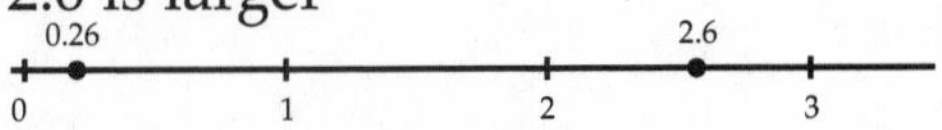

2.20 0.175 is smaller
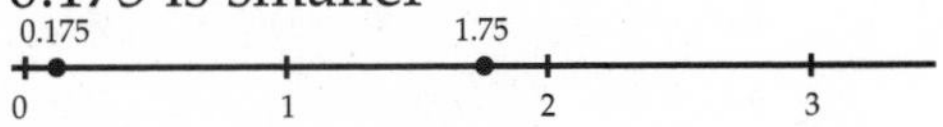

2.21 $100 \div 5 = 20$
$\frac{1}{5} = \frac{1 \times 20}{5 \times 20} = \frac{20}{100} = 0.20$ or 0.2

2.22 $100 \div 25 = 4$
$\frac{8}{25} = \frac{8 \times 4}{25 \times 4} = \frac{32}{100} = 0.32$

2.23 $\frac{3}{200} = \frac{3 \times 5}{200 \times 5} = \frac{15}{1,000} = 0.015$

2.24 $\frac{7}{500} = \frac{7 \times 2}{500 \times 2} = \frac{14}{1,000} = 0.014$

2.25 $100 \div 4 = 25$
$\frac{1}{4} = \frac{1 \times 25}{4 \times 25} = \frac{25}{100} = 0.25$
$100 \div 2 = 50$
$\frac{1}{2} = \frac{1 \times 50}{2 \times 50} = \frac{50}{100} = 0.50$ or 0.5

2.26 $100 \div 5 = 20$
$\frac{3}{5} = \frac{3 \times 20}{5 \times 20} = \frac{60}{100} = 0.60$ or 0.6
$100 \div 4 = 25$
$\frac{3}{4} = \frac{3 \times 25}{4 \times 25} = \frac{75}{100} = 0.75$

2.27 $0.5 = \frac{5}{10} = \frac{5 \div 5}{10 \div 5} = \frac{1}{2}$

2.28 $0.15 = \frac{15}{100} = \frac{15 \div 5}{100 \div 5} = \frac{3}{20}$

2.29 $\frac{3}{1,000}$

2.30 $0.0018 = \frac{18}{10,000} = \frac{18 \div 2}{10,000 \div 2} = \frac{9}{5,000}$

2.31 $1.25 = 1\frac{25}{100} = 1\frac{25 \div 25}{100 \div 25} = 1\frac{1}{4}$

2.32 $5.625 = 5\frac{625}{1,000} = 5\frac{625 \div 125}{1,000 \div 125} = 5\frac{5}{8}$

2.33 $0.44\frac{4}{9}$

2.34 0.375

2.35 $\frac{18}{24} = \frac{18 \div 6}{24 \div 6} = \frac{3}{4} = 0.75$

2.36 $\frac{16}{48} = \frac{16 \div 16}{48 \div 16} = \frac{1}{3} = 0.33\frac{1}{3}$

2.37 $\frac{4}{5}$

2.38 $\frac{2}{9}$

2.39 2.75

2.40 $3.33\frac{1}{3}$

2.41 $1\frac{7}{8}$

2.42 $4\frac{1}{6}$

2.43 5 out of 100 or 0.05 or $\frac{5}{100}$

2.44 18 out of 100 or 0.18 or $\frac{18}{100}$

2.45 27%

2.46 53%

2.47 31%

2.48 85%

2.49 Examples:
%, hundredths, out of 100, part of 100, in each 100, divided by 100, per 100

2.50 Drop the percent sign and move the decimal point two places to the left.
11% = 0.11

2.51 Drop the percent sign and move the decimal point two places to the left.
33% = 0.33

2.52 Move the decimal point two places to the right and add the percent sign.
0.27 = 27%

2.53 Move the decimal point two places to the right and add the percent sign.
0.61 = 61%

2.54 Drop the percent sign and move the decimal point two places to the left.
29.8% = 0.298

2.55 Drop the percent sign and move the decimal point two places to the left.
2.6% = 0.026

2.56 Move the decimal point two places to the right and add the percent sign.
0.157 = 15.7%

2.57 Move the decimal point two places to the right and add the percent sign.
0.807 = 80.7%

2.58 Drop the percent sign and write 100 as the denominator. $17\% = \frac{17}{100}$

2.59 Drop the percent sign and write 100 as the denominator. $71\% = \frac{71}{100}$

2.60 $\frac{1}{6} = 0.16\frac{2}{3} = 16\frac{2}{3}\%$

2.61 $\frac{4}{9} = 0.44\frac{4}{9} = 44\frac{4}{9}\%$

2.62 $28\% = \frac{28}{100} = \frac{28 \div 4}{100 \div 4} = \frac{7}{25}$

2.63 $85\% = \frac{85}{100} = \frac{85 \div 5}{100 \div 5} = \frac{17}{20}$

2.64 Move the decimal point two places to the right and add the percent sign.
0.141 = 14.1%

2.65 Move the decimal point two places to the right and add the percent sign.
0.571 = 57.1%

III. SECTION THREE

3.1 9:11 or $\frac{9}{11}$

3.2 27:71 or $\frac{27}{71}$

3.3 8:12 = 2:3 or $\frac{2}{3}$

3.4 15:36 = 5:12 or $\frac{5}{12}$

3.5 Change dimes to nickels.
7 dimes = 14 nickels
The ratio is 5:14 or $\frac{5}{14}$.

3.6 Change yards to inches.
2 yards = 2 x 36 = 72 inches
$\frac{18}{72} = \frac{18 \div 18}{72 \div 18} = \frac{1}{4}$ or 1:4

3.7 Change days to hours.
7 days = 7 x 24 = 168 hours
$\frac{168}{6} = \frac{168 \div}{6} = \frac{28}{1}$ or 28:1

3.8 Change 1 mile to feet.
1 mile = 5,280 feet.
$\frac{5{,}280}{150} = \frac{5{,}280 \div 30}{150 \div 30} = \frac{176}{5}$ or 176:5

3.9 $\frac{37}{74} = \frac{37 \div 37}{74 \div 37} = \frac{1}{2}$ or 1:2

$\frac{1}{2} = 50\%$

3.10 $\frac{8}{32} = \frac{8 \div 8}{32 \div 8} = \frac{1}{4}$ or 1:4

$\frac{1}{4} = 25\%$

3.11 $\frac{2}{3}$; $\frac{5}{7}$; $\frac{8}{1}$; $\frac{27}{31}$; $\frac{15}{100} = \frac{15 \div 5}{100 \div 5} = \frac{3}{20}$

3.12 2:3, 5:6, 9:7, 2:1, 36:53

3.13 $\frac{110}{125} = \frac{110 \div 5}{125 \div 5} = \frac{22}{25}$ or 22:25

3.14 $\frac{15}{20} = \frac{15 \div 5}{20 \div 5} = \frac{3}{4}$ or 3:4

3.15 The means are 6 and 10; the extremes are 5 and 12.
6 x 10 = 60 and 5 x 12 = 60;
6 x 10 = 5 x 12 is true.

3.16 The means are 9 and 14; the extremes are 7 and 18.
9 x 14 = 126 and 7 x 18 = 126;
9 x 14 = 7 x 18 is true.

3.17 5 x 6 = 3 x ?
30 = 3 x ?
30 ÷ 3 = ?
10 = ?

3.18 9 x 16 = 8 x ?
144 = 8 x ?
144 ÷ 8 = ?
18 = ?

3.19 $\frac{5}{7} = \frac{10}{14}$; 7 x 10 = 70 and 5 x 14 = 70;
7 x 10 = 5 x 14 is true

3.20 $\frac{6}{11} = \frac{12}{22}$; 11 x 12 = 132 and 6 x 22 = 132; 11 x 12 = 6 x 22 is true

3.21 2:3 = 39:?
3 x 39 = 2 x ?
117 = 2 x ?
117 ÷ 2 = ?
$58.50 = ?

3.22 5:25 = 35:?
1:5 = 35:?
5 x 35 = 1 x ?
175 = ?
175 minutes

3.23 2:100 = 6:?
1:50 = 6:?
50 x 6 = 1 x ?
300 = ?
300 pounds

3.24 3:50 = 9:?
50 x 9 = 3 x ?
450 = 3 x ?
450 ÷ 3 = ?
150 = ?
150¢ or $1.50

3.25 $\frac{DE}{AB} = \frac{EF}{BC} = \frac{DF}{AC}$

3.26 $\frac{EG}{HJ} = \frac{GF}{JI} = \frac{EF}{HI}$

3.27 $\frac{DE}{AB} = \frac{DF}{AC}$

$\frac{9}{6} = \frac{12}{AC}$

6 x 12 = 9 x AC
72 = 9 x AC
72 ÷ 9 = AC
8 = AC

3.28 $\frac{GF}{JI} = \frac{EF}{HI}$

$\frac{4}{5} = \frac{8}{HI}$

$5 \times 8 = 4 \times HI$

$40 = 4 \times HI$

$40 \div 4 = HI$

$10 = HI$

3.29 Example:

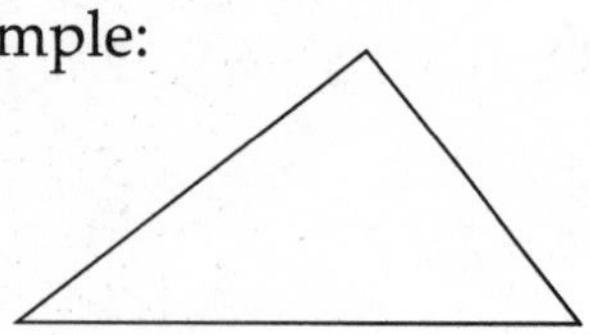

3.30 Example:

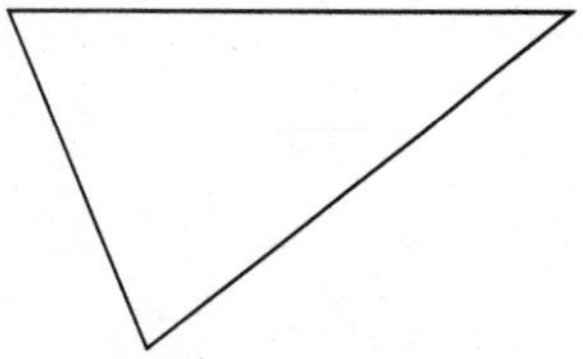

3.31 $AB:EF = 6:3 = 2:1$

Each side of rectangle *ABCD* is twice as long as the corresponding side of rectangle *EFGH*.

$GH = EF = 3$

$EH = FG = \frac{1}{2}BC = \frac{1}{2} \times 14 = 7$

$FG = \frac{1}{2}BC = \frac{1}{2} \times 14 = 7$

$CD = AB = 6$

$AD = BC = 14$

3.32 $QR = PQ = 2$

$RS = PQ = 2$

$PS = PQ = 2$

$UV = TU = 5$

$VW = TU = 5$

$TW = TU = 5$

3.33 30:3 = ?:4

10:1 = ?:4

1 x ? = 10 x 4

? = 40 feet

3.34 5:50 = 6:?

1:10 = 6:?

10 x 6 = 1 x ?

60 = ?

60 feet

3.35 1:1 = ?:8

1 x ? = 1 x 8

? = 8 cm

1:1 = ?:7

1 x ? = 1 x 7

? = 7 cm

1:1 = ?:5

1 x ? = 1 x 5

? = 5 cm

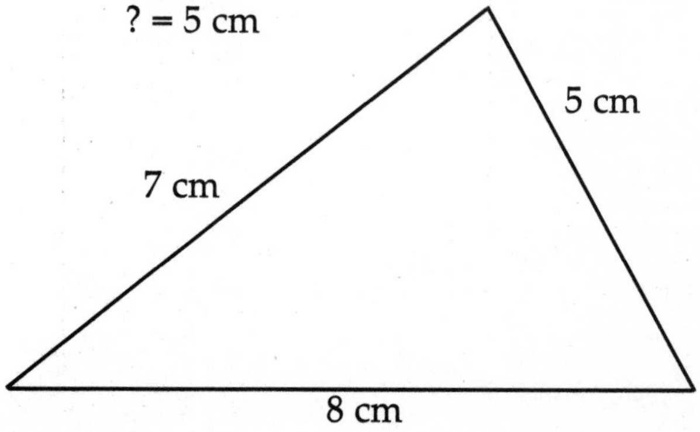

3.36 1:10 = L:50

10 x L = 1 x 50

10 x L = 50

L = 50 ÷ 10

L = 5 cm

1:10 = W:30

10 x W = 1 x 30

10 x W = 30

W = 30 ÷ 10

W = 3 cm

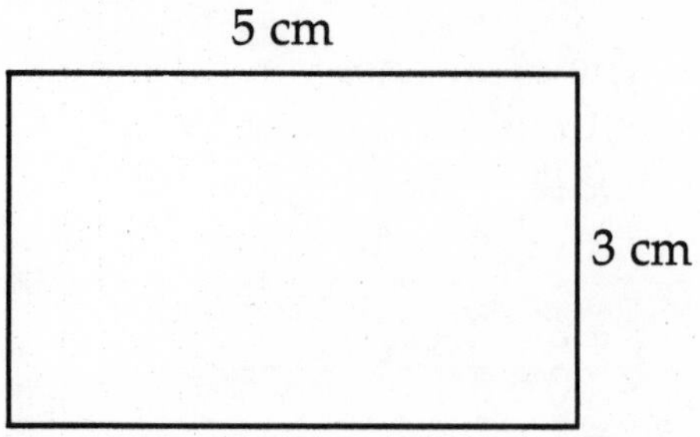

3.37
$1:200 = ?:1,000$
$200 \times ? = 1 \times 1,000$
$200 \times ? = 1,000$
$? = 1,000 \div 200$
$? = 5$ inches

3.38
$1:50 = 8:?$
$50 \times 8 = 1 \times ?$
$400 = ?$
400 miles

3.39
$1:8 = 12:L$
$8 \times 12 = 1 \times L$
$96 = L$

$1:8 = 12:W$
$8 \times 8 = 1 \times W$
$64 = W$
96 feet by 64 feet

3.40
$1:5 = 36:?$
$5 \times 36 = 1 \times ?$
$180 = ?$

$1:5 = 41:?$
$5 \times 41 = 1 \times ?$
$205 = ?$

$1:5 = 27:?$
$5 \times 27 = 1 \times ?$
$135 = ?$
180 m, 205 m, 135 m

3.41
a. 10^2
b. 10^4

3.42 10^6

3.43
a. 10^0
b. 10^{-2}

3.44
a. 10^{-3}
b. 10^{-5}

3.45 186×10^{-3}

3.46 $8,746 \times 10^{-4}$

3.47 2.1×0.001 or 0.0021

3.48 8.142×0.0001 or 0.0008142

3.49
1 cm = 10 mm, so
8 cm = $8 \times 10 = 80$ mm

3.50
1 dm = 10 cm, so
14 dm = $14 \times 10 = 140$ cm

3.51
1 mm = 0.001 m, so
9 mm = $9 \times 0.001 = 0.009$ m

3.52
1 mm = 0.01 dm, so
28 mm = $28 \times 0.01 = 0.28$ dm

3.53
1 m = 100 cm, so
13.8 m = $13.8 \times 100 = 1,380$ cm

3.54
1 dm = 100 mm, so
47.5 dm = $47.5 \times 100 = 4,750$ mm

3.55
1 dm = 0.1 m, so
7.64 dm = $7.64 \times 0.1 = 0.764$ m

3.56
1 mm = 0.001 m, so
31.76 mm = $31.76 \times 0.001 = 0.03176$ m

3.57
1 km = 1,000,000 mm, so
1.8 km = $1.8 \times 1,000,000 =$
1,800,000 mm

3.58
1 mm = 0.000001 km, so
1,461 mm = $1,461 \times 0.000001 =$
0.001461 km

3.59 $\frac{1}{4} = 0.25$ or 0.250

3.60 $\frac{3}{5} = 0.6$ or 0.60

3.61 $1\frac{1}{2}$

3.62 $3\frac{1}{8}$

3.63 $\frac{3}{1} = 3$ or 3.0

$\frac{1}{2} = 0.5$ or 0.50

$\frac{0}{5} = 0$ or 0.0

3.64 $\frac{5}{1} = 5$ or 5.0

$\frac{1}{5} = 0.2$ or 0.20

$\frac{0}{7} = 0$ or 0.0

$\frac{1}{16} = 0.0625$ or 0.06250

3.65 $\frac{1}{9} = 0.11\frac{1}{9} = 0.111\ldots = 0.\overline{1}$

3.66 $\frac{4}{9} = 0.44\frac{4}{9} = 0.444\ldots = 0.\overline{4}$

3.67 $\frac{1}{6} = 0.16\frac{2}{3} = 0.1666\ldots = 0.1\overline{6}$

3.68 $\frac{5}{6} = 0.83\frac{1}{3} = 0.8333\ldots = 0.8\overline{3}$

3.69 $\frac{3}{5} = 0.6$ or 0.60

3.70 $\frac{1}{8} = 0.125$ or 0.1250

3.71 Drop the % sign and move the decimal point two places to the left.
0.4% = 0.004

3.72 Drop the % sign and move the decimal point two places to the left.
0.05% = 0.0005

3.73 Move the decimal point two places to the right and add the % sign.
0.007 = 0.7%

3.74 Move the decimal point two places to the right and add the % sign.
0.0008 = 0.08%

3.75 $\frac{1}{2} = 0.5$; $\frac{1}{2}\% = 0.5\%$

0.5% = 0.005

3.76 $\frac{2}{5} = 0.4\%$; $\frac{2}{5}\% = 0.4\%$

0.4% = 0.004

3.77 0.0025 = 0.25%
$0.25 = \frac{1}{4}$; $0.25\% = \frac{1}{4}\%$

3.78 0.00125 = 0.125%
$0.125 = \frac{1}{8}$; $0.125\% = \frac{1}{8}\%$

3.79 1 mill = 0.1¢, so
5 mills = 5 x 0.1 = 0.5¢

3.80 1 mill = 0.1¢, so
27 mills = 27 x 0.1 = 2.7¢

3.81 Drop the % sign and move the decimal point to the left two places.
400% = 4.00 or 4

3.82 Drop the % sign and move the decimal point to the left two places.
146% = 1.46

3.83 Move the decimal point two places to the right and add the % sign.
2.9 = 290%

3.84 Move the decimal point two places to the right and add the % sign.
8.61 = 861%

3.85 A whole = 100%;
127% – 100% = 27%.

3.86 A whole = 100%;
186% – 100% = 86%.

3.87 $\frac{12}{6} = \frac{12 \div 6}{6 \div 6} = \frac{2}{1}$

$\frac{2}{1} = 2 = 200\%$

3.88 $\frac{27}{9} = \frac{27 \div 9}{9 \div 9} = \frac{3}{1}$

$\frac{3}{1} = 3 = 300\%$

3.89 $\frac{25}{10} = \frac{25 \div 5}{10 \div 5} = \frac{5}{2}$ or $2\frac{1}{2}$

$2\frac{1}{2} = 2.5 = 250\%$

3.90 $\frac{15}{10} = \frac{15 \div 5}{10 \div 5} = \frac{3}{2}$ or $1\frac{1}{2}$

$1\frac{1}{2} = 1.5 = 150\%$

3.91 6¢ out of every dollar goes for tax

3.92 4¢ out of every dollar goes for tax

3.93 $50 – $40 = $10
$\frac{10}{50} = \frac{10 \div 10}{50 \div 10} = \frac{1}{5} = 0.2 = 20\%$

3.94 $100 – $75 = $25
$\frac{25}{100} = \frac{25 \div 25}{100 \div 25} = \frac{1}{4} = 0.25 = 25\%$

3.95 12¢ = $0.12
$\frac{0.12}{3} = 0.04 = 4\%$ or

$3 = 300¢
$\frac{12}{300} = \frac{12 \div 3}{300 \div 3} = \frac{4}{100} = 0.04 = 4\%$

3.96 $\frac{2}{20} = \frac{2 \div 2}{20 \div 2} = \frac{1}{10} = 0.1 = 10\%$

3.97 $\frac{6}{7}$

3.98 $\frac{6}{12}$ or $\frac{6 \div 6}{12 \div 6} = \frac{1}{2}$

3.99 $\frac{1}{4} + \frac{3}{8} = \frac{2}{8} + \frac{3}{8} = \frac{5}{8}$

3.100 $\frac{1}{6} + \frac{7}{12} = \frac{2}{12} + \frac{7}{12} = \frac{9}{12}$ or

$\frac{9 \div 3}{12 \div 3} = \frac{3}{4}$

3.101 $\frac{14}{15}$

3.102 $\frac{24}{18}$ or $\frac{24 \div 6}{18 \div 6} = \frac{4}{3}$ or $1\frac{1}{3}$

3.103 $\frac{1}{2} + \frac{1}{3} + \frac{1}{6} + \frac{1}{12} =$

$\frac{6}{12} + \frac{4}{12} + \frac{2}{12} + \frac{1}{12} = \frac{13}{12}$ or $1\frac{1}{12}$

3.104 $\frac{2}{5} + \frac{3}{10} + \frac{7}{20} + \frac{9}{40} =$

$\frac{16}{40} + \frac{12}{40} + \frac{14}{40} + \frac{9}{40} = \frac{51}{40}$ or $1\frac{11}{40}$

3.105
$$\begin{array}{r} 0.2 \\ +\,0.7 \\ \hline 0.9 \end{array}$$

3.106
$$\begin{array}{r} 0.31 \\ +\,0.54 \\ \hline 0.85 \end{array}$$

3.107
$$\begin{array}{r} 0.08 \\ +\,2.6 \\ \hline 2.68 \end{array}$$

3.108
$$\begin{array}{r} 5.17 \\ +\,2.019 \\ \hline 7.189 \end{array}$$

3.109
$$\begin{array}{r} 1.2 \\ 2.61 \\ 3.05 \\ +\,5 \\ \hline 11.86 \end{array}$$

3.110
$$\begin{array}{r} 8.1 \\ 4.08 \\ 0.514 \\ +\,2 \\ \hline 14.694 \end{array}$$

3.111 $\frac{1}{4} = 0.25$

$$\begin{array}{r} 0.5 \\ 0.27 \\ \frac{3}{5} = \underline{0.6} \\ 1.62 \end{array}$$

3.112 $\frac{1}{8} = 0.125$

$$\begin{array}{r} \frac{3}{2} = 1.5 \\ 4.15 \\ \frac{3}{4} = \underline{0.75} \\ 6.525 \end{array}$$

The solutions to Problems 3.113 through 3.132 are shown for students who do not have access to calculators.

3.113
$$\begin{array}{r} 8.6 \\ 4.1 \\ \underline{+\ 8.8} \\ 21.5 \end{array}$$

3.114
$$\begin{array}{r} 16.4 \\ 47.6 \\ \underline{+\ 0.115} \\ 64.115 \end{array}$$

3.115
$$\begin{array}{r} 81.114 \\ 16.1 \\ \underline{+\ 9.8} \\ 107.014 \end{array}$$

3.116
$$\begin{array}{r} 5 \\ 27.1189 \\ \underline{+\ 109.8} \\ 141.9189 \end{array}$$

3.117
$$\begin{array}{r} 41.81 \\ \underline{-\ 34.78} \\ 7.03 \end{array}$$

3.118
$$\begin{array}{r} 101.6 \\ \underline{-\ 69.8} \\ 31.8 \end{array}$$

3.119
$$\begin{array}{r} 8.1114 \\ \underline{-\ 4.8} \\ 3.3114 \end{array}$$

3.120
$$\begin{array}{r} 327.010 \\ \underline{-\ 218.962} \\ 108.048 \end{array}$$

3.121
$$\begin{array}{r} 6.23 \\ \underline{\times\ 5.1} \\ 623 \\ \underline{3115\ } \\ 31.773 \end{array}$$

3.122
$$\begin{array}{r} 15.6 \\ \underline{\times\ 41.9} \\ 1404 \\ 156\ \\ \underline{624\ \ } \\ 653.64 \end{array}$$

3.123
$$\begin{array}{r} 0.066 \\ \underline{\times\ 7.14} \\ 0264 \\ 066\ \\ \underline{0462\ \ } \\ 0.47124 \end{array}$$

3.124
$$\begin{array}{r} 147.816 \\ \underline{2.68} \\ 1182528 \\ 886896\ \\ \underline{295632\ \ } \\ 396.14688 \end{array}$$

3.125
$$\begin{array}{r} 407.2 \\ 2\overline{)814.4} \\ \underline{8\ \ \ \ } \\ 014 \\ \underline{14} \\ 04 \\ \underline{4} \\ 0 \end{array}$$

3.126

5.415
4)21.660
20
16
16
06
4
20
20
0

3.127

22.326923
52)1161.000000
104
121
104
170
156
140
104
360
312
480
468
120
104
160
156
4

3.128

5.396
775)4181.900
3875
3069
2325
7440
6975
4650
4650
0

3.129

$\frac{6}{11}$ = 11)6.0000 = 0.5454
55
50
44
60
55
50
44
6

3.130

$\frac{22}{7}$ = 7)22.0000000 = 3.1428571
21
10
7
30
28
20
14
60
56
40
35
50
49
10
7
3

3.131

$\frac{4}{7}$ = 7)4.0000000 = 0.5714285
35
50
49
10
7
30
28
20
14
60
56
40
35
5

$3\frac{4}{7} = 3.5714285$

3.132

$\frac{23}{25}$ = 25)23.00 = 0.92
225
50
50
0

$15\frac{23}{25} = 15.92$

3.133 $4 \times 4 = 16$ or $4^2 = 16$, so
$\sqrt{16} = 4$. 4 is a rational number.

3.134 $5 \times 5 = 25$ or $5^2 = 25$, so
$\sqrt{25} = 5$. 5 is a rational number.

3.135 $? \times ? = 7$ or $?^2 = 7$, so
$\sqrt{7}$ is an irrational number.

3.136 $? \times ? = 11$ or $?^2 = 11$, so
$\sqrt{11}$ is an irrational number.

3.137 $\sqrt{3}$

3.138 $\sqrt{64} = 8$

I. SECTION ONE

1.1 $\frac{2}{5}$

1.2 $\frac{5}{7}$

1.3 $\frac{7}{13}$

1.4 $\frac{10}{11}$

1.5 $\frac{5}{7}$

1.6 $\frac{13}{17}$

1.7 $\frac{6}{11}$

1.8 $\frac{23}{49}$

1.9 $\frac{6}{8} = \frac{3}{4}$

Note: This problem and some following problems have answers that can be simplified. Count correct with either answer.

1.10 $\frac{11}{13}$

1.11 $\frac{4}{5}$

1.12 $\frac{12}{12} = 1$

1.13 $\frac{32}{33}$

1.14 $\frac{20}{32} = \frac{20}{32} \div \frac{4}{4} = \frac{5}{8}$

1.15 $\frac{14}{17}$

1.16 $\frac{10}{16} = \frac{10}{16} \div \frac{2}{2} = \frac{5}{8}$

1.17 $\frac{7}{14} = \frac{7}{14} \div \frac{7}{7} = \frac{1}{2}$

1.18 $\frac{10}{11}$

1.19 $\frac{20}{23}$

1.20 $\frac{22}{24} = \frac{22}{24} \div \frac{2}{2} = \frac{11}{12}$

1.21 $\frac{2}{6} = \frac{2}{6} \div \frac{2}{2} = \frac{1}{3}$

1.22 $\frac{4}{8} = \frac{4}{8} \div \frac{4}{4} = \frac{1}{2}$

1.23 $\frac{9}{13}$

1.24 $\frac{8}{16} = \frac{8}{16} \div \frac{8}{8} = \frac{1}{2}$

1.25 $\frac{12}{36} = \frac{12}{36} \div \frac{12}{12} = \frac{1}{3}$

1.26 $\frac{18}{72} = \frac{18}{72} \div \frac{18}{18} = \frac{1}{4}$

1.27 $\frac{10}{19}$

1.28 $\frac{9}{108} = \frac{9}{108} \div \frac{9}{9} = \frac{1}{12}$

1.29 $\frac{6}{8} = \frac{6}{8} \div \frac{2}{2} = \frac{3}{4}$

1.30 $\frac{14}{16} = \frac{14}{16} \div \frac{2}{2} = \frac{7}{8}$

1.31 $\frac{25}{32}$

1.32 $\frac{36}{144} = \frac{36}{144} \div \frac{36}{36} = \frac{1}{4}$

1.33 $\frac{17}{23}$

1.34 $\frac{80}{150} = \frac{80}{150} \div \frac{10}{10} = \frac{8}{15}$

1.35 $\frac{86}{216} = \frac{86}{216} \div \frac{2}{2} = \frac{43}{108}$

1.36 $\frac{75}{175} = \frac{75}{175} \div \frac{25}{25} = \frac{3}{7}$

1.37 $\frac{30}{180} = \frac{30}{180} \div \frac{30}{30} = \frac{1}{6}$

1.38 $\frac{102}{180} = \frac{102}{180} \div \frac{6}{6} = \frac{17}{30}$

1.39 $\frac{15}{15} = \frac{15}{15} \div \frac{15}{15} = \frac{1}{1} = 1$

1.40 $\frac{26}{44} = \frac{26}{44} \div \frac{2}{2} = \frac{13}{22}$

1.41 $\frac{72}{77}$

1.42 $\frac{44}{48} = \frac{44}{48} \div \frac{4}{4} = \frac{11}{12}$

1.43 $\frac{20}{25} = \frac{20}{25} \div \frac{5}{5} = \frac{4}{5}$

1.44 $\frac{14}{21} = \frac{14}{21} \div \frac{7}{7} = \frac{2}{3}$

1.45 $\frac{1}{2} = \frac{2}{4}$; $\frac{1}{4} = \frac{1}{4}$; sum: $\frac{3}{4}$

1.46 $\frac{1}{3} = \frac{2}{6}$; $\frac{1}{2} = \frac{3}{6}$; sum: $\frac{5}{6}$

1.47 $\frac{1}{4} = \frac{2}{8}$; $\frac{1}{8} = \frac{1}{8}$; sum: $\frac{3}{8}$

1.48 $\frac{2}{3} = \frac{4}{6}$; $\frac{1}{6} = \frac{1}{6}$; sum: $\frac{5}{6}$

1.49 $\frac{3}{4} = \frac{15}{20}$; $\frac{1}{5} = \frac{4}{20}$; sum: $\frac{19}{20}$

1.50 $\frac{1}{5} = \frac{2}{10}$ or $\frac{1}{5}$; $\frac{2}{10} = \frac{2}{10}$ or $\frac{1}{5}$; sum: $\frac{4}{10} = \frac{2}{5}$

1.51 $\frac{7}{8} = \frac{63}{72}$; $\frac{1}{9} = \frac{8}{72}$; sum: $\frac{71}{72}$

1.52 $\frac{3}{5} = \frac{18}{30}$; $\frac{1}{6} = \frac{5}{30}$; sum: $\frac{23}{30}$

1.53 $\frac{1}{3} = \frac{4}{12}$; $\frac{1}{4} = \frac{3}{12}$; sum: $\frac{7}{12}$

1.54 $\frac{1}{4} = \frac{3}{12}$ or $\frac{1}{4}$; $\frac{6}{12} = \frac{6}{12}$ or $\frac{2}{4}$; sum: $\frac{9}{12} = \frac{3}{4}$

1.55 $\frac{1}{7} = \frac{2}{14}$; $\frac{3}{14} = \frac{3}{14}$; sum: $\frac{5}{14}$

1.56

$$25 = 5 \cdot 5$$
$$15 = 3 \cdot 5$$
$$\text{LCD} = 3 \cdot 5 \cdot 5 = 75$$
$$\frac{7}{25} = \frac{21}{75}$$
$$\frac{4}{15} = \frac{20}{75}$$
$$\frac{41}{75}$$

1.57

$$25 = 5 \cdot 5$$
$$35 = 5 \cdot 7$$
$$\text{LCD} = 5 \cdot 5 \cdot 7 = 175$$
$$\frac{18}{25} = \frac{126}{175}$$
$$\frac{9}{35} = \frac{45}{175}$$
$$\frac{171}{175}$$

1.58

$$\frac{1}{4} = \frac{3}{12}$$
$$\frac{1}{3} = \frac{4}{12}$$
$$\frac{7}{12}$$

1.59

$$\frac{1}{5} = \frac{3}{15}$$
$$\frac{2}{3} = \frac{10}{15}$$
$$\frac{13}{15}$$

1.60

$$\frac{1}{6} = \frac{4}{24}$$
$$\frac{1}{8} = \frac{3}{24}$$
$$\frac{7}{24}$$

1.61

$$\frac{1}{4} = \frac{4}{16}$$
$$\frac{7}{16} = \frac{7}{16}$$
$$\frac{11}{16}$$

1.62

$$\frac{1}{3} = \frac{5}{15}$$
$$\frac{3}{5} = \frac{9}{15}$$
$$\frac{14}{15}$$

1.63

$$\frac{4}{7} = \frac{24}{42}$$
$$\frac{1}{6} = \frac{7}{42}$$
$$\frac{31}{42}$$

1.64

$$20 = 2 \cdot 2 \cdot 5$$
$$15 = 3 \cdot 5$$
$$\text{LCD} = 2 \cdot 2 \cdot 3 \cdot 5 = 60$$
$$\frac{3}{20} = \frac{9}{60}$$
$$\frac{7}{15} = \frac{28}{60}$$
$$\frac{37}{60}$$

1.65

$$48 = 2 \cdot 2 \cdot 2 \cdot 2 \cdot 3$$
$$60 = 2 \cdot 2 \cdot 3 \cdot 5$$
$$\text{LCD} = 2 \cdot 2 \cdot 2 \cdot 2 \cdot 3 \cdot 5 = 240$$
$$\frac{5}{48} = \frac{25}{240}$$
$$\frac{8}{60} = \frac{32}{240}$$
$$\frac{57}{240} = \frac{19}{80}$$

1.66

$$\frac{1}{8} = \frac{3}{24}$$
$$\frac{1}{3} = \frac{8}{24}$$
$$\frac{1}{4} = \frac{6}{24}$$
$$\frac{17}{24}$$

1.67

$$\frac{1}{2} = \frac{5}{10}$$
$$\frac{1}{5} = \frac{2}{10}$$
$$\frac{1}{10} = \frac{1}{10}$$
$$\frac{8}{10} = \frac{4}{5}$$

1.68
$8 = 2 \cdot 2 \cdot 2$
$10 = 2 \cdot 5$
$16 = 2 \cdot 2 \cdot 2 \cdot 2$
$\text{LCD} = 2 \cdot 2 \cdot 2 \cdot 2 \cdot 5 = 80$
$\frac{1}{8} = \frac{10}{80}$
$\frac{3}{10} = \frac{24}{80}$
$\frac{3}{16} = \frac{15}{80}$
$\frac{49}{80}$

1.69
$\frac{1}{7} = \frac{4}{28}$
$\frac{1}{4} = \frac{7}{28}$
$\frac{1}{2} = \frac{14}{28}$
$\frac{25}{28}$

1.70
$\frac{1}{5} = \frac{14}{70}$
$\frac{1}{2} = \frac{35}{70}$
$\frac{2}{7} = \frac{20}{70}$
$\frac{69}{70}$

1.71
$\frac{1}{4} = \frac{10}{40}$
$\frac{1}{5} = \frac{8}{40}$
$\frac{3}{8} = \frac{15}{40}$
$\frac{33}{40}$

1.72
$\frac{1}{3} = \frac{4}{12}$
$\frac{1}{4} = \frac{3}{12}$
$\frac{1}{6} = \frac{2}{12}$
$\frac{9}{12} = \frac{3}{4}$

1.73
$8 = 2 \cdot 2 \cdot 2$
$6 = 2 \cdot 3$
$5 = 5$
$\text{LCD} = 2 \cdot 2 \cdot 2 \cdot 3 \cdot 5 = 120$
$\frac{1}{8} = \frac{15}{120}$
$\frac{1}{6} = \frac{20}{120}$
$\frac{1}{5} = \frac{24}{120}$
$\frac{59}{120}$

1.74
$\frac{1}{6} = \frac{4}{24}$
$\frac{3}{8} = \frac{9}{24}$
$\frac{1}{12} = \frac{2}{24}$
$\frac{15}{24} = \frac{15}{24} \div \frac{3}{3} = \frac{5}{8}$

1.75
$15 = 3 \cdot 5$
$3 = 3$
$4 = 2 \cdot 2$
$\text{LCD} = 2 \cdot 2 \cdot 3 \cdot 5 = 60$
$\frac{4}{15} = \frac{16}{60}$
$\frac{1}{3} = \frac{20}{60}$
$\frac{1}{4} = \frac{15}{60}$
$\frac{51}{60} = \frac{51}{60} \div \frac{3}{3} = \frac{17}{20}$

1.76
$21 = 3 \cdot 7$
$6 = 2 \cdot 3$
$14 = 2 \cdot 7$
$\text{LCD} = 2 \cdot 3 \cdot 7 = 42$
$\frac{2}{21} = \frac{4}{42}$
$\frac{1}{6} = \frac{7}{42}$
$\frac{1}{14} = \frac{3}{42}$
$\frac{14}{42} = \frac{14}{42} \div \frac{14}{14} = \frac{1}{3}$

1.77
$5\frac{1}{3}$

1.78 $4\frac{2}{3}$

1.79 $21\frac{4}{5}$

1.80 $7\frac{12}{16} = 7\frac{12 \div 4}{16 \div 4} = 7\frac{3}{4}$ inches

1.81
$$2\frac{3}{4} = 2\frac{9}{12}$$
$$3\frac{1}{6} = 3\frac{2}{12}$$
$$5\frac{11}{12}$$

1.82 $28\frac{5}{7}$

1.83
$$7\frac{3}{8} = 7\frac{9}{24}$$
$$2\frac{5}{12} = 2\frac{10}{24}$$
$$9\frac{19}{24}$$

1.84
$$15\frac{7}{16} = 15\frac{7}{16}$$
$$29\frac{3}{8} = 29\frac{6}{16}$$
$$44\frac{13}{16}$$

1.85
$$422\frac{3}{8} = 422\frac{15}{40}$$
$$359\frac{2}{5} = 359\frac{16}{40}$$
$$781\frac{31}{40}$$

1.86
$$16 = 2 \cdot 2 \cdot 2 \cdot 2$$
$$12 = 2 \cdot 2 \cdot 3$$
$$\text{LCD} = 2 \cdot 2 \cdot 2 \cdot 2 \cdot 3 = 48$$
$$4\frac{3}{16} = 4\frac{9}{48}$$
$$3\frac{5}{12} = 3\frac{20}{48}$$
$$7\frac{29}{48}$$

1.87
$$15 = 3 \cdot 5$$
$$6 = 2 \cdot 3$$
$$\text{LCD} = 2 \cdot 3 \cdot 5 = 30$$
$$25\frac{4}{15} = 25\frac{8}{30}$$
$$17\frac{1}{6} = 17\frac{5}{30}$$
$$42\frac{13}{30}$$

1.88
$$21 = 3 \cdot 7$$
$$6 = 2 \cdot 3$$
$$\text{LCD} = 2 \cdot 3 \cdot 7 = 42$$
$$66\frac{8}{21} = 66\frac{16}{42}$$
$$75\frac{1}{6} = 75\frac{7}{42}$$
$$141\frac{23}{42}$$

1.89
$$72\frac{5}{16} = 72\frac{45}{144}$$
$$85\frac{5}{9} = 85\frac{80}{144}$$
$$157\frac{125}{144}$$

1.90
$$14 = 2 \cdot 7$$
$$6 = 2 \cdot 3$$
$$\text{LCD} = 2 \cdot 3 \cdot 7 = 42$$
$$848\frac{1}{14} = 848\frac{3}{42}$$
$$157\frac{5}{6} = 157\frac{35}{42}$$
$$1{,}005\frac{38}{42} = 1{,}005\frac{38 \div 2}{42 \div 2} = 1{,}005\frac{19}{21}$$

1.91
$$12 = 2 \cdot 2 \cdot 3$$
$$32 = 2 \cdot 2 \cdot 2 \cdot 2 \cdot 2$$
$$\text{LCD} = 2 \cdot 2 \cdot 2 \cdot 2 \cdot 2 \cdot 3 = 96$$
$$654\frac{3}{12} = 654\frac{24}{96}$$
$$767\frac{1}{32} = 767\frac{3}{96}$$
$$1{,}421\frac{27}{96} = 1{,}421\frac{27 \div 3}{96 \div 3} = 1{,}421\frac{9}{32}$$

1.92
$$742\frac{4}{5} = 742\frac{24}{30}$$
$$627\frac{1}{30} = 627\frac{1}{30}$$
$$1{,}369\frac{25}{30} = 1{,}369\frac{25 \div 5}{30 \div 5} = 1{,}369\frac{5}{6}$$

1.93 $941\frac{28}{32} = 941\frac{28 \div 4}{32 \div 4} = 941\frac{7}{8}$

1.94
$25 = 5 \cdot 5$
$30 = 2 \cdot 3 \cdot 5$
$\text{LCD} = 2 \cdot 3 \cdot 5 \cdot 5 = 150$
$752\frac{8}{25} = 752\frac{48}{150}$
$627\frac{11}{30} = 627\frac{55}{150}$
$1{,}379\frac{103}{150}$

1.95
$49 = 7 \cdot 7$
$14 = 2 \cdot 7$
$\text{LCD} = 2 \cdot 7 \cdot 7 = 98$
$42\frac{8}{49} = 42\frac{16}{98}$
$24\frac{5}{14} = 24\frac{35}{98}$
$66\frac{51}{98}$

1.96 $3\frac{8}{8} = 4$

1.97
$4\frac{4}{5} = 4\frac{24}{30}$
$13\frac{5}{6} = 13\frac{25}{30}$
$17\frac{49}{30} = 18\frac{19}{30}$

1.98 $13\frac{12}{8} = 13\frac{12 \div 4}{8 \div 4} = 13\frac{3}{2} = 14\frac{1}{2}$

1.99
$4\frac{4}{5} = 4\frac{28}{35}$
$3\frac{6}{7} = 3\frac{30}{35}$
$1\frac{3}{5} = 1\frac{21}{35}$
$8\frac{79}{35} = 10\frac{9}{35}$

1.100
$357\frac{4}{5} = 357\frac{84}{105}$
$265\frac{6}{7} = 265\frac{90}{105}$
$152\frac{2}{3} = 152\frac{70}{105}$
$774\frac{244}{105} = 776\frac{34}{105}$

1.101
$75\frac{4}{5} = 75\frac{64}{80}$
$62\frac{5}{16} = 62\frac{25}{80}$
$137\frac{89}{80} = 138\frac{9}{80}$

1.102
$42\frac{5}{7} = 42\frac{15}{21}$
$69\frac{17}{21} = 69\frac{17}{21}$
$111\frac{32}{21} = 112\frac{11}{21}$

1.103
$189\frac{7}{8} = 189\frac{21}{24}$
$234\frac{5}{6} = 234\frac{20}{24}$
$423\frac{41}{24} = 424\frac{17}{24}$

1.104
$69\frac{7}{9} = 69\frac{14}{18}$
$92\frac{5}{6} = 92\frac{15}{18}$
$161\frac{29}{18} = 162\frac{11}{18}$

1.105
$75\frac{8}{9} = 75\frac{128}{144}$
$26\frac{9}{16} = 26\frac{81}{144}$
$101\frac{209}{144} = 102\frac{65}{144}$

1.106
$14 = 2 \cdot 7$
$6 = 2 \cdot 3$
$\text{LCD} = 2 \cdot 3 \cdot 7 = 42$
$342\frac{9}{14} = 342\frac{27}{42}$
$267\frac{5}{6} = 267\frac{35}{42}$
$609\frac{62}{42} = 609\frac{62 \div 2}{42 \div 2} =$
$609\frac{31}{21} = 610\frac{10}{21}$

1.107
$25 = 5 \cdot 5$
$15 = 3 \cdot 5$
$75 = 3 \cdot 5 \cdot 5$
$\text{LCD} = 3 \cdot 5 \cdot 5 = 75$

1.107 cont.

$45\frac{8}{25} = 45\frac{24}{75}$

$64\frac{8}{15} = 64\frac{40}{75}$

$76\frac{8}{75} = 76\frac{8}{75}$

$185\frac{72}{75} = 185\frac{72 \div 3}{75 \div 3} = 185\frac{24}{25}$

1.108 $14 = 2 \cdot 7$
$6 = 2 \cdot 3$
$21 = 3 \cdot 7$
$\text{LCD} = 2 \cdot 3 \cdot 7 = 42$

$76\frac{1}{14} = 76\frac{3}{42}$

$57\frac{1}{6} = 57\frac{7}{42}$

$7\frac{20}{21} = 7\frac{40}{42}$

$140\frac{50}{42} = 140\frac{50 \div 2}{42 \div 2} =$

$140\frac{25}{21} = 141\frac{4}{21}$

1.109 $9 = 3 \cdot 3$
$6 = 2 \cdot 3$
$16 = 2 \cdot 2 \cdot 2 \cdot 2$
$\text{LCD} = 2 \cdot 2 \cdot 2 \cdot 2 \cdot 3 \cdot 3 = 144$

$85\frac{8}{9} = 85\frac{128}{144}$

$275\frac{5}{6} = 275\frac{120}{144}$

$67\frac{15}{16} = 67\frac{135}{144}$

$427\frac{383}{144} = 429\frac{95}{144}$

1.110 $12 = 12$

$\frac{3}{4} = \frac{3}{4}$

$+\ \frac{1}{2} = \frac{3}{4}$

$12\frac{5}{4} = 13\frac{1}{4}$ inches

1.111 $\frac{5}{8} = \frac{5}{8}$

$+\ \frac{1}{2} = \frac{4}{8}$

$\frac{9}{8} = 1\frac{1}{8}$ inches

1.112 $2\frac{3}{4} = 2\frac{6}{8}$

$2\frac{3}{8} = 2\frac{3}{8}$

$4\frac{9}{8} = 5\frac{1}{8}$ cups

1.113 $2\frac{17}{32} = 2\frac{17}{32}$

$1\frac{5}{8} = 1\frac{20}{32}$

$+\ 2\frac{3}{4} = 2\frac{24}{32}$

$5\frac{61}{32} = 6\frac{29}{32}$ inches

1.114 $3\frac{3}{8} = 3\frac{12}{32}$

$4\frac{11}{32} = 4\frac{11}{32}$

$+\ 2\frac{3}{4} = 2\frac{24}{32}$

$9\frac{47}{32} = 10\frac{15}{32}$ inches

1.115 $1\frac{3}{4} = 1\frac{6}{8}$

$\frac{5}{8} = \frac{5}{8}$

$\frac{1}{8} = \frac{1}{8}$

$+\ \frac{1}{4} = \frac{2}{8}$

$1\frac{14}{8} = 1\frac{14 \div 2}{8 \div 2} = 1\frac{7}{4} = 2\frac{3}{4}$"

1.116 $14\frac{3}{5} = 14\frac{12}{20}$

$14\frac{3}{5} = 14\frac{12}{20}$

$11\frac{1}{4} = 11\frac{5}{20}$

$+\ 11\frac{1}{4} = 11\frac{5}{20}$

$50\frac{34}{20} = 50\frac{34 \div 2}{20 \div 2} =$

$50\frac{17}{10} = 51\frac{7}{10}$ cm

1.117 $7\frac{3}{10} + 11\frac{9}{10} = 18\frac{12}{10} =$

$18\frac{12 \div 2}{10 \div 2} = 18\frac{6}{5} = 19\frac{1}{5}$ meters

1.118 $2\frac{7}{8} = 2\frac{7}{8}$

$+\,2\frac{1}{4} = 2\frac{2}{8}$

$4\frac{9}{8} = 5\frac{1}{8}$ yards

1.119 Outside length: $17\frac{1}{8} = 17\frac{1}{8}$

$\frac{1}{4} = \frac{2}{8}$

$+\,\frac{1}{4} = \frac{2}{8}$

$17\frac{5}{8}$ in.

Outside width: $12\frac{7}{8} = 12\frac{7}{8}$

$\frac{1}{4} = \frac{2}{8}$

$+\,\frac{1}{4} = \frac{2}{8}$

$12\frac{11}{8} = 13\frac{3}{8}$ in.

II. SECTION TWO

2.1 $\frac{2}{4} = \frac{1}{2}$

2.2 $\frac{2}{8} = \frac{1}{4}$

2.3 $\frac{3}{7}$

2.4 $\frac{6}{16} = \frac{3}{8}$

2.5 $\frac{6}{15} = \frac{2}{5}$

2.6 $\frac{2}{12} = \frac{1}{6}$

2.7 $\frac{6}{24} = \frac{1}{4}$

2.8 $\frac{8}{32} = \frac{1}{4}$

2.9 0

2.10 $\frac{3}{13}$

2.11 $\frac{2}{5}$

2.12 $\frac{2}{12} = \frac{1}{6}$

2.13 $\frac{3}{33} = \frac{1}{11}$

2.14 $\frac{6}{32} = \frac{3}{16}$

2.15 $\frac{8}{17}$

2.16 $\frac{4}{16} = \frac{1}{4}$

2.17 $\frac{7}{14} = \frac{1}{2}$

2.18 $\frac{2}{11}$

2.19 $\frac{4}{23}$

2.20 $\frac{12}{24} = \frac{1}{2}$

2.21 $\frac{7}{8} = \frac{14}{16}$

$\frac{5}{16} = \frac{5}{16}$

$\frac{9}{16}$

2.22 $\frac{4}{5} = \frac{28}{35}$

$\frac{4}{7} = \frac{20}{35}$

$\frac{8}{35}$

2.23
$$25 = 5 \cdot 5$$
$$10 = 2 \cdot 5$$
$$\text{LCD} = 2 \cdot 5 \cdot 5 = 50$$
$$\frac{12}{25} = \frac{24}{50}$$
$$-\frac{3}{10} = -\frac{15}{50}$$
$$\frac{9}{50}$$

2.24
$$\frac{5}{7} = \frac{20}{28}$$
$$-\frac{1}{4} = -\frac{7}{28}$$
$$\frac{13}{28}$$

2.25
$$\frac{11}{18} = \frac{11}{18}$$
$$-\frac{5}{9} = -\frac{10}{18}$$
$$\frac{1}{18}$$

2.26
$$\frac{5}{6} = \frac{20}{24}$$
$$-\frac{3}{8} = -\frac{9}{24}$$
$$\frac{11}{24}$$

2.27
$$\frac{14}{17} = \frac{56}{68}$$
$$-\frac{3}{4} = -\frac{51}{68}$$
$$\frac{5}{68}$$

2.28
$$\frac{8}{9} = \frac{16}{18}$$
$$-\frac{5}{6} = -\frac{15}{18}$$
$$\frac{1}{18}$$

2.29
$$15 = 3 \cdot 5$$
$$25 = 5 \cdot 5$$
$$\text{LCD} = 3 \cdot 5 \cdot 5 = 75$$
$$\frac{11}{15} = \frac{55}{75}$$
$$-\frac{12}{25} = -\frac{36}{75}$$
$$\frac{19}{75}$$

2.30
$$16 = 2 \cdot 2 \cdot 2 \cdot 2$$
$$6 = 2 \cdot 3$$
$$\text{LCD} = 2 \cdot 2 \cdot 2 \cdot 2 \cdot 3 = 48$$
$$\frac{15}{16} = \frac{45}{48}$$
$$-\frac{5}{6} = -\frac{40}{48}$$
$$\frac{5}{48}$$

2.31
$$18 = 2 \cdot 3 \cdot 3$$
$$12 = 2 \cdot 2 \cdot 3$$
$$\text{LCD} = 2 \cdot 2 \cdot 3 \cdot 3 = 36$$
$$\frac{11}{18} = \frac{22}{36}$$
$$-\frac{5}{12} = -\frac{15}{36}$$
$$\frac{7}{36}$$

2.32
$$\frac{9}{16} = \frac{81}{144}$$
$$-\frac{4}{9} = -\frac{64}{144}$$
$$\frac{17}{144}$$

2.33
$$\frac{4}{5} = \frac{24}{30}$$
$$-\frac{1}{6} = -\frac{5}{30}$$
$$\frac{19}{30}$$

2.34
$$\frac{8}{11} = \frac{40}{55}$$
$$-\frac{3}{5} = -\frac{33}{55}$$
$$\frac{7}{55}$$

2.35
$$\frac{11}{13} = \frac{66}{78}$$
$$-\frac{1}{6} = -\frac{13}{78}$$
$$\frac{53}{78}$$

2.36
$$\frac{1}{7} = \frac{11}{77}$$
$$-\frac{1}{11} = -\frac{7}{77}$$
$$\frac{4}{77}$$

2.37
$$\frac{23}{25} = \frac{138}{150}$$
$$\frac{5}{6} = \frac{125}{150}$$
$$\frac{13}{150}$$

2.38
$$\frac{7}{17} = \frac{14}{34}$$
$$\frac{12}{34} = \frac{12}{34}$$
$$\frac{2}{34} = \frac{1}{17}$$

2.39
$$\frac{7}{8} = \frac{63}{72}$$
$$\frac{5}{9} = \frac{40}{72}$$
$$\frac{23}{72}$$

2.40
$$\frac{5}{8} = \frac{5}{8}$$
$$\frac{10}{16} = \frac{5}{8}$$
$$0$$

2.41 $7\frac{3}{5}$

2.42
$$5\frac{1}{4} = 4 + \frac{4}{4} + \frac{1}{4} = 4\frac{5}{4}$$
$$2\frac{3}{4} = \quad 2\frac{3}{4}$$
$$2\frac{2}{4} = 2\frac{1}{2}$$

2.43
$$14\frac{6}{8} = 14\frac{18}{24} = 13\frac{42}{24}$$
$$9\frac{5}{6} = \quad 9\frac{20}{24}$$
$$4\frac{22}{24} = 4\frac{11}{12}$$

2.44
$$7\frac{5}{8} = 7\frac{25}{40}$$
$$5\frac{1}{5} = 5\frac{8}{40}$$
$$2\frac{17}{40}$$

2.45
$$23 = 22\frac{8}{8}$$
$$17\frac{7}{8} = 17\frac{7}{8}$$
$$5\frac{1}{8}$$

2.46
$$152\frac{3}{7} = 152\frac{12}{28} = 151\frac{40}{28}$$
$$61\frac{23}{28} = \quad 61\frac{23}{28}$$
$$90\frac{17}{28}$$

2.47
$$214\frac{7}{16} = 214\frac{7}{16}$$
$$75\frac{3}{8} = 75\frac{6}{16}$$
$$139\frac{1}{16}$$

2.48
$$245\frac{4}{9} = 245\frac{16}{36} = 244\frac{52}{36}$$
$$167\frac{11}{12} = \quad 167\frac{33}{36}$$
$$77\frac{19}{36}$$

2.49
$$5\frac{4}{5} = 5\frac{24}{30}$$
$$2\frac{1}{6} = 2\frac{5}{30}$$
$$3\frac{19}{30}$$

2.50
$$13\frac{1}{3} = 12\frac{4}{3}$$
$$11\frac{2}{3} = 11\frac{2}{3}$$
$$1\frac{2}{3}$$

2.51
$$14\frac{1}{5} = 14\frac{4}{20} = 13\frac{24}{20}$$
$$6\frac{3}{4} = \quad 6\frac{15}{20}$$
$$7\frac{9}{20}$$

2.52

$$756\frac{4}{5} = 756\frac{16}{20}$$

$$\underline{275\frac{3}{4}} = \underline{275\frac{15}{20}}$$

$$481\frac{1}{20}$$

2.53

$$347\frac{2}{9} = 347\frac{32}{144} = 346\frac{176}{144}$$

$$\underline{98\frac{7}{16}} = \qquad \underline{98\frac{63}{144}}$$

$$248\frac{113}{144}$$

2.54

$$246\frac{3}{15} = 246\frac{15}{75} = 245\frac{90}{75}$$

$$\underline{78\frac{17}{25}} = \qquad \underline{78\frac{51}{75}}$$

$$167\frac{39}{75} = 167\frac{13}{25}$$

2.55

$$972\frac{3}{4} = 972\frac{9}{12}$$

$$\underline{279\frac{2}{3}} = \underline{279\frac{8}{12}}$$

$$693\frac{1}{12}$$

2.56

$$981\frac{5}{9} = 981\frac{80}{144} = 980\frac{224}{144}$$

$$\underline{93\frac{13}{16}} = \qquad \underline{93\frac{117}{144}}$$

$$887\frac{107}{144}$$

2.57

$$488\frac{8}{21} = 488\frac{16}{42} = 487\frac{58}{42}$$

$$\underline{266\frac{8}{14}} = \qquad \underline{266\frac{24}{42}}$$

$$221\frac{34}{42} = 221\frac{17}{21}$$

2.58

$$324\frac{5}{8} = 324\frac{10}{16} = 323\frac{26}{16}$$

$$\underline{97\frac{13}{16}} = \qquad \underline{97\frac{13}{16}}$$

$$226\frac{13}{16}$$

2.59

$$14 = 2 \cdot 7$$
$$6 = 2 \cdot 3$$
$$\text{LCD} = 2 \cdot 3 \cdot 7 = 42$$

$$375\frac{5}{14} = 375\frac{15}{42} =$$

$$\underline{89\frac{5}{6}} = \underline{89\frac{35}{42}}$$

$$374 + \frac{42}{42} + \frac{15}{42} = 374\frac{57}{42}$$

$$\underline{89\frac{35}{42}}$$

$$285\frac{22}{42} = 285\frac{11}{21}$$

2.60

$$25 = 5 \cdot 5$$
$$20 = 2 \cdot 2 \cdot 5$$
$$\text{LCD} = 2 \cdot 2 \cdot 5 \cdot 5 = 100$$

$$927\frac{7}{25} = 927\frac{28}{100} =$$

$$\underline{471\frac{13}{20}} = \underline{471\frac{65}{100}} =$$

$$926 + \frac{100}{100} + \frac{28}{100} = 926\frac{128}{100}$$

$$\underline{471\frac{65}{100}}$$

$$455\frac{63}{100}$$

2.61

$$15 = 3 \cdot 5$$
$$40 = 2 \cdot 2 \cdot 2 \cdot 5$$
$$\text{LCD} = 2 \cdot 2 \cdot 2 \cdot 3 \cdot 5 = 120$$

$$495\frac{7}{15} = 495\frac{56}{120} =$$

$$\underline{191\frac{33}{40}} = \underline{191\frac{99}{120}} =$$

$$494 + \frac{120}{120} + \frac{56}{120} = 494\frac{176}{120}$$

$$\underline{191\frac{99}{120}}$$

$$303\frac{77}{120}$$

2.62

$$49 = 7 \cdot 7$$
$$28 = 2 \cdot 2 \cdot 7$$
$$\text{LCD} = 2 \cdot 2 \cdot 7 \cdot 7 = 196$$

$$415\frac{27}{49} = 415\frac{108}{196} =$$

$$\underline{297\frac{25}{28}} = \underline{297\frac{175}{196}} =$$

$$414 + \frac{196}{196} + \frac{108}{196} = 414\frac{304}{196}$$

$$\underline{297\frac{175}{196}}$$

$$117\frac{129}{196}$$

2.63 $25\frac{2}{8} = 25\frac{1}{4}$

Check: $23\frac{3}{8} = 23\frac{3}{8}$

$+\ 25\frac{1}{4} = 25\frac{2}{8}$

$48\frac{5}{8}$

2.64 $245\frac{1}{4} = 245\frac{2}{8} = 244\frac{10}{8}$

$77\frac{5}{8} = \quad 77\frac{5}{8}$

$167\frac{5}{8}$

Check: $77\frac{5}{8}$

$+\ 167\frac{5}{8}$

$244\frac{10}{8} = 244\frac{5}{4} = 244\frac{1}{4}$

2.65 $47\frac{7}{16} = 47\frac{35}{80} = 46\frac{115}{80}$

$9\frac{4}{5} = 9\frac{64}{80} = 9\frac{64}{80}$

$37\frac{51}{80}$

2.66 $14 = 2 \cdot 7$

$21 = 3 \cdot 7$

$\text{LCD} = 2 \cdot 3 \cdot 7 = 42$

$66\frac{5}{14} = 66\frac{15}{42} = 65\frac{57}{42}$

$44\frac{19}{21} = 44\frac{38}{42} = 44\frac{38}{42}$

$21\frac{19}{42}$

2.67 $41\frac{3}{4} = 41\frac{12}{16} = 40\frac{28}{16}$

$20\frac{15}{16} = \quad 20\frac{15}{16}$

$20\frac{13}{16}$

2.68 $15 = 3 \cdot 5$

$25 = 5 \cdot 5$

$\text{LCD} = 3 \cdot 5 \cdot 5 = 75$

$65\frac{7}{15} = 65\frac{35}{75} = 64\frac{110}{75}$

$9\frac{13}{25} = 9\frac{39}{75} = 9\frac{39}{75}$

$55\frac{71}{75}$

2.69 $35\frac{1}{6} = 35\frac{1}{6} = 34\frac{7}{6}$

$33\frac{1}{2} = 33\frac{3}{6} = 33\frac{3}{6}$

$1\frac{4}{6} = 1\frac{2}{3}$

2.70 $74\frac{5}{6} = 74\frac{20}{24}$

$22\frac{1}{8} = 22\frac{3}{24}$

$52\frac{17}{24}$

2.71 $18\frac{1}{6} = 18\frac{1}{6} = 17\frac{7}{6}$

$8\frac{2}{3} = 8\frac{4}{6} = 8\frac{4}{6}$

$9\frac{3}{6} = 9\frac{1}{2}$

2.72 $16\frac{1}{4} = 16\frac{3}{12} = 15\frac{15}{12}$

$12\frac{2}{3} = 12\frac{8}{12} = 12\frac{8}{12}$

$3\frac{7}{12}$

2.73 $25 = 5 \cdot 5$

$15 = 3 \cdot 5$

$\text{LCD} = 3 \cdot 5 \cdot 5 = 75$

$7\frac{7}{25} = 7\frac{21}{75} = 6\frac{96}{75}$

$3\frac{12}{15} = 3\frac{60}{75} = 3\frac{60}{75}$

$3\frac{36}{75} = 3\frac{12}{25}$

2.74 $82\frac{9}{10} = 82\frac{99}{110} = 81\frac{209}{110}$

$9\frac{10}{11} = 9\frac{100}{110} = 9\frac{100}{110}$

$72\frac{109}{110}$

2.75
$$48\tfrac{3}{4} = 48\tfrac{15}{20} = 47\tfrac{35}{20}$$
$$\underline{31\tfrac{9}{10}} = \underline{31\tfrac{18}{20}} = \underline{31\tfrac{18}{20}}$$
$$16\tfrac{17}{20}$$

2.76
$$37\tfrac{1}{6} = 37\tfrac{4}{24} = 36\tfrac{28}{24}$$
$$\underline{8\tfrac{3}{8}} = \underline{8\tfrac{9}{24}} = \underline{8\tfrac{9}{24}}$$
$$28\tfrac{19}{24}$$

2.77
$$6 = 2 \cdot 3$$
$$16 = 2 \cdot 2 \cdot 2 \cdot 2$$
$$\text{LCD} = 2 \cdot 2 \cdot 2 \cdot 2 \cdot 3 = 48$$
$$18\tfrac{5}{6} = 18\tfrac{40}{48} = 17\tfrac{88}{48}$$
$$\underline{7\tfrac{15}{16}} = \underline{7\tfrac{45}{48}} = \underline{7\tfrac{45}{48}}$$
$$10\tfrac{43}{48}$$

2.78
$$23\tfrac{3}{8} = 23\tfrac{9}{24} = 22\tfrac{33}{24}$$
$$\underline{12\tfrac{5}{12}} = \underline{12\tfrac{10}{24}} = \underline{12\tfrac{10}{24}}$$
$$10\tfrac{23}{24}$$

2.79
$$12 = 2 \cdot 2 \cdot 3$$
$$10 = 2 \cdot 5$$
$$\text{LCD} = 2 \cdot 2 \cdot 3 \cdot 5 = 60$$
$$24\tfrac{7}{12} = 24\tfrac{35}{60} = 23\tfrac{95}{60}$$
$$\underline{15\tfrac{7}{10}} = \underline{15\tfrac{42}{60}} = \underline{15\tfrac{42}{60}}$$
$$8\tfrac{53}{60}$$

2.80
$$7\tfrac{1}{16} = 7\tfrac{3}{48} = 6\tfrac{51}{48}$$
$$\underline{3\tfrac{7}{48}} = \underline{3\tfrac{7}{48}}$$
$$3\tfrac{44}{48} = 3\tfrac{11}{12}$$

2.81
$$17 = 16\tfrac{16}{16}$$
$$\underline{7\tfrac{5}{16}} = \underline{7\tfrac{5}{16}}$$
$$9\tfrac{11}{16}$$

2.82
$$8\tfrac{3}{4} = 8\tfrac{12}{16} = 7\tfrac{28}{16}$$
$$\underline{2\tfrac{15}{16}} = \underline{2\tfrac{15}{16}}$$
$$5\tfrac{13}{16}$$

2.83
$$1\tfrac{1}{8} = 1\tfrac{4}{32} = \tfrac{36}{32}$$
$$\underline{\tfrac{11}{32}} \qquad \underline{\tfrac{11}{32}}$$
$$\tfrac{25}{32}$$

2.84
$$7 = 6\tfrac{25}{25}$$
$$\underline{3\tfrac{4}{25}} = \underline{3\tfrac{4}{25}}$$
$$3\tfrac{21}{25}$$

2.85
$$18\tfrac{4}{5} = 18\tfrac{32}{40} = 17\tfrac{72}{40}$$
$$\underline{6\tfrac{7}{8}} = \underline{6\tfrac{35}{40}} = \underline{6\tfrac{35}{40}}$$
$$11\tfrac{37}{40}$$

2.86
$$10 = 2 \cdot 5$$
$$8 = 2 \cdot 2 \cdot 2$$
$$\text{LCD} = 2 \cdot 2 \cdot 2 \cdot 5 = 40$$
$$27\tfrac{9}{10} = 27\tfrac{36}{40}$$
$$\underline{18\tfrac{5}{8}} = \underline{18\tfrac{25}{40}}$$
$$9\tfrac{11}{40}$$

2.87
$$75\tfrac{1}{16} = 75\tfrac{9}{144} = 74\tfrac{153}{144}$$
$$\underline{62\tfrac{1}{9}} = \underline{62\tfrac{16}{144}} = \underline{62\tfrac{16}{144}}$$
$$12\tfrac{137}{144}$$

2.88

$15 = 3 \cdot 5$
$25 = 5 \cdot 5$
$\text{LCD} = 3 \cdot 5 \cdot 5 = 75$

$$
\begin{array}{rcccl}
87\frac{4}{15} & = & 87\frac{20}{75} & = & 86\frac{95}{75} \\
22\frac{12}{25} & = & 22\frac{36}{75} & = & 22\frac{36}{75} \\
\hline
 & & & & 64\frac{59}{75}
\end{array}
$$

2.89

$$
\begin{array}{rcccl}
75\frac{3}{8} & = & 75\frac{27}{72} & = & 74\frac{99}{72} \\
70\frac{5}{9} & = & 70\frac{40}{72} & = & 70\frac{40}{72} \\
\hline
 & & & & 4\frac{59}{72}
\end{array}
$$

2.90

$$
\begin{array}{rcl}
3\frac{4}{5} & = & 3\frac{20}{25} \\
1\frac{20}{25} & = & 1\frac{20}{25} \\
\hline
 & & 2
\end{array}
$$

2.91

$$
\begin{array}{rcl}
4\frac{5}{8} & = & 4\frac{10}{16} \\
-\,2\frac{7}{16} & = & 2\frac{7}{16} \\
\hline
 & & 2\frac{3}{16}
\end{array}
$$
centimeters

2.92

$$
\begin{array}{rcccl}
27\frac{1}{3} & = & 27\frac{8}{24} & = & 26\frac{32}{24} \\
-\,17\frac{5}{8} & = & 17\frac{15}{24} & = & 17\frac{15}{24} \\
\hline
 & & & & 9\frac{17}{24}
\end{array}
$$
kilometers

2.93 You have two choices.

Two-Step Method: Get a common denominator for the addition step only, then get a common denominator for the subtraction step.

$$
\begin{array}{rcl}
6\frac{7}{12} & = & 6\frac{7}{12} \\
+\,2\frac{5}{6} & = & 2\frac{10}{12} \\
\hline
 & & 8\frac{17}{12}
\end{array}
\qquad
\begin{array}{rcl}
8\frac{17}{12} & = & 8\frac{34}{24} \\
-\,5\frac{7}{8} & = & 5\frac{21}{24} \\
\hline
 & & 3\frac{13}{24}
\end{array}
$$

One-Denominator Method: Find a common denominator for all 3 fractions.

$$
\begin{array}{rcl}
6\frac{7}{12} & = & 6\frac{14}{24} \\
+\,2\frac{5}{6} & = & 2\frac{20}{24} \\
\hline
 & & 8\frac{34}{24}
\end{array}
\qquad
\begin{array}{rcl}
8\frac{34}{24} & = & 8\frac{34}{24} \\
-\,5\frac{7}{8} & = & 5\frac{21}{24} \\
\hline
 & & 3\frac{13}{24}
\end{array}
$$

Note that the sums ($8\frac{17}{12}$ and $8\frac{34}{24}$) were not simplified. If they had been simplified, more work would have resulted. Wait until the final answer is reached before simplifying.

III. SECTION THREE

3.1 0.61

3.2 1.114

3.3 48.8618

3.4 81.42

3.5 419.087

3.6 677.184

3.7 168.13

3.8 99.754

3.9 779.864

3.10 12,902.794

3.11 1,070.023

3.12 346.3098

3.13 0.41

3.14 20.73

3.15 12.09

3.16 82.65

3.17 17.75 kilometers

3.18 $4.56 Do you have the ($)?

3.19 18.725

3.20 387.06 Check: 387.06
+ 65.04
452.10

3.21 20.27 Check: 20.27
+ 4.73
25.00

3.22 55.85 Check: 55.85
+ 16.15
72.00

3.23 247.47 Check: 247.47
+ 127.57
375.04

3.24 $6.55 Check $ 6.55
+ 3.45
$10.00

3.25 5.456 Check: 37
+ 5.456
42.456

3.26 52.52

3.27 85.66

3.28 557.08

3.29 67,493.2

3.30 414.865

3.31 209.2575

3.32 215.069

3.33 5,116.756

3.34 4.5
+ 6.7
11.2
− 3.3
7.9 mm

3.35 17,978.63

3.36 652.75
+ 25.1
677.85
− 356.777
321.073

3.37 975.74
+ 672.13
1,647.87
− 597.47
1,050.40

3.38 475.4
+ 77.9
553.300
− 327.543
225.757

3.39 47,972
+ 6,543.4
54,515.40
− 29,647.39
24,868.01

3.40 $ 75.04
+ 253.37
328.41
− 197.75
$130.66

The solutions to Problems 3.41 through 3.50 are given for students who do not have access to calculators.

3.41 4.57
x 322
914
914
1371
147.154
x 21.7
1030078
147154
294308
3,193.2418

3.42 65.23
x 25.6
39138
32615
13046
1,669.888

3.42 cont.

```
       52.184
32 )1,669.888
     160
      69
      64
       58
       32
       268
       256
        128
        128
          0
```

```
         11.0656
642 )7104.1152
     642
      684
      642
       4211
       3852
        3595
        3210
         3852
         3852
            0
```

3.43

```
       2.98
742 )2211.16
     1484
      7271
      6678
       5936
       5936
          0
```

```
     2.98
   x 5.27
     2086
     596
   1490
  15.7046
```

3.44

```
     7.25
   x 19.4
     2900
    6525
    725
  140.650
```

```
     140.65
      x 314
      56260
     14065
    42195
    44,164.10 = 44,164.1
```

3.45

```
   2.87616
  x   2.47
   2013312
  1150464
  575232
  7.1041152
```

3.46

```
           148.7
1,645 )244611.5
        1645
         8011
         6580
         14311
         13160
          11515
          11515
              0
```

```
     148.7
   x  4.52
      2974
     7435
    5948
   672.124
```

3.47

```
    1,451
   x 0.24
     5804
    2902
    348.24
   x 0.16
   208944
   34824
   55.7184
```

3.48

```
     43.52
   x 14.75
     21760
    30464
   17408
   4352
   641.9200
 +  45.2527
   687.1727
```

3.49

```
                 61,046
  1,256 )76,673,776
          7536
           1313
           1256
             5777
             5024
              7536
              7536
                 0
```

```
   61,046
 - 16,523
   44,523
```

3.50

```
     48
   x 17
    336
    48
    816 cans
 x 0.39
   7344
  2448
$318.24
```

3.51

```
      459. 36
     - 15.47
a.    443.89
     + 62.45
b.    506.34
     - 75.14
c.    431.20
    - 125.47
d.    305.73
```

IV. SECTION FOUR

4.1 740

4.2 680

4.3 830

4.4 7,700

4.5 9,900

4.6 6,800

4.7 7,548,400

4.8 7,548,400

4.9 7,548,000

4.10 7,550,000

4.11 7,500,000

4.12 8,000,000

4.13 a. 27.453
b. 27.45
c. 27.5
d. 27

4.14 a. 6.205
b. 6.20
c. 6.2
d. 6

4.15 a. 452.992
b. 452.99
c. 453.0
d. 453

4.16 2.7

4.17 2.8

4.18 2.0

4.19 9,752.3

4.20 760.0

4.21 a. 67.44
b. 67.435
c. 67

4.22 a. 43.00
b. 42.998
c. 43

4.23 a. 67,543.78
b. 67,543.782
c. 67,544

4.24 $\frac{6}{8} = \frac{3}{4}$

4.25 $\frac{8}{16} = \frac{1}{2}$

4.26
$$17\frac{4}{7} = 17\frac{16}{28}$$
$$+\ 21\frac{3}{4} = 21\frac{21}{28}$$
$$38\frac{37}{28} = 39\frac{9}{28}$$

4.27
$$7\frac{3}{8} = 6\frac{11}{8}$$
$$-\ 4\frac{5}{8} = 4\frac{5}{8}$$
$$2\frac{6}{8} = 2\frac{3}{4}$$

4.28
$$475\frac{5}{9} = 475\frac{80}{144}$$
$$+\ 237\frac{7}{16} = 237\frac{63}{144}$$
$$712\frac{143}{144}$$

4.29
$8 = 2 \cdot 2 \cdot 2$
$10 = 2 \cdot 5$
$\text{LCD} = 2 \cdot 2 \cdot 2 \cdot 5 = 40$
$$7\frac{1}{8} = 7\frac{5}{40} = 6\frac{45}{40}$$
$$-\ 3\frac{3}{10} = 3\frac{12}{40} = 3\frac{12}{40}$$
$$3\frac{33}{40}$$

4.30
$16 = 2 \cdot 2 \cdot 2 \cdot 2$
$6 = 2 \cdot 3$
$\text{LCD} = 2 \cdot 2 \cdot 2 \cdot 2 \cdot 3 = 48$
$$27\frac{3}{16} = 27\frac{9}{48} = 26\frac{57}{48}$$
$$-\ 19\frac{5}{6} = 19\frac{40}{48} = 19\frac{40}{48}$$
$$7\frac{17}{48}$$

4.31
$14 = 2 \cdot 7$
$12 = 2 \cdot 2 \cdot 3$
$\text{LCD} = 2 \cdot 2 \cdot 3 \cdot 7 = 84$
$$363\frac{9}{14} = 36\frac{54}{84} = 362\frac{138}{84}$$
$$-\ 166\frac{11}{12} = 166\frac{77}{84} = 166\frac{77}{84}$$
$$196\frac{61}{84}$$

4.32
$$477\frac{6}{7} = 477\frac{48}{56}$$
$$+\ 392\frac{7}{8} = 392\frac{49}{56}$$
$$869\frac{97}{56} = 870\frac{41}{56}$$

4.33
$$711\frac{1}{7} = 711\frac{4}{28} = 710\frac{32}{28}$$
$$-\ 428\frac{7}{28} = \qquad 428\frac{7}{28}$$
$$282\frac{25}{28}$$

4.34
$$24\frac{1}{4} = 24\frac{4}{16}$$
$$62\frac{7}{8} = 62\frac{14}{16}$$
$$+\ 75\frac{15}{16} = 75\frac{15}{16}$$
$$161\frac{33}{16} = 163\frac{1}{16}$$

4.35
$9 = 3 \cdot 3$
$6 = 2 \cdot 3$
$16 = 2 \cdot 2 \cdot 2 \cdot 2$
$\text{LCD} = 2 \cdot 2 \cdot 2 \cdot 2 \cdot 3 \cdot 3 = 144$
$$42\frac{7}{9} = 42\frac{112}{144}$$
$$67\frac{5}{6} = 67\frac{120}{144}$$
$$+\ 79\frac{7}{16} = 79\frac{63}{144}$$
$$188\frac{295}{144} = 190\frac{7}{144}$$

4.36
$25 = 5 \cdot 5$
$75 = 3 \cdot 5 \cdot 5$
$15 = 3 \cdot 5$
$\text{LCD} = 3 \cdot 5 \cdot 5 = 75$
$$63\frac{14}{25} = 63\frac{42}{75}$$
$$82\frac{67}{75} = 82\frac{67}{75}$$
$$+\ 91\frac{11}{15} = 91\frac{55}{75}$$
$$236\frac{164}{75} = 238\frac{14}{75}$$

4.37
$$\begin{array}{rcrcr} 16\frac{3}{8} & = & 16\frac{15}{40} & = & 15\frac{55}{40} \\ -7\frac{3}{5} & = & 7\frac{24}{40} & = & 7\frac{24}{40} \\ \hline & & & & 8\frac{31}{40} \end{array}$$

4.38
$$\begin{array}{rcrcr} 77\frac{3}{4} & = & 77\frac{6}{8} & = & 76\frac{14}{8} \\ -16\frac{7}{8} & = & & & 16\frac{7}{8} \\ \hline & & & & 60\frac{7}{8} \end{array}$$

4.39
$$\begin{array}{rcrcr} 477\frac{3}{8} & = & 477\frac{6}{16} & = & 476\frac{22}{16} \\ -284\frac{10}{16} & = & & & 284\frac{10}{16} \\ \hline & & & & 192\frac{12}{16} = 192\frac{3}{4} \end{array}$$

4.40
$$\begin{array}{rcr} 38\frac{5}{6} & = & 38\frac{10}{12} \\ -19\frac{5}{12} & = & 19\frac{5}{12} \\ \hline & & 19\frac{5}{12} \end{array}$$

4.41 5,000

4.42 4,540

4.43 4,537.4

4.44 4,537.40

4.45 17,000

4.46 17,400

4.47 17,396.2

4.48 17,396.22

4.49 64.17

4.50 510.72

4.51 817.777

4.52
$$\begin{array}{r} 65.43 \\ +\ 65.98 \\ \hline 131.41 \\ -\ 47.65 \\ \hline 83.76 \end{array}$$

4.53
$$\begin{array}{r} 471.92 \\ +\ 67.354 \\ \hline 539.274 \\ -\ 276.7 \\ \hline 262.574 \end{array}$$

4.54
$$\begin{array}{r} 345.1 \\ +\ 98.17 \\ \hline 443.27 \\ -\ 237.45 \\ \hline 205.82 \end{array}$$

4.55
$$\begin{array}{r} 0.4783 \\ +\ 1.3275 \\ \hline 1.8058 \\ -\ 0.564 \\ \hline 1.2418 \end{array}$$

4.56 2.0829

4.57
$$\begin{array}{r} 7.4032 \\ +\ 67.31 \\ \hline 74.7132 \\ -\ 38.307 \\ \hline 36.4062 \end{array}$$

4.58
$$\begin{array}{r} 6.41 \\ +\ 77.9103 \\ \hline 84.3203 \\ -\ 16.7014 \\ \hline 67.6189 \end{array}$$

4.59 $7.96 - 4.32 = 3.64$

4.60
$$\begin{array}{rcrcr} 7\frac{1}{8} & = & 7\frac{1}{8} & = & 6\frac{9}{8} \\ -3\frac{3}{4} & = & 3\frac{6}{8} & = & 3\frac{6}{8} \\ \hline & & & & 3\frac{3}{8} \end{array}$$

4.61

$$\begin{array}{rr} 34\frac{3}{8} = & 34\frac{9}{24} \\ -\ 7\frac{1}{6} = & 7\frac{4}{24} \\ \hline & 27\frac{5}{24} \end{array}$$

4.62 $170.134 + 67.43 = 237.564$

4.63 $0.965 + 0.417 = 1.382$

4.64

$$\begin{array}{rr} 7\frac{3}{28} = & 7\frac{3}{28} \\ +\ 4\frac{1}{7} = & 4\frac{4}{28} \\ \hline & 11\frac{7}{28} = 11\frac{1}{4} \end{array}$$

I. SECTION ONE

1.1 $\frac{1}{8}$

1.2 $\frac{4}{15}$

1.3 $\frac{15}{30} = \frac{15 \div 15}{30 \div 15} = \frac{1}{2}$

1.4 $\frac{6}{35}$

1.5 $\frac{6}{6} = 1$

1.6 $\frac{6}{60} = \frac{6 \div 6}{60 \div 6} = \frac{1}{10}$

1.7 $\frac{15}{128}$

1.8 $\frac{18}{60} = \frac{18 \div 6}{60 \div 6} = \frac{3}{10}$

1.9 $\frac{\cancel{4}^{1}}{5} \times \frac{3}{\cancel{8}_{2}} = \frac{3}{10}$

1.10 $\frac{4}{_{5}\cancel{15}} \times \frac{\cancel{3}^{1}}{5} = \frac{4}{25}$

1.11 $\frac{\cancel{8}^{1}}{_{1}\cancel{11}} \times \frac{\cancel{33}^{3}}{\cancel{40}_{5}} = \frac{3}{5}$

1.12 $\frac{5}{7} \times \frac{\cancel{3}^{1}}{\cancel{6}_{2}} = \frac{5}{14}$

1.13 $\frac{\cancel{7}^{1}}{_{2}\cancel{10}} \times \frac{\cancel{5}^{1}}{\cancel{49}_{7}} = \frac{1}{14}$

1.14 $\frac{9}{20}$

1.15 $\frac{\cancel{4}^{1}}{_{1}\cancel{5}} \times \frac{\cancel{5}^{1}}{\cancel{12}_{\cancel{3}_{1}}} \times \frac{\cancel{3}^{1}}{11} = \frac{1}{11}$

1.16 $\frac{\cancel{4}^{1}}{5} \times \frac{3}{\cancel{16}_{4}} = \frac{3}{20}$

1.17 $\frac{\cancel{7}^{1}}{_{3}\cancel{15}} \times \frac{\cancel{10}}{\cancel{21}_{3}} = \frac{2}{9}$

1.18 $\frac{^{5}\cancel{25}}{_{2}\cancel{36}} \times \frac{\cancel{18}^{1}}{\cancel{35}_{7}} = \frac{5}{14}$

1.19 $\frac{3}{_{1}\cancel{4}} \times \frac{\cancel{16}^{\cancel{4}^{1}}}{\cancel{25}_{5}} \times \frac{\cancel{5}^{1}}{\cancel{8}_{2}} = \frac{3}{10}$

1.20 $\frac{^{1}\cancel{5}}{_{2}\cancel{8}} \times \frac{\cancel{4}^{1}}{\cancel{35}_{\cancel{7}_{1}}} \times \frac{\cancel{7}^{1}}{12} = \frac{1}{24}$

1.21 $\frac{1}{4} \times \frac{1}{8} = \frac{1}{32}$ inch

1.22 $2\frac{1}{4} \times \frac{1}{2} = \frac{9}{4} \times \frac{1}{2} = \frac{9}{8} = 1\frac{1}{8}$ cups

1.23 $3\frac{1}{3} \times 4 = \frac{10}{3} \times \frac{4}{1} = \frac{40}{3} = 13\frac{1}{3}$ ft.

1.24 Example:

$4 = \frac{4 \times 5}{5} = \frac{20}{5}$

1.25 Example:

$3 = \frac{3 \times 7}{7} = \frac{21}{7}$

1.26 Example:

$6 = \frac{6 \times 3}{3} = \frac{18}{3}$

1.27 $4\frac{1}{5} = \frac{(4 \times 5) + 1}{5} = \frac{20 + 1}{5} = \frac{21}{5}$

1.28 $3\frac{4}{7} = \frac{(3 \times 7) + 4}{7} = \frac{21 + 4}{7} = \frac{25}{7}$

1.29 $6\frac{2}{3} = \frac{(6 \times 3) + 2}{3} = \frac{18 + 2}{3} = \frac{20}{3}$

1.30 $5\frac{7}{8} = \frac{(5 \times 8) + 7}{8} = \frac{40 + 7}{8} = \frac{47}{8}$

1.31 $6\frac{5}{16} = \frac{(6 \times 16) + 5}{16} = \frac{96 + 5}{16} = \frac{101}{16}$

1.32 $2\frac{11}{32} = \frac{(2 \times 32) + 11}{32} = \frac{64 + 11}{32} = \frac{75}{32}$

1.33 $\frac{3}{1} \times \frac{5}{4} = \frac{15}{4} = 3\frac{3}{4}$

1.34 $\frac{\overset{3}{\cancel{6}}}{1} \times \frac{9}{\underset{1}{\cancel{2}}} = \frac{27}{1} = 27$

1.35 $\frac{\overset{2}{\cancel{14}}}{\underset{1}{\cancel{3}}} \times \frac{\overset{20}{\cancel{60}}}{\underset{1}{\cancel{7}}} = \frac{40}{1} = 40$

1.36 $\frac{\overset{10}{\cancel{50}}}{\underset{1}{\cancel{3}}} \times \frac{\overset{4}{\cancel{12}}}{\underset{1}{\cancel{5}}} = \frac{40}{1} = 40$

1.37 $\frac{13}{4} \times \frac{27}{5} = \frac{351}{20} =$

$$\begin{array}{r} 17 \\ 20\overline{)351} \\ \underline{20} \\ 151 \\ \underline{140} \\ 11 \end{array} = 17\frac{11}{20}$$

1.38 $\frac{\overset{1}{\cancel{8}}}{\underset{1}{\cancel{11}}} \times \frac{\overset{1}{\cancel{11}}}{\underset{1}{\cancel{8}}} = \frac{1}{1} = 1$

1.39 $\frac{539}{8} \times \frac{41}{4} = \frac{22,099}{32} =$

$$\begin{array}{r} 690 \\ 32\overline{)22,099} \\ \underline{19\,2} \\ 289 \\ \underline{288} \\ 19 \end{array} = 690\frac{19}{32}$$

1.40 $\frac{53}{\underset{2}{\cancel{8}}} \times \frac{\overset{9}{\cancel{36}}}{5} = \frac{477}{10} = 47\frac{7}{10}$

1.41 $\frac{\overset{49}{\cancel{147}}}{16} \times \frac{35}{\underset{1}{\cancel{3}}} = \frac{1,715}{16} =$

$$\begin{array}{r} 107 \\ 16,\overline{)1,715} \\ \underline{1\,6} \\ 115 \\ \underline{112} \\ 3 \end{array} = 107\frac{3}{16}$$

1.42 $\frac{\overset{31}{\cancel{186}}}{25} \times \frac{103}{\underset{2}{\cancel{12}}} = \frac{3,193}{50} =$

$$\begin{array}{r} 63 \\ 50\overline{)3,193} \\ \underline{3\,00} \\ 193 \\ \underline{150} \\ 43 \end{array} = 63\frac{43}{50}$$

1.43 $\frac{63}{\underset{1}{\cancel{5}}} \times \frac{\overset{9}{\cancel{45}}}{8} = \frac{567}{8} =$

$$\begin{array}{r} 70 \\ 8\overline{)567} \\ \underline{56} \\ 07 \end{array} = 70\frac{7}{8}$$

1.44 $\frac{115}{\underset{4}{\cancel{16}}} \times \frac{\overset{1}{\cancel{4}}}{1} = \frac{115}{4} =$

$$\begin{array}{r} 28 \\ 4\overline{)115} \\ \underline{8} \\ 35 \\ \underline{32} \\ 3 \end{array} = 28\frac{3}{4}$$

1.45 $\frac{\overset{11}{\cancel{44}}}{5} \times \frac{1}{\underset{1}{\cancel{4}}} = \frac{11}{5} = 2\frac{1}{5}$

1.46 $\frac{\overset{21}{\cancel{63}}}{5} \times \frac{107}{\underset{2}{\cancel{6}}} = \frac{2,247}{10} = 224\frac{7}{10}$

1.47 $\frac{\overset{1}{\cancel{7}}}{\underset{1}{\cancel{16}}} \times \frac{\overset{1}{\cancel{16}}}{\underset{1}{\cancel{7}}} = \frac{1}{1} = 1$

1.48 $\frac{\overset{3}{\cancel{36}}}{\underset{1}{\cancel{5}}} \times \frac{\overset{13}{\cancel{65}}}{\underset{1}{\cancel{12}}} = \frac{39}{1} = 39$

1.49 $\frac{\overset{16}{\cancel{48}}}{\underset{1}{\cancel{5}}} \times \frac{\overset{1}{\cancel{5}}}{\underset{1}{\cancel{3}}} = \frac{16}{1} = 16$

1.50 $\frac{25}{\underset{1}{\cancel{3}}} \times \frac{\overset{21}{\cancel{63}}}{8} = \frac{525}{8} =$

$$\begin{array}{r} 65 \\ 8\overline{)525} \\ \underline{48} \\ 45 \\ \underline{40} \\ 5 \end{array} = 65\frac{5}{8}$$

1.51 $\frac{51}{\underset{1}{\cancel{4}}} \times \frac{\overset{16}{\cancel{64}}}{5} = \frac{816}{5} =$

$$\begin{array}{r} 163 \\ 5\overline{)816} \\ \underline{5} \\ 31 \\ \underline{30} \\ 16 \\ \underline{15} \\ 1 \end{array} = 163\frac{1}{5}$$

1.52 $\frac{52}{3} \times \frac{7}{1} = \frac{364}{3} = 121\frac{1}{3}$

1.53 $\frac{\cancel{136}^{68}}{3} \times \frac{1}{2} = \frac{68}{3} = 22\frac{2}{3}$

1.54 $\frac{56}{255}$

1.55 $\frac{\cancel{5}^{1}}{{}_{2}\cancel{6}} \times \frac{\cancel{3}^{1}}{4} \times \frac{73}{\cancel{10}_{2}} = \frac{73}{16} = 4\frac{9}{16}$

1.56 $\frac{\cancel{55}^{11}}{8} \times \frac{19}{\cancel{5}_{1}} \times \frac{13}{3} = \frac{2{,}717}{24} =$

```
     113
24)2,717 = 113 5/24
   24
    31
    24
     77
     72
      5
```

1.57 $\frac{\cancel{14}^{7}}{{}_{1}\cancel{5}} \times \frac{\cancel{51}^{17}}{\cancel{16}_{8}} \times \frac{\cancel{35}^{7}}{\cancel{6}_{2}} = \frac{833}{16} =$

```
    52
16)833 = 52 1/16
   80
    33
    32
     1
```

1.58 $0 \times \frac{3}{4} = 0$

1.59 $\frac{1}{2} \times \frac{3}{4} = \frac{3}{8}$

1.60 $\frac{3}{4} \times \frac{3}{4} = \frac{9}{16}$

1.61 $1 \times \frac{3}{4} = \frac{3}{4}$

1.62 $1\frac{1}{3} \times \frac{3}{4} = \frac{\cancel{4}^{1}}{{}_{1}\cancel{3}} \times \frac{\cancel{3}^{1}}{\cancel{4}_{1}} = \frac{1}{1} = 1$

1.63 a. $0 \times \frac{1}{2} = 0$

b. $0 + \frac{1}{4} = \frac{1}{4}$

1.64 a. $\frac{1}{2} \times \frac{1}{2} = \frac{1}{4}$

b. $\frac{1}{4} + \frac{1}{4} = \frac{2}{4} = \frac{1}{2}$

1.65 a. $\frac{1}{3} \times \frac{1}{2} = \frac{1}{6}$

b. $\frac{1}{6} + \frac{1}{4} = \frac{2}{12} + \frac{3}{12} = \frac{5}{12}$

1.66 a. $1\frac{1}{2} \times \frac{1}{2} = \frac{3}{2} \times \frac{1}{2} = \frac{3}{4}$

b. $\frac{3}{4} + \frac{1}{4} = \frac{4}{4} = 1$

1.67
```
   0.3 (1)
 x 0.2 (1)
  0.06 (2)
```

1.68
```
    1.4 (1)
 x 0.32 (2)
     28
    42
  0.448 (3)
```

1.69
```
$ 3.25 (2)
    x 7 (0)
$22.75 (2)
```

1.70
```
   452.3 (1)
  x 7.37 (2)
   31661
  13569
 31661
3333.451 (3)
```

1.71
```
  23.456 (3)
  x 6.66 (2)
  140736
 140736
140736
156.21696 (5)
```

1.72
```
     465 (0)
  x 0.32 (2)
     930
   1395
  148.80 (2)
```

1.73
36.75 (2)
x 60.6 (1)
22050
22050
2,227.050 (3)

1.74
14.34 (2)
x 3.17 (2)
10038
1434
4302
45.4578 (4)

1.75
0.3472 (4)
x 22 (0)
6944
6944
7.6384 (4)

1.76
7,456 (0)
x 0.0014 (4)
29824
7456
10.4384 (4)

1.77
63.4 (2)
x 7.07 (2)
4438
4438
44.8238 (4)

1.78
4,050 (0)
x 23.2 (1)
8100
12150
8100
93,960.0 (1)

1.79
3,400 (0)
x 0.071 (3)
3400
23800
241.400 (3)

1.80
2.375
x 12
4750
2375
28.500 = 28.5 cm

1.81
\$0.47 x 5 = \$2.35
\$0.99 x 2 = 1.98
4.75
\$9.08

1.82
0 x 0.45 = 0
0 + 2.17 = 2.17

1.83
0.45
x 0.2
0.090
+ 2.17
2.26

1.84
0.45
x 0.5
0225
+ 2.17
2.395

1.85
1 x 0.45 = 0.45
0.45 + 2.17 = 2.62

1.86
0.45
x 1.7
315
45
0.765
+ 2.17
2.935

1.87
2 x 2.7 = 5.4
5.4 – 3.4 = 2.0

1.88
2.5
x 2.7
175
50
6.75
– 3.4
3.35

1.89
$$\begin{array}{r} 3.2 \\ \times\ 2.7 \\ \hline 224 \\ 64 \\ \hline 8.64 \\ -\ 3.4 \\ \hline 5.24 \end{array}$$

1.90
$$\begin{array}{r} 4.7 \\ \times\ 2.7 \\ \hline 329 \\ 94 \\ \hline 12.69 \\ -3.4 \\ \hline 9.29 \end{array}$$

1.91
$$\begin{array}{r} 9.6 \\ \times\ 2.7 \\ \hline 672 \\ 192 \\ \hline 25.92 \\ -3.4 \\ \hline 22.52 \end{array}$$

II. SECTION TWO

2.1 $\frac{3}{2}$ or $1\frac{1}{2}$

2.2 $\frac{32}{7}$ or $4\frac{4}{7}$

2.3 $\frac{25}{14}$ or $1\frac{11}{14}$

2.4 $\frac{845}{77}$ or $10\frac{75}{77}$

2.5 $\frac{20}{19}$ or $1\frac{1}{19}$

2.6
a. $\frac{17}{5}$
b. $\frac{5}{17}$

2.7
a. $\frac{23}{3}$
b. $\frac{3}{23}$

2.8
a. $\frac{81}{5}$
b. $\frac{5}{81}$

2.9
a. $\frac{1,081}{16}$
b. $\frac{16}{1,081}$

2.10 $\frac{2}{3} \times \frac{\cancel{6}^{3}}{{}_{2}\cancel{4}} = \frac{{}^{1}\cancel{2}}{{}_{1}\cancel{3}} \times \frac{\cancel{3}^{1}}{\cancel{2}_{1}} = \frac{1}{1} = 1$

2.11 $\frac{31}{5} \div \frac{30}{7} = \frac{31}{5} \times \frac{7}{30} = \frac{217}{150} = 1\frac{67}{150}$

2.12 $\frac{22}{5} \times \frac{3}{2} = \frac{{}^{11}\cancel{22}}{5} \times \frac{3}{\cancel{2}_{1}} = \frac{33}{5} = 6\frac{3}{5}$

2.13 $\frac{7}{16} \div \frac{21}{8} = \frac{{}^{1}\cancel{7}}{{}_{2}\cancel{16}} \times \frac{\cancel{8}^{1}}{\cancel{21}_{3}} = \frac{1}{6}$

2.14 $\frac{7}{8} \times \frac{1}{3} = \frac{7}{24}$

2.15 $8 \times \frac{2}{1} = \frac{16}{1} = 16$

2.16 $\frac{9}{2} \div 7 = \frac{9}{2} \times \frac{1}{7} = \frac{9}{14}$

2.17 $\frac{14}{3} \div 2 = \frac{{}^{7}\cancel{14}}{3} \times \frac{1}{\cancel{2}_{1}} = \frac{7}{3} = 2\frac{1}{3}$

2.18 $\frac{5}{3} \div 5 = \frac{{}^{1}\cancel{5}}{3} \times \frac{1}{\cancel{5}_{1}} = \frac{1}{3}$

2.19 $\frac{5}{2} \div \frac{5}{6} = \frac{{}^{1}\cancel{5}}{{}_{1}\cancel{2}} \times \frac{\cancel{6}^{3}}{\cancel{5}_{1}} = \frac{3}{1} = 3$

2.20 $\frac{21}{8} \div \frac{3}{5} = \frac{{}^{7}\cancel{21}}{8} \times \frac{5}{\cancel{3}_{1}} = \frac{35}{8} = 4\frac{3}{8}$

2.21 $\frac{15}{8} \div \frac{3}{2} = \frac{{}^{5}\cancel{15}}{{}_{4}\cancel{8}} \times \frac{\cancel{2}^{1}}{\cancel{3}_{1}} = \frac{5}{4} = 1\frac{1}{4}$

2.22 $\frac{7}{8} \div \frac{7}{4} = \frac{{}^{1}\cancel{7}}{{}_{2}\cancel{8}} \times \frac{\cancel{4}^{1}}{\cancel{7}_{1}} = \frac{1}{2}$

2.23 $\frac{3}{4} \div \frac{8}{5} = \frac{3}{4} \times \frac{5}{8} = \frac{15}{32}$

2.24 $\frac{34}{3} \div \frac{17}{6} = \frac{\overset{2}{\cancel{34}}}{\underset{1}{\cancel{3}}} \times \frac{\overset{2}{\cancel{6}}}{\underset{1}{\cancel{17}}} = \frac{4}{1} = 4$

2.25 $\frac{9}{2} \div \frac{3}{5} = \frac{\overset{3}{\cancel{9}}}{2} \times \frac{5}{\underset{1}{\cancel{3}}} = \frac{15}{2} = 7\frac{1}{2}$

2.26 $\frac{25}{4} \div 8 = \frac{25}{4} \times \frac{1}{8} = \frac{25}{32}$

2.27 $\frac{8}{3} \div \frac{15}{2} = \frac{8}{3} \times \frac{2}{15} = \frac{16}{45}$

2.28 $\frac{15}{4} \div \frac{22}{5} = \frac{15}{4} \times \frac{5}{22} = \frac{75}{88}$

2.29 $\frac{13}{4} \div \frac{5}{6} = \frac{13}{\underset{2}{\cancel{4}}} \times \frac{\overset{3}{\cancel{6}}}{5} = \frac{39}{10} = 3\frac{9}{10}$

2.30 $\frac{8}{3} \div \frac{9}{16} = \frac{8}{3} \times \frac{16}{9} = \frac{128}{27} = 4\frac{20}{27}$

2.31 $\frac{15}{8} \div \frac{4}{3} = \frac{15}{8} \times \frac{3}{4} = \frac{45}{32} = 1\frac{13}{32}$

2.32 $\frac{7}{8} \div \frac{15}{4} = \frac{7}{\underset{2}{\cancel{8}}} \times \frac{\overset{1}{\cancel{4}}}{15} = \frac{7}{30}$

2.33 $\frac{13}{16} \div \frac{39}{32} = \frac{\overset{1}{\cancel{13}}}{\underset{1}{\cancel{16}}} \times \frac{\overset{2}{\cancel{32}}}{\underset{3}{\cancel{39}}} = \frac{2}{3}$

2.34 $\frac{1}{4} \div \frac{67}{16} = \frac{1}{\underset{1}{\cancel{4}}} \times \frac{\overset{4}{\cancel{16}}}{67} = \frac{4}{67}$

2.35 $\frac{33}{16} \div \frac{11}{8} = \frac{\overset{3}{\cancel{33}}}{\underset{2}{\cancel{16}}} \times \frac{\overset{1}{\cancel{8}}}{\underset{1}{\cancel{11}}} = \frac{3}{2} = 1\frac{1}{2}$

2.36 $\frac{5}{3} \div \frac{23}{16} = \frac{5}{3} \times \frac{16}{23} = \frac{80}{69} = 1\frac{11}{69}$

2.37 $\frac{8}{3} \div \frac{16}{3} = \frac{\overset{1}{\cancel{8}}}{\underset{1}{\cancel{3}}} \times \frac{\overset{1}{\cancel{3}}}{\underset{2}{\cancel{16}}} = \frac{1}{2}$

2.38 $\frac{27}{8} \div \frac{9}{2} = \frac{\overset{3}{\cancel{27}}}{\underset{4}{\cancel{8}}} \times \frac{\overset{1}{\cancel{2}}}{\underset{1}{\cancel{9}}} = \frac{3}{4}$

2.39 $\frac{63}{16} \div \frac{21}{8} = \frac{\overset{3}{\cancel{63}}}{\underset{2}{\cancel{16}}} \times \frac{\overset{1}{\cancel{8}}}{\underset{1}{\cancel{21}}} = \frac{3}{2} = 1\frac{1}{2}$

2.40 $\frac{85}{8} \div \frac{17}{4} = \frac{\overset{5}{\cancel{85}}}{\underset{2}{\cancel{8}}} \times \frac{\overset{1}{\cancel{4}}}{\underset{1}{\cancel{17}}} = \frac{5}{2} = 2\frac{1}{2}$

2.41 $\frac{22}{5} \div \frac{11}{4} = \frac{\overset{2}{\cancel{22}}}{5} \times \frac{4}{\underset{1}{\cancel{11}}} = \frac{8}{5} = 1\frac{3}{5}$

2.42 $\frac{5}{8} \times \frac{1}{2} = \frac{5}{16}$

2.43 $\frac{100}{3} \div 100 = \frac{\overset{1}{\cancel{100}}}{3} \times \frac{1}{\underset{1}{\cancel{100}}} = \frac{1}{3}$

2.44 $1 \div \frac{8}{3} = 1 \times \frac{3}{8} = \frac{3}{8}$

2.45 $\frac{5}{\underset{2}{\cancel{16}}} \times \frac{\overset{1}{\cancel{8}}}{3} = \frac{5}{6}$

2.46 $\frac{2}{3} \div \frac{5}{8} = \frac{2}{3} \times \frac{8}{5} = \frac{16}{15} = 1\frac{1}{15}$

2.47 $4\frac{5}{8} \div 3\frac{1}{4} =$

$\frac{37}{8} \div \frac{13}{4} = \frac{37}{\underset{2}{\cancel{8}}} \times \frac{\overset{1}{\cancel{4}}}{13} = \frac{37}{26} = 1\frac{11}{26}$

2.48 $(\frac{9}{8} \div \frac{9}{4}) \div \frac{14}{3} =$

$(\frac{\overset{1}{\cancel{9}}}{\underset{2}{\cancel{8}}} \times \frac{\overset{1}{\cancel{4}}}{\underset{1}{\cancel{9}}}) \div \frac{14}{3} =$

$\frac{1}{2} \div \frac{14}{3} = \frac{1}{2} \times \frac{3}{14} = \frac{3}{28}$

2.49 $(\frac{19}{\underset{1}{\cancel{5}}} \times \frac{\overset{5}{\cancel{25}}}{6}) \div \frac{10}{3} =$

$\frac{95}{6} \div \frac{10}{3} = \frac{\overset{19}{\cancel{95}}}{\underset{2}{\cancel{6}}} \times \frac{\overset{1}{\cancel{3}}}{\underset{2}{\cancel{10}}} = \frac{19}{4} = 4\frac{3}{4}$

2.50
```
         3 2.4
4.5)1 4 5.8.0
   ^1 3 5  ^
      1 0 8
        9 0
      1 8 0
      1 8 0
          0
```

Check:
```
  32.4
 x 4.5
  1620
 1296
145.80
```

2.51
```
          0.456
0.23)0.10.488
        92
        128
        115
         138
         138
           0
```

Check:
```
   0.456
  x 0.23
    1368
    912
 0.10488
```

2.52
```
           42.5
14.4)612.0.0
      576
       360
       288
        720
        720
          0
```

Check:
```
   42.5
 x 14.4
   1700
  1700
  425
 612.00
```

2.53
```
       0.471
143)67.353
     57 2
     10 15
     10 01
        143
        143
          0
```

Check:
```
  0.471
  x 143
   1413
  1884
  471
 67.353
```

2.54
```
            14.2
0.071)1.008.2
        71
        298
        284
         142
         142
           0
```

Check:
```
    14.2
 x 0.071
     142
    994
  1.0082
```

2.55
```
          70.43
41.2)2901.7.16
     2884
      1771
      1648
       1236
       1236
          0
```

Check:
```
     70.43
     x 412
     14086
     7043
   28172
 2,901.716
```

2.56
```
           71.71
6.43)461.09.53
      450 1
       10 99
        6 43
        4 565
        4 501
           643
           643
             0
```

2.57
```
             1.943
0.524)1.018.132
         524
         4941
         4716
          2253
          2096
           1572
           1572
              0
```

2.58
```
              0.0417
0.0422)0.0017.5974
          1688
           717
           422
          2954
          2954
             0
```

2.59
```
         47400
42.2)2000280.0.
     1688
      3122
      2954
       1688
       1688
          0
```

2.60
```
        4.708
2.75)12.94.700
     1100
      1947
      1925
        2200
        2200
           0
```

2.61
```
      12.43
197)2,448.71
    1 97
      478
      394
       847
       788
        591
        591
          0
```

2.62
```
    9.166→9.17
96)880.000
   864
    160
     96
     640
     576
      640
      576
```

2.63
```
  100.00
4)400.00
  4
  0
```

2.64
```
    83.333 → 83.33
84)7000.000
   672
    280
    252
     280
     252
      280
      252
       280
       252
```

2.65
```
    7.837 → 7.84
37)290.000
   259
    310
    296
     140
     111
      290
      259
```

2.66
```
    59.189 → 59.19
37)2190.000
   185
    340
    333
      70
      37
      330
      296
       340
       333
```

2.67
```
     22.222 → 22.22
153)3400.000
    306
     340
     306
      340
      306
       340
       306
        340
        306
```

2.68

6.185 → 6.19
221)1367.000
1326
410
221
1890
1768
1220
1105

2.69

2.380 → 2.38
3)7.141
6
11
9
24
24
0

2.70

74.774 → 74.77
333)24900.000
2331
1590
1332
2580
2331
2490
2331
1590
1332

2.71

4,166.666 → 4,166.67
6)25000.000
24
10
6
40
36
40
36
40
36
40
36
40
36

2.72

4,330.769 → 4,330.77
13)56300.000
52
43
39
40
39
100
91
90
78
120
117

2.73

27.038 → 27.04
314)8490.000
628
2210
2198
1200
942
2580
2512

2.74

3,772.151 → 3,772.15
79)298000.000
237
610
553
570
553
170
158
120
79
410
395
150
79

2.75

1.628 → 1.63
592)964.000
592
3720
3552
1680
1184
4960
4736

2.76

30,000.00
7)210000.00
21
0

2.77

47.1428 → 47.143
7)330.0000
28
50
49
10
7
30
28
20
14
60
56

2.78

7.6428 → 7.643
56)428.0000
392
360
336
240
224
160
112
480
448

2.79

8.6445 → 8.645
723)6250.0000
5784
4660
4338
3220
2892
3280
2892
3880
3615

2.80

14.9517 → 14.952
2381)35600.0000
2381
11790
9524
22660
21429
12310
11905
4050
2381
16690
16667

2.81

9.5457 → 9.546
4007)38250.0000
36063
21870
20035
18350
16028
23220
20035
31850
28049

2.82

0.0181 → 0.018
54)0.9800
54
440
432
80
54

2.83

12.2463 → 12.246
69)845.0000
69
155
138
170
138
320
276
440
414
260
207

2.84

1.6423 → 1.642
26)42.7000
26
167
156
110
104
60
52
80
78

2.85

12.1272 → 12.127
503)6100.0000
503
1070
1006
640
503
1370
1006
3640
3521
1190
1005

2.86

0.0066 → 0.007
243)1.6200
1458
1620
1458

2.87

0.0032 → 0.003
7462)24.3200
22386
19340
14924

2.88

3.400
753)2560.200
2259
3012
3012
0

2.89

$10.745 → $10.75
7)$75.220
7
052
49
32
28
40
35

2.90

$ 5.320 → $5.32
16)$85.130
80
51
48
33
32
10

2.91

$ 177.431 → $177.43
42)$7452.110
42
325
294
312
294
181
168
131
126
50
42

2.92

$17\frac{3}{4} \div \frac{1}{4} =$

$\frac{71}{\not{4}_1} \times \frac{\not{4}^1}{1} = \frac{71}{1} = 71$ pieces

2.93

$12\frac{1}{3} \times \frac{1}{2} = \frac{37}{3} \times \frac{1}{2} = \frac{37}{6} = 6\frac{1}{6}$ inches

2.94

$742\frac{1}{6} \div 62\frac{1}{2} = \frac{4,453}{6} \div \frac{125}{2} =$

$\frac{4,453}{\not{6}_3} \times \frac{\not{2}^1}{125} = \frac{4,453}{375} =$

$11\frac{328}{375}$ cubic feet
375)4,453
375
703
375
328

2.95

$\frac{\not{13,400}^{1,675}}{1} \times \frac{1}{\not{8}_1} = \frac{1,675}{1} = \$1,675$

2.96

$\frac{\not{39,600}^{13,200}}{1} \times \frac{1}{\not{3}_1} = \frac{13,200}{1} = \$13,200$

2.97

$7\frac{3}{4} \times 749 =$

$\frac{31}{4} \times \frac{749}{1} = \frac{23,219}{4} =$

5804.75¢ = $58.05
4)23,219.00
20
32
32
019
16
30
28
20
20
0

2.98 $N = 8 \div 0.01$

$$0.01\overline{)8.00} = 800$$
8
0

$N = 800$

2.99 $0.06 \times N = 42$
$N = 42 \div 0.06$

$$0.06\overline{)42.00} = 700$$
42
0

$N = 700$

2.100 $0.25 \times N = 15$
$N = 15 \div 0.25$

$$0.25\overline{)15.00} = 60$$
15 0
0

$N = 60$

2.101 Balls : $2.75 x 12
550
275
$33.00

Bats: $2.95 x 7
$20.65
+ 33.00
Total: $53.65

$$9\overline{)\$53.650} = \$5.961 \rightarrow \$5.96$$
45
86
81
55
54
10
9

2.102 3 dozen = 36 pencils

$0.06 or 6¢ per pencil

$$36\overline{)\$2.16}$$
2 16
0

2.103

$$0.96\overline{)7.44,00} = 7.75$$ lbs., $7\frac{3}{4}$ lbs., or 7 lbs., 12 oz
6 72
720
672
480
480
0

2.104 20
x 18
160
20
360 miles

2.105 a. $0 \times \frac{2}{3} = 0$

b. $0 + \frac{1}{6} = \frac{1}{6}$

2.106 a. $1 \times \frac{2}{3} = \frac{2}{3}$

b. $\frac{2}{3} + \frac{1}{6} = \frac{4}{6} + \frac{1}{6} = \frac{5}{6}$

2.107 a. $\frac{3}{4} \times \frac{2}{3} =$

$\frac{\cancel{3}^1}{{}_2\cancel{4}} \times \frac{\cancel{2}^1}{\cancel{3}_1} = \frac{1}{2}$

b. $\frac{1}{2} + \frac{1}{6} = \frac{3}{6} + \frac{1}{6} = \frac{4}{6} = \frac{2}{3}$

2.108 a. $\frac{7}{8} \times \frac{2}{3} =$

$\frac{7}{{}_4\cancel{8}} \times \frac{\cancel{2}^1}{3} = \frac{7}{12}$

b. $\frac{7}{12} + \frac{1}{6} = \frac{7}{12} + \frac{2}{12} = \frac{9}{12} = \frac{3}{4}$

2.109 a. $1 \times \frac{3}{2} = \frac{3}{2} = 1\frac{1}{2}$

b. $\frac{3}{2} + \frac{1}{4} = \frac{6}{4} + \frac{1}{4} = \frac{7}{4} = 1\frac{3}{4}$

2.110 a. $\frac{2}{3} \times \frac{3}{2} = \frac{\cancel{2}^1}{_1\cancel{3}} \times \frac{\cancel{3}^1}{\cancel{2}_1} = \frac{1}{1} = 1$

b. $1 + \frac{1}{4} = 1\frac{1}{4}$

2.111 a. $\frac{3}{4} \times \frac{3}{2} = \frac{9}{8} = 1\frac{1}{8}$

b. $\frac{9}{8} + \frac{1}{4} = \frac{9}{8} + \frac{2}{8} = \frac{11}{8} = 1\frac{3}{8}$

2.112 a. $\frac{5}{6} \times \frac{3}{2} =$

$\frac{5}{_2\cancel{6}} \times \frac{\cancel{3}^1}{2} = \frac{5}{4} = 1\frac{1}{4}$

b. $\frac{5}{4} + \frac{1}{4} = \frac{6}{4} + \frac{3}{2} = 1\frac{1}{2}$

2.113 a. $0 \times 7.4 = 0$

b. $0 + 2.37 = 2.37$

2.114 a. $1 \times 7.4 = 7.4$

b. $7.4 + 2.37 = 9.77$

2.115 a.

```
   2.43
  x 7.4
    972
  1701
 17.982
```

b.

```
 17.982
 + 2.37
 20.352
```

2.116 a.

```
   3.27
  x 7.4
   1308
  2289
 24.198
```

2.116 (cont.)

b.

```
 24.198
 + 2.37
 26.568
```

2.117 a.

```
   6.85
  x 7.4
   2740
  4795
 50.690 = 50.69
```

b.

```
 50.69
 + 2.37
 53.06
```

2.118 a.

```
   8.14
  x 7.4
   3256
  5698
 60.236
```

b.

```
 60.236
 + 2.37
 62.606
```

2.119 a.

```
      9.3023 → 9.302
 43)400.0000
    387
     130
     129
       100
        86
       140
       129
```

b.

```
 9.302
 - 7.14
 2.162
```

2.120

```
     11.0465 → 11.047
43)475.0000
   43
    45
    43
     200
     172
      280
      258
       220
       215
```

b.
```
  11.047
 – 7.14
  3.907
```

2.121

a.
```
     11.9302 → 11.930
43)513.0000
   43
    83
    43
    400
    387
     130
     129
      100
       86
```

b.
```
  11.930
 – 7.14
  4.790
```

2.122

a.
```
      38.1511 → 38.151
43)1640.5000
   129
    350
    344
      65
      43
      220
      215
        50
        43
         70
         43
```

2.122 (cont.)

b.
```
  38.151
 – 7.14
  31.011
```

2.123

```
      51.8604 → 51.860
43)2230.0000
   215
     80
     43
     370
     344
      260
      258
        200
        172
```

b.
```
  51.860
 – 7.14
  44.720
```

2.124

a.
```
      63.1279 → 63.128
43)2714.5000
   258
    134
    129
      55
      43
      120
       86
       340
       301
        390
        387
```

b.
```
  63.128
 – 7.14
  55.988
```

2.125
a. $60 - 32 = 28$
b. $28 \times 5 = 140$
c. $140.0 \div 9 = 15.55 \rightarrow 15.6$

$$\begin{array}{r} 15.55 \\ 9\overline{)140.0} \\ \underline{9} \\ 50 \\ \underline{45} \\ 50 \\ \underline{45} \end{array}$$

d. 16

2.126
a. $50 - 32 = 18$
b. $18 \times 5 = 90$
c. $90 \div 9 = 10.0$
d. 10

2.127
a. $40 - 32 = 8$
b. $8 \times 5 = 40$
c.

$$\begin{array}{r} 4.4 \\ 9\overline{)40.0} \\ \underline{36} \\ 40 \\ \underline{36} \end{array}$$

d. 4

2.128
a. $90 - 32 = 58$
b. $58 \times 5 = 290$
c.

$$\begin{array}{r} 32.2 \\ 9\overline{)290.0} \\ \underline{27} \\ 20 \\ \underline{18} \\ 20 \end{array}$$

d. 32

III. SECTION THREE

3.1 $0.45 \times \frac{100}{100} = \frac{45}{100} = 45\%$

3.2 $0.67 \times \frac{100}{100} = \frac{67}{100} = 67\%$

3.3 $0.27 \times \frac{100}{100} = \frac{27}{100} = 27\%$

3.4 $0.032 \times \frac{100}{100} = \frac{3.2}{100} = 3.2\%$

3.5 $0.042 \times \frac{100}{100} = \frac{4.2}{100} = 4.2\%$

3.6 $4.17 \times \frac{100}{100} = \frac{417}{100} = 417\%$

3.7 $0.003 \times \frac{100}{100} = \frac{0.3}{100} = 0.3\%$

3.8 $66.3 \times \frac{100}{100} = \frac{6,630}{100} = 6,630\%$

3.9 $1.45 \times \frac{100}{100} = \frac{145}{100} = 145\%$

3.10 $0.0004 \times \frac{100}{100} = \frac{0.04}{100} = 0.04\%$

3.11 $1.0 \times \frac{100}{100} = \frac{100}{100} = 100\%$

3.12 $\frac{1}{4} = 0.25 = 25\%$

3.13 $\frac{1}{2} = 0.5 = 50\%$

3.14 $\frac{7}{8} = 0.875 = 87.5\%$

3.15 $\frac{3}{2} = 1\frac{1}{2} = 1.5 = 150\%$

3.16 $\frac{1}{16} = 0.0625 = 6.25\%$

3.17 $\frac{9}{4} = 2\frac{1}{4} = 2.25 = 225\%$

3.18 $\frac{17}{17} = 1 = 100\%$

3.19 The percent times the base is equal to the number or, $\% \times B = N$.

3.20 $\frac{10}{100} \times 100 = 0.1 \times 100 = 10$

3.21 $\frac{150}{100} \times 4 = 1.5 \times 4 = 6$

3.22 $\frac{15}{100} \times 3 = 0.15 \times 3 = 0.45$

3.23 $0.001 \times 700 = 0.7$

3.24
$$\begin{array}{r} 0.75 \\ \times\ 16 \\ \hline 450 \\ 75\ \ \\ \hline 12.00 \end{array} = 12$$

3.25 $0.0002 \times 400 = 0.08$

3.26
$$\begin{array}{r} 0.72 \\ \times\ 45 \\ \hline 360 \\ 288\ \ \\ \hline 32.40 \end{array} = 32.4$$

3.27 $0.22 \times 4 = 0.88$

3.28 $0.048 \times 6 = 0.288$

3.29
$$\begin{array}{r} 0.88 \\ \times\ 25 \\ \hline 440 \\ 176\ \ \\ \hline 22.00 \end{array} = 22 \text{ problems correct}$$

3.30 $0.015 \times B = 45$

$B = 45 \div 0.015$

$$\begin{array}{r} 3{,}000 \\ 15\overline{)45000} \\ 45 \\ \hline 0 \end{array}$$

$B = 3{,}000$

3.31 $1.25 \times B = 55$

$B = 55 \div 1.25$

$$\begin{array}{r} 44 \\ 125\overline{)5500} \\ 500\ \\ \hline 500 \\ 500 \\ \hline 0 \end{array}$$

$B = 44$

3.32 $0.025 \times B = 25$

$B = 25 \div 0.025$

$$\begin{array}{r} 1{,}000 \\ 25\overline{)25000} \\ 25 \\ \hline 0 \end{array}$$

$B = 1{,}000$

3.33 $0.001 \times B = 2.4$

$B = 2.4 \div 0.001$

$$\begin{array}{r} 2{,}400 \\ 1\overline{)2400} \\ 2 \\ \hline 04 \\ 4 \\ \hline 0 \end{array}$$

$B = 2{,}400$

3.34 $0.75 \times B = 125$

$B = 125 \div 0.75$

$$\begin{array}{r} 166 \\ 75\overline{)12500} \\ 75\ \ \\ \hline 500 \\ 450 \\ \hline 500 \\ 450 \\ \hline 50 \end{array} = 166\frac{50}{75}$$

$B = 166\frac{50}{75} = 166\frac{2}{3}$

3.35 $1 \times B = 7$

$B = 7 \div 1$

$B = 7$

3.36
$6 = 0.75 \times B$
$6 \div 0.75 = B$
$$\begin{array}{r} 8 \\ 75\overline{)600} \\ \underline{600} \\ 0 \end{array}$$
$B = 8$

3.37
$77 = 0.11 \times B$
$77 \div 0.11 = B$
$$\begin{array}{r} 700 \\ 11\overline{)7700} \\ \underline{77} \\ 0 \end{array}$$
$B = 700$

3.38
$8 = R \times 16$
$8 \div 16 = R$
$0.5 = R$
$50\% = R$

3.39
$14 = R \times 7$
$14 \div 7 = R$
$2 = R$
$200\% = R$

3.40
$12 = R \times 4$
$12 \div 4 = R$
$3 = R$
$300\% = R$

3.41
$4 = R \times 5$
$4 \div 5 = R$
$0.8 = R$
$80\% = R$

3.42
$R \times 12 = 6$
$R = 6 \div 12$
$R = 0.5$
$R = 50\%$

3.43
$32 = R \times 16$
$32 \div 16 = R$
$2 = R$
$200\% = R$

3.44
$21 \div 25 = R$
$0.84 = R$
$84\% = R$

3.45
$R \times 2.50 = 1.50$
$R = 1.50 \div 2.50$
$R = 0.6$
$R = 60\%$

3.46
a. $175 - 140 = 35$
b. $\frac{35}{175}$
c. $$\begin{array}{r} 0.2 \\ 175\overline{)35.0} \\ \underline{35\,0} \\ 0 \end{array}$$
d. $0.2 = 20\%$
e. decrease

3.47
a. $45 - 25 = 20$
b. $\frac{20}{25}$
c. $$\begin{array}{r} 0.8 \\ 25\overline{)20.0} \\ \underline{20\,0} \\ 0 \end{array}$$
d. $0.8 = 80\%$
e. increase

3.48
a. $70 - 35 = 35$
b. $\frac{35}{35}$
c. $35 \div 35 = 1$
d. $1 = 100\%$
e. increase

3.49

a. $340 - 170 = 170$

b. $\frac{170}{340}$

c. $$\begin{array}{r} 0.5 \\ 340\overline{)170.0} \\ \underline{1700} \\ 0 \end{array}$$

d. $0.5 = 50\%$

e. decrease

3.50

a. $20 - 10 = 10$

b. $\frac{10}{10}$

c. $10 \div 10 = 1$

d. $1 = 100\%$

e. increase

3.51

a. $110 - 100 = 10$

b. $\frac{10}{100}$

c. $10 \div 100 = 0.1$

d. $0.1 = 10\%$

e. increase

3.52

a. $210 - 200 = 10$

b. $\frac{10}{200}$

c. $10 \div 200 = 0.05$

d. $0.05 = 5\%$

e. increase

3.53

$30 - 24 = 6$

$\frac{6}{24} = \frac{1}{4} = 0.25 = 25\%$ increase

3.54

$49 - 35 = 14$

$\frac{14}{35} = \frac{2}{5} = 0.4 = 40\%$ increase

3.55

$0.14 \times 76 = N$

$$\begin{array}{r} 0.14 \\ \underline{\times\ 76} \\ 84 \\ \underline{98\ \ } \\ 10.64 = N \end{array}$$

3.56

$0.45 \times B = 9$

$B = 9 \div 0.45$

$$\begin{array}{r} 20 \\ 45\overline{)900} \\ \underline{90\ } \\ 0 \end{array}$$

$B = 20$

3.57

$R \times 100 = 25$

$R = 25 \div 100$

$R = 0.25$

$R = 25\%$

3.58

$0.72 \times 25 = N$

$$\begin{array}{r} 0.72 \\ \underline{\times\ 25} \\ 360 \\ \underline{144\ \ } \\ 18.00 \end{array}$$

$N = 18$

3.59

$25 - 17 = 8$

$\frac{8}{25} = 0.32 = 32\%$ decrease

3.60

$33 - 25 = 8$

$\frac{8}{25} = 0.32 = 32\%$ increase

3.61

$0.27 \times B = 1.62$

$B = 1.62 \div 0.27$

$$\begin{array}{r} 6 \\ 27\overline{)162} \\ \underline{162} \\ 0 \end{array}$$

$B = 6$

3.62 $1.45 \times 5.7 = N$

$$\begin{array}{r} 1.45 \\ \times\ 5.7 \\ \hline 1015 \\ 725 \\ \hline 8.265 \end{array} = N$$

3.63 $6 = R \times 24$
$6 \div 24 = R$
$0.25 = R$
$25\% = R$

3.64 $R \times 16 = 12$
$R = 12 \div 16$
$R = 0.75$
$R = 75\%$

3.65 $120 - 90 = 30$
$\frac{30}{120} = \frac{1}{4} = 0.25 = 25\%$ decrease

3.66 $60 - 40 = 20$
$\frac{20}{40} = \frac{1}{2} = 0.5 = 50\%$ increase

3.67 $0.0004 \times 75 = 0.03$

3.68 $0.27 \times B = 5.4$
$B = 5.4 \div 0.27$

$$\begin{array}{r} 20 \\ 27\overline{)540} \\ 54 \\ \hline 0 \end{array}$$

$B = 20$

3.69 $R \times 7 = 2.1$
$R = 2.1 \div 7$
$R = 0.3$
$R = 30\%$

3.70 $42 = 0.21 \times B$
$42 \div 0.21 = B$
$200 = B$

3.71 $N = 1.75 \times 24$

$$\begin{array}{r} 1.75 \\ \times\ 24 \\ \hline 700 \\ 350 \\ \hline 42.00 \end{array}$$

$N = 42$

3.72 $6 = R \times 0.3$
$6 \div 0.3 = R$
$20 = R$
$2{,}000\% = R$

3.73 $0.4 \times B = 48$
$B = 48 \div 0.4$
$B = 120$

3.74 $N = 0.17 \times 2.2$

$$\begin{array}{r} 0.17 \\ \times\ 2.2 \\ \hline 34 \\ 34 \\ \hline 0.374 \end{array}$$

$N = 0.374$

3.75 $R \times 70 = 35$
$R = 35 \div 70$
$R = 0.5$
$R = 50\%$

3.76 $4 \times 16 = N$
$64 = N$

3.77 $15 - 12 = 3$
$\frac{3}{12} = \frac{1}{4} = 0.25 = 25\%$ increase

3.78 $150 - 120 = 30$
$\frac{30}{120} = \frac{1}{4} = 0.25 = 25\%$ increase

3.79

$42 = 1.75 \times B$
$42 \div 1.75 = B$
$24 = B$
$175\overline{)4200}$
350
700
700
0

3.80

$0.75 \times 72 = N$
0.75
x 72
150
525
54.00
$N = 54$

3.81

$72 = 0.75 \times B$
$72 \div 0.75 = B$
$96 = B$
$75\overline{)7200}$
675
450
450
0

3.82

$72 = R \times 75$
$72 \div 75 = R$
$0.96 = R$
$75\overline{)72.00}$
675
450
450
0

3.83

$N = 0.77 \times 3$
$N = 2.31$

3.84

$0.24 \times 34 = N$
0.24
x 34
96
72
$8.16 = N$

3.85

$4 - 3 = 1$

$\frac{1}{4} = 0.25 = 25\%$ decrease

3.86

$175 - 140 = 35$

$\frac{35}{175} = \frac{1}{5} = 0.2 = 20\%$ decrease

3.87

$\$4.37 \times 0.05 = \$0.2185 =$
$0.22 or 22¢

3.88

$\frac{27}{45} = \frac{3}{5} = 0.6 = 60\%$

3.89

a. $\$200 \times 0.1 = \$20$ cut
$\$200 = \$20 = \$180$ per week
b. $\$180 \times 0.1 = \$18$ raise
$\$180 + \$18 = \$198$ per week
c. The raise was smaller than the cut because the base was lower.

3.90

$\frac{36}{0.8} = \$45$

3.91

$120 \times 12 = \$1{,}440$ saved per year

$\frac{1{,}440}{14{,}400} = \frac{1}{10} = 0.1 = 10\%$

3.92

$\$4{,}500 - \$4{,}050 = \$450$

$\frac{450}{4500} = \frac{1}{10} = 0.1 = 10\%$ discount

3.93

$\$24{,}000 \times 0.07 = \$1{,}680$ commission

3.94

$\$4{,}900 \div 0.07 = \$70{,}000$ selling price

3.95

$\$20{,}000 \times 0.09 = \$1{,}800$ interest per year

$\$1{,}800 \div 12 = \$150$ for the first month's interest

3.96 $0.09 \times \$575 = \$51.75$

3.97

$$\begin{array}{r} 125 \text{ tons of ore} \\ 0.16\overline{)2000} \\ \underline{16} \\ 40 \\ \underline{32} \\ 80 \\ \underline{80} \\ 0 \end{array}$$

3.98 Base = $220 \div 0.55 = 400$ total
$400 - 220 = 180$ boys

3.99 $9 \times 0.07 = 0.63$

3.100 $24 \times 0.07 = 1.68$

3.101 $0.03 \times 0.07 = 0.0021$

3.102 $4{,}400 \times 0.07 = 308$

3.103

$$\begin{array}{r} 12 \\ \underline{\times\ 0.15} \\ 60 \\ \underline{12} \\ 1.80 \end{array} = 1.8$$

3.104

$$\begin{array}{r} 35 \\ \underline{\times\ 0.15} \\ 175 \\ \underline{35} \\ 5.25 \end{array}$$

3.105

$$\begin{array}{r} 0.004 \\ \underline{\times\ 0.15} \\ 0.00060 \end{array} = 0.0006$$

3.106

$$\begin{array}{r} 2{,}700 \\ \underline{\times\ 0.15} \\ 13500 \\ \underline{2700} \\ 405.00 \end{array} = 405$$

I. SECTION ONE

1.1 $S = 22 + 56 + 75 + 30 + 38 + 50$
$= 271$
$m = \frac{271}{6} = 45\frac{1}{6} \rightarrow 45$

1.2 $S = 12 + 18 + 17 + 6 + 25 + 18 + 22$
$= 118$
$m = \frac{118}{7} = 16\frac{6}{7} \rightarrow 17$

1.3 $S = 35 + 44 + 77 + 68 + 43 + 55$
$= 322$
$m = \frac{322}{6} = 53\frac{2}{3} \rightarrow 54$

1.4 $S = 17 + 7 + 2 + 9 + 8 + 22 + 65$
$= 130$
$m = \frac{130}{7} = 18\frac{4}{7} \rightarrow 19$

1.5 $S = 29 + 73 + 41 + 55 + 76 + 81 + 92$
$= 447$
$m = \frac{447}{7} = 63\frac{6}{7} \rightarrow 64$

1.6 $S = 14 + 24 + 28 + 34 + 18 + 26 + 30$
$= 174$
$m = \frac{174}{7} = 24\frac{6}{7} \rightarrow 25$

1.7 $S = 13 + 26 + 29 + 38 + 62 + 73 + 55$
$= 296$
$m = \frac{296}{7} = 42\frac{2}{7} \rightarrow 42$

1.8 $S = 19 + 28 + 17 + 39 + 22 + 44 + 56$
$= 225$
$m = \frac{225}{7} = 32\frac{1}{7} \rightarrow 32$

1.9 $S = 3.2 + 7.5 + 9.8 + 11.5 + 2.9 + 3.5$
$= 38.4$
$m = \frac{38.4}{6} = 6.4$

1.10 $S = 16.25 + 17.96 + 3.58 + 4.61 + 5.23$
$= 47.63$
$m = \frac{47.63}{5} = 9.52 \rightarrow 9.5$

1.11 $\frac{S}{6} = 17$
$S = 17 \times 6 = 102$
$13 + 19 + 21 + 14 + 16 = 83$
$102 - 83 = 19$, the missing number

1.12 $\frac{S}{8} = 20$
$S = 20 \times 8 = 160$
$22 + 19 + 26 + 17 + 15 + 28 + 30 = 157$
$160 - 157 = 3$, the missing number

1.13 $\frac{S}{5} = 80$
$S = 80 \times 5 = 400$
$85 + 90 + 95 + 70 = 340$
$400 - 340 = 60$, the missing number

1.14 16, 18, 19, 20, 22, 22, 36
median = 20

1.15 24, 28, 39, 46, 50, 68, 73
median = 46

1.16 1, 2, 5, 6, 7, 9, 15, 22
median $= \frac{6+7}{2} = \frac{13}{2} = 6.5$

1.17 28, 29, 39, 39, 46, 48, 56, 68
median $= \frac{39+46}{2} = \frac{85}{2} = 42.5$

1.18 25, 29, 30, 31, 48, 56, 56
median = 31

1.19 36, 62, 65, 80, 80, 85, 90, 90

median = $\frac{80 + 80}{2} = \frac{160}{2} = 80$

1.20 a. mean = $\frac{1{,}215}{15} = 81$

b. 62, 65, 65, 66, 70, 75, 75,

85, 88, 88, 90, 92, 96, 98, 100

median = 85

1.21 a. mean = $\frac{2{,}260}{15} = 150\,\frac{2}{3} \rightarrow 151$

b. 100, 105, 110, 115, 125,

149, 153, 156, 162, 162,

165, 175, 180, 198, 205

median = 156

1.22 mean = $\frac{690}{7} = 98\,\frac{4}{7} \rightarrow 99°$

To find the median, reorder the temperatures from least to greatest as 96, 96, 97, 98, 100, 105; median = 98°.

1.23 average = $\frac{56}{150}$ =

```
  0.3733  0.373
150)56.0000
    450
    1100
    1050
      500
      450
       500
       450
```

1.24 average = $\frac{253}{15} = 16\,\frac{13}{15} \rightarrow$ 17 points per game

1.25 average = $\frac{60}{102}$ =

```
  0.5882 → 0.588
102)60.0000
    510
    900
    816
     840
     816
      240
      204
```

1.26 average = $\frac{2{,}270}{11} = 206\,\frac{4}{11} \rightarrow 206$

To find the median, reorder the numbers from least to greatest as 165, 175, 180, 180, 190, 195, 195, 225, 245, 255, 265; median = 195.

1.27 2

1.28 20

1.29 none or all

1.30 80

1.31 mean = $\frac{296}{14} = 21\,\frac{1}{7} \rightarrow 21$

To find the median, reorder the numbers from least to greatest as 16, 16, 17, 18, 18, 18, 19, 20, 22, 22, 22, 28, 30, 30;

median = $\frac{19 + 20}{2} = \frac{39}{2} = 19.5$

mode = 18 and 22

1.32 mean = $\frac{1{,}539}{18} = 85\,\frac{1}{2} \rightarrow 86$

To find the median, reorder the numbers from least to greatest as 60, 70, 75, 78, 80, 80, 83, 85, 87, 90, 90, 90, 90, 93, 95, 95, 98, 100;

median = $\frac{87 + 90}{2} = \frac{177}{2} = 88.5$

mode = 90

1.33 $40 - 2 = 38$

1.34 $91 - 38 = 53$

1.35 $10.1 - 1.1 = 9$

1.36 $1.33 - 0.01 = 1.32$

1.37 mean $\frac{28}{6} = 4\frac{2}{3} \rightarrow 5$

Number	Mean	Diff.
1	5	4
3	5	2
4	5	1
5	5	0
6	5	1
9	5	4
28	30	12

average spread = $\frac{12}{6} = 2$

1.38 mean = $\frac{135}{6} = 22.5 \rightarrow 23$

Number	Mean	Diff.
15	23	8
20	23	3
20	23	3
25	23	2
25	23	2
30	23	7
135	138	25

average spread = $\frac{25}{6} = 4\frac{1}{6} \rightarrow 4$

1.39 mean = $\frac{125}{6} = 20\frac{5}{6} \rightarrow 21$

Number	Mean	Diff.
3	21	18
12	21	9
15	21	6
25	21	4
30	21	9
40	21	19
125	126	65

average spread = $\frac{65}{6} = 10\frac{5}{6} \rightarrow 11$

1.40 mean = $\frac{530}{7} = 75\frac{5}{7} \rightarrow 76$

Number	Mean	Diff.
50	76	26
60	76	16
70	76	6
80	76	4
80	76	4
90	76	14
100	76	24
530	532	94

average spread = $\frac{94}{7} = 13\frac{3}{7} \rightarrow 13$

1.41 a. mean = $\frac{356}{30} = 11\frac{13}{15} \rightarrow 12$

b. 4, 4, 5, 6, 7, 8, 8, 8, 8,
9, 9, 10, 10, 11, 12, 12,
12, 12, 13, 15, 15, 15, 16,
17, 17, 17, 18, 19, 19, 20
median = $\frac{12 + 12}{2}$ $\frac{24}{2} = 12$

c. 8 and 12

d. $20 - 4 = 16$

e.

Score	Mean	Diff.
4	12	8
4	12	8
5	12	7
6	12	6
7	12	5
8	12	4
8	12	4
8	12	4
8	12	4
9	12	3
9	12	3
10	12	2
10	12	2
11	12	1
12	12	0
12	12	0
12	12	0
12	12	0
13	12	1
15	12	3
15	12	3
15	12	3
16	12	4
17	12	5
17	12	5
17	12	5
18	12	6
19	12	7
19	12	7
20	12	8
356	360	118

1.41 (cont.)

average spread = $\frac{118}{30} = 3\frac{14}{15} \to 4$

1.42

a. mean = $\frac{2,457}{30} = 81\frac{9}{10} \to 82$

b. 60, 65, 68, 72, 73, 73, 73, 75, 75, 75, 76, 77, 80, 80, 80, 82, 82, 85, 86, 87, 90, 90, 91, 92, 92, 93, 94, 95, 96, 100

median = $\frac{80 + 82}{2} = \frac{162}{2} = 81$

c. 73, 75, and 80

d. 100 – 60 = 40

e.

Score	Mean	Diff.
60	82	22
65	82	17
68	82	14
72	82	10
73	82	9
73	82	9
73	82	9
75	82	7
75	82	7
75	82	7
76	82	6
77	82	5
80	82	2
80	82	2
80	82	2
82	81	0
82	82	0
85	82	3
86	82	4
87	82	5
90	82	8
90	82	8
91	82	9
92	82	10
92	82	10
93	82	11
94	82	12
95	82	13
96	82	14
100	82	18
2,457	2,460	253

1.42 (cont.)

average spread = $\frac{253}{30} = 8\frac{13}{30} \to 8$

1.43

mean = $\frac{610}{50} = 12\frac{1}{5} \to 12$

E	F	E x F	D	D x F
2	2	4	10	20
3	1	3	9	9
4	1	4	8	8
5	1	5	7	7
6	0	0	6	0
7	2	14	5	10
8	4	32	4	16
9	3	27	3	9
10	3	30	2	6
11	5	55	1	5
12	5	60	0	0
13	4	52	1	4
14	0	0	2	0
15	6	90	3	18
16	3	48	4	12
17	2	34	5	10
18	3	54	6	18
19	2	38	7	14
20	3	60	8	24
	50	610		190

median = 12 (the twenty-fifth and twenty-sixth numbers are both 12)

mode = 15

spread = 20 – 2 = 18

average spread = $\frac{190}{50} = 3\frac{4}{5} \to 4$

1.44 mean = $\frac{605}{50} = 12\frac{1}{10} \rightarrow 12$

I	E	F	E x F	D	D x F
1–5	3	5	3 x 5 = 15	12 – 3 = 9	45
6–10	8	12	8 x 12 = 96	12 – 8 = 4	48
11–15	13	20	13 x 20 = 260	13 – 12 = 1	20
16–20	18	13	18 x 13 = 234	18 – 12 = 6	78
I = Interval		50	605		191

median interval: the twenty-fifth and twenty-sixth numbers are in the 11–15 interval; therefore, the median interval is 11–15.

mode interval 11–15

spread = 20 – 2 = 18

average spread = $\frac{191}{50} = 3\frac{41}{50} \rightarrow 4$

1.45 mean = $\frac{3,190}{50} = 63\frac{4}{5} \rightarrow 64$

I	E	F	E x F	D	D x F
10-19	15	2	15 x 2 = 30	64 – 15 = 49	98
20-29	25	4	25 x 4 = 100	64 – 25 = 39	156
30-39	35	5	35 x 5 = 175	64 – 35 = 29	145
40-49	45	5	45 x 5 = 225	64 – 45 = 19	95
50-59	55	4	55 x 4 = 220	64 – 55 = 9	36
60-69	65	6	65 x 6 = 390	65 – 64 = 1	6
70-79	75	9	75 x 9 = 675	75 – 64 = 11	99
80-89	85	5	85 x 5 = 425	85 – 64 = 21	105
90-100	95	10	95 x 10 = 950	95 – 64 = 31	310
		50	3,190		1,050

median interval: the twenty-fifth and twenty-sixth numbers are in the 60-69 interval; therefore, the medial interval is 60-69.

mode interval = 90-100

spread = 100 – 10 = 90

average spread = $\frac{1,050}{50} = 21$

II. SECTION TWO

2.1

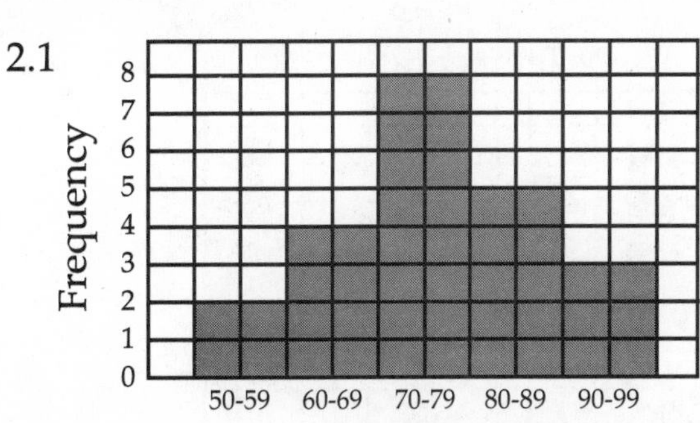

2.2

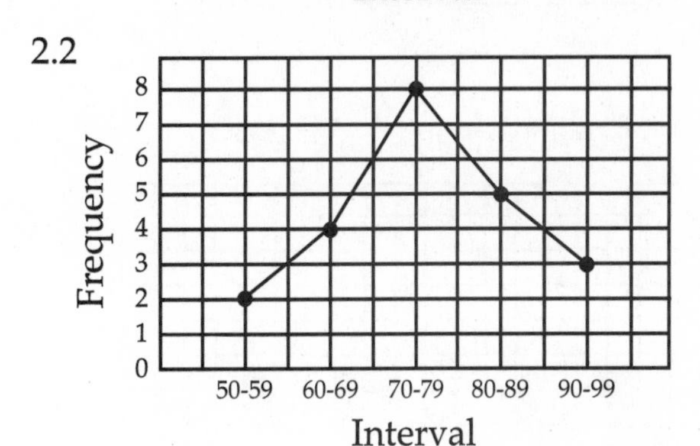

2.3

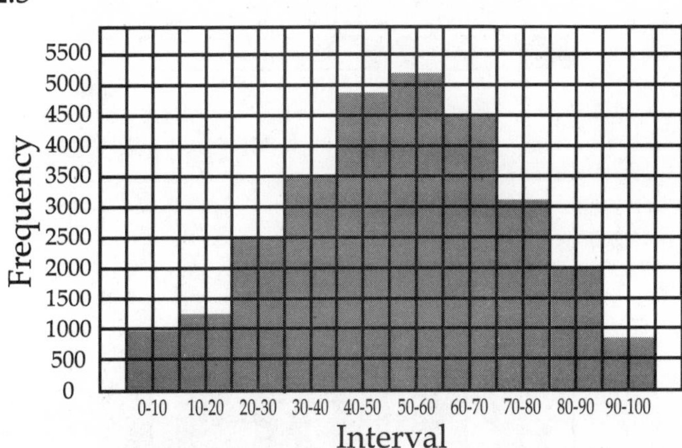

2.4

Interval	Frequency
60-69	10
70-79	16
80-89	14
90-99	10

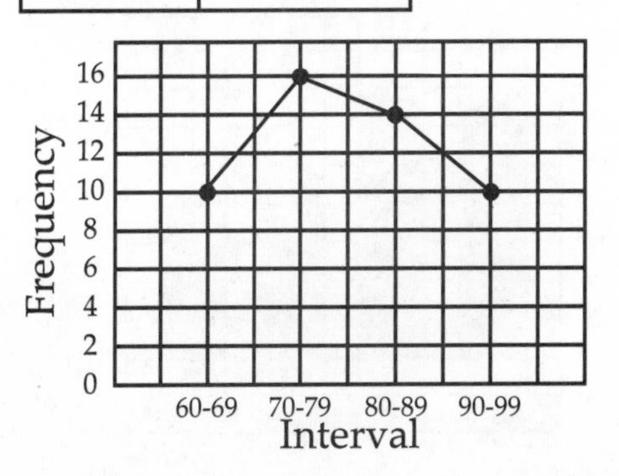

2.5 teacher check

2.6 Daily Temperature

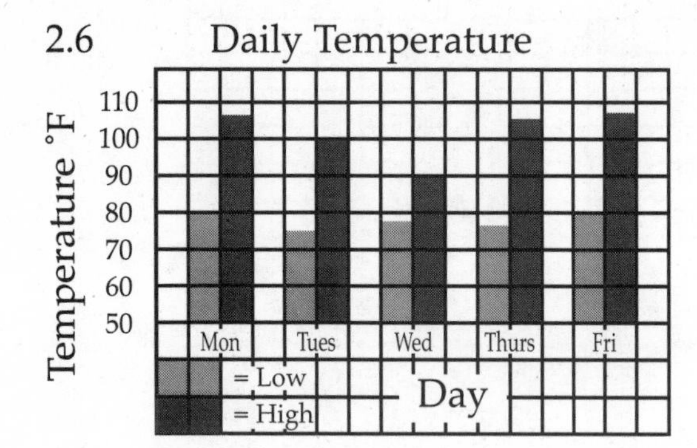

2.7 Scores in Ball Game

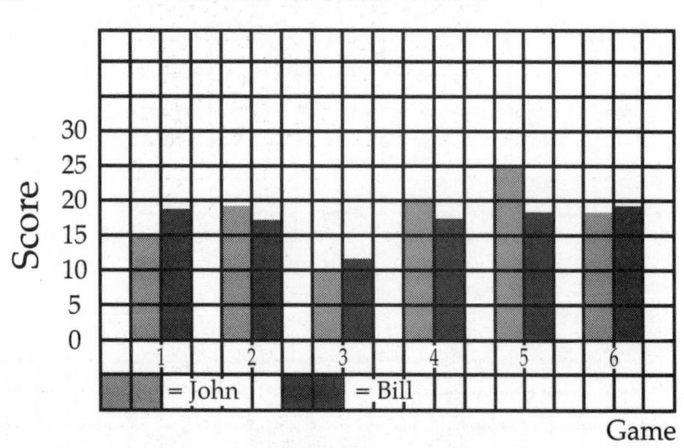

2.8 Sales Boat Company

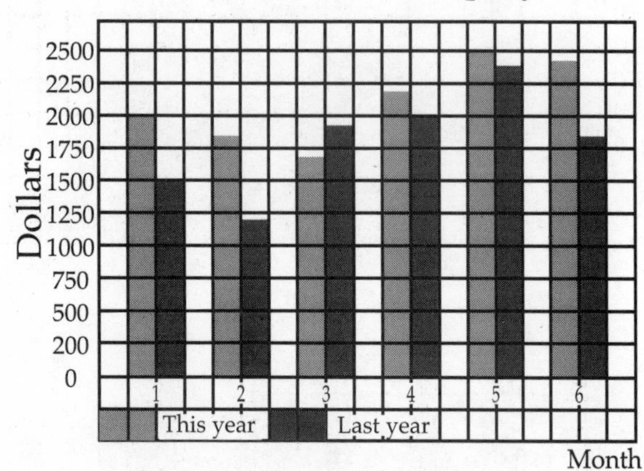

2.9 Enrollment, Murphy High

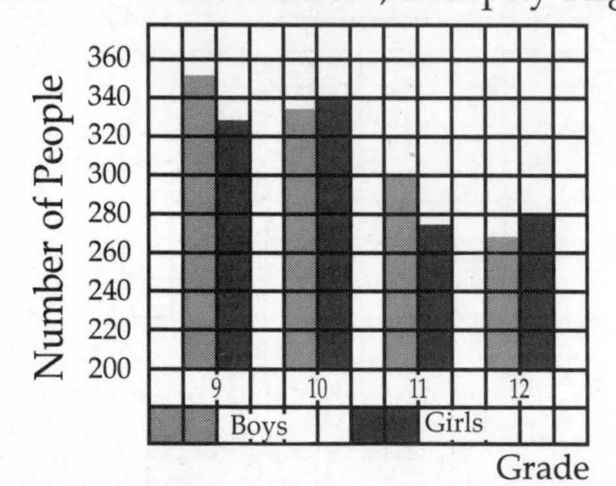

2.10 2,471

2.11 average = $\frac{350 + 327 + 300 + 275}{4}$ = $\frac{1,252}{4}$ = 313

2.12 average = $\frac{100}{6} = 16\frac{2}{3} \rightarrow 17$

2.13 teacher check

2.14 Scores in Ball Game

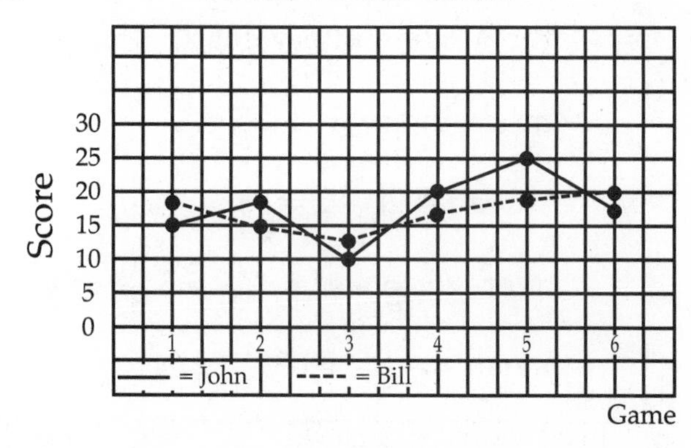

2.15 Sales Boat Company

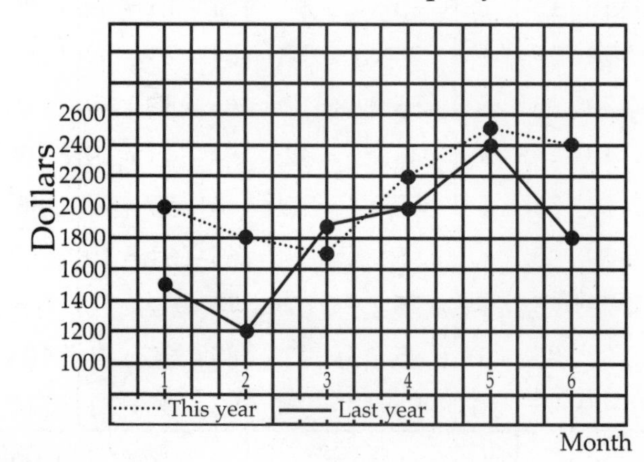

2.16 teacher check

2.17 a. $f(n) = 2 + 3 = 5$

b. $f(n) = 3 + 3 = 6$

c. $f(n) = 4 + 3 = 7$

2.18 a. $f(n) = 8 - 2 = 6$
b. $f(n) = 8 - 3 = 5$
c. $f(n) = 8 - 4 = 4$
d. $f(n) = 8 - 5 = 3$

2.19 a. $f(n) = 2 \times 2 + 2$
$= 4 + 2$
$= 6$
b. $f(n) = 2 \times 4 + 2$
$= 8 + 2$
$= 10$
c. $f(n) = 2 \times 6 + 2$
$= 12 + 2$
$= 14$
d. $f(n) = 2 \times 8 + 2$
$= 16 + 2$
$= 18$

2.20 a. $f(n) = 5 \times 3 - 5$
$= 15 - 5$
$= 10$
b. $f(n) = 5 \times 5 - 5$
$= 25 - 5$
$= 20$
c. $f(n) = 5 \times 7 - 5$
$= 35 - 5$
$= 30$
d. $f(n) = 5 \times 9 - 5$
$= 45 - 5$
$= 40$

2.21 $f(n) = 2 \times n$

2.22 $f(n) = 2 \times n - 3$

2.23

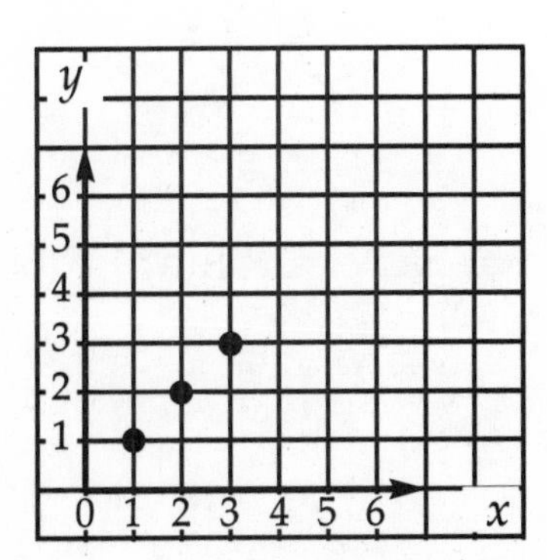

2.24

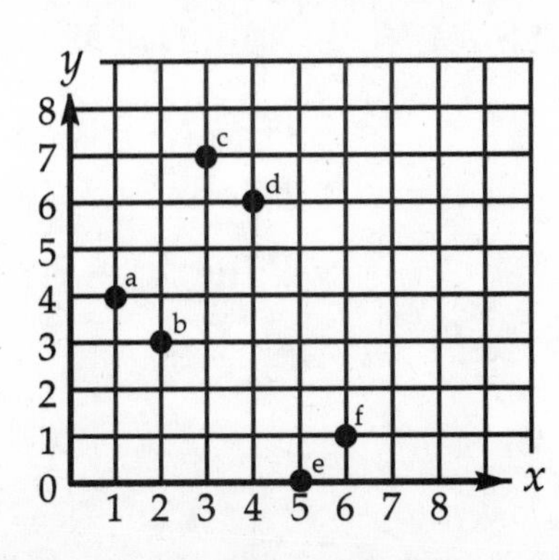

2.25 A (1, 2)
B (3, 1)
C (4, 4)
D (6, 6)
E (8, 4)

2.26 a. $f(n) = 2 \times 1 - 1$
$= 2 - 1$
$= 1$
(1, 1)
$f(n) = 2 \times 2 - 1$
$= 4 - 1$
$= 3$
(2, 3)
$f(n) = 2 \times 3 - 1$
$= 6 - 1$
$= 5$
(3, 5)

b.

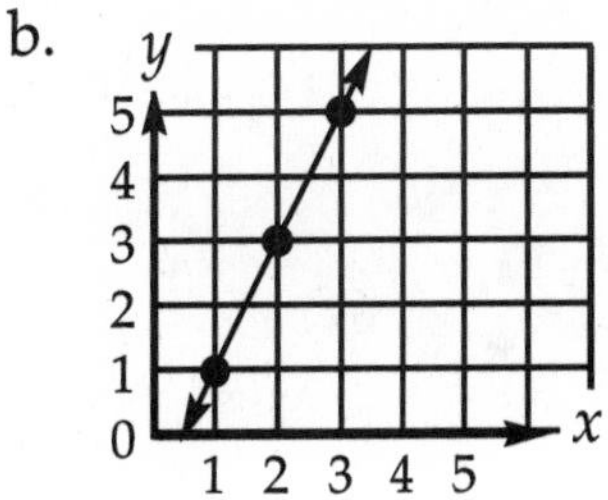

2.27 a. (1, 1), (2, 4), (3, 7)
b. $f(n) = 3 \times n - 2$

2.28 a. (0, 5), (1, 4), (2, 3), (3, 2), (4, 1), (5, 0)
b. $f(n) = 5 - n$

III. SECTION THREE

3.1 a. $\frac{2}{5}$
b. $\frac{3}{5}$

3.2 a. $\frac{7}{10}$
b. $\frac{3}{10}$

3.3 sample space = {a, e, i, o, u}
$P(\text{vowel}) = \frac{5}{26}$

3.4 sample space = {1, 3, 5, 7, 9}
$P(\text{odd}) = \frac{5}{10} = \frac{1}{2}$

3.5 sample space = {3, 6, 9, 12, 15, 18}
$P(3n) = \frac{6}{20} = \frac{3}{10}$

3.6 $\frac{4}{8} = \frac{1}{2}$

3.7 $\frac{2}{8} = \frac{1}{4}$

3.8 $\frac{2}{8} = \frac{1}{4}$

3.9 $\frac{3}{8}$

3.10 $\frac{1}{8}$

3.11 $\frac{1}{8}$

3.12 $\frac{1}{8}$

3.13 sample space = {2, 4, 6, 8, 10, 12, 14, 16, 18}
$P(\text{even number}) = \frac{10}{20} = \frac{1}{2}$

3.14 sample space = {1, 3, 5, 7, 9, 11, 13, 15, 17, 19}
$P(\text{odd number}) = \frac{10}{20} = \frac{1}{2}$

3.15 sample space = {5, 10, 15, 20}
$P(\text{multiple of 5}) = \frac{4}{20} = \frac{1}{5}$

3.16 sample space = {10, 11, 12, 13, 14, 15, 16, 17, 18, 19, 20}
$P(\text{2-digit number}) = \frac{11}{20}$

3.17 sample space = {10, 20}
$P(\text{number with zero}) = \frac{2}{20} = \frac{1}{10}$

3.18 sample space = {5, 14}
$P(\text{sum of digits} = 5) = \frac{2}{20} = \frac{1}{10}$

3.19 sample space = {2, 3, 5, 7, 11, 13, 17, 19}
$P(\text{prime number} = \frac{8}{20} = \frac{2}{5}$

3.20 sample space = {1, 4, 6, 8, 9, 10, 12, 14, 15, 16, 18, 20}
$P(\text{not prime}) = \frac{12}{20} = \frac{3}{5}$

3.21 $\frac{1}{5}$

3.22 $\frac{1}{5}$

3.23 $\frac{1}{5}$

3.24 $\frac{1}{5}$

3.25 $\frac{1}{5}$

3.26 $\frac{3}{5}$

3.27 $\frac{2}{5}$

3.28 $P(\text{John or Jim}) = P(\text{John}) + P(\text{Jim})$
$= \frac{1}{5} + \frac{1}{5}$
$= \frac{2}{5}$

3.29 $P(\text{Sally or Sue}) = P(\text{Sally}) + P(\text{Sue})$
$= \frac{1}{5} + \frac{1}{5}$
$= \frac{2}{5}$

3.30 $P(\text{Jim or Sally}) = P(\text{Jim}) + P(\text{Sally})$
$= \frac{1}{5} + \frac{1}{5}$
$= \frac{2}{5}$

3.31 $P(\text{not Sue}) = P(\text{John or Jim or Jerry or Sally})$
$= P(\text{John}) + P(\text{Jim}) + P(\text{Jerry}) + P(\text{Sally})$
$= \frac{1}{5} + \frac{1}{5} + \frac{1}{5} + \frac{1}{5}$
$= \frac{4}{5}$

3.32 P(not Jim or Jerry) = P(John or Sue or Sally)
= P (John) + P(Sue) + P(Sally)
$= \frac{1}{5} + \frac{1}{5} + \frac{1}{5}$
$= \frac{3}{5}$

3.33 $P(\text{even}) = \frac{25}{50} = \frac{1}{2}$

3.34 $P(\text{odd}) = \frac{25}{50} = \frac{1}{2}$

3.35 P(10 or odd) = P(10) + P(odd)
$= \frac{1}{50} + \frac{25}{50}$
$= \frac{26}{50}$
$= \frac{13}{25}$

3.36 sample space = {1, 2, 3, 4, 5}
$P(\text{less than } 6) = \frac{5}{50} = \frac{1}{10}$

3.37 Let A = event of drawing a number < 10 and
B = event of drawing a number > 40
sample space A = {1, 2, 3, 4, 5, 6, 7, 8, 9}

sample space B = {41, 42, 43, 44, 45, 46, 47, 48, 49, 50}

$P(A) = \frac{9}{50}$

$P(B) = \frac{10}{50}$

P (A or B) = $P(A) + P(B)$
$= \frac{9}{50} + \frac{10}{50} = \frac{19}{50}$

3.38 P (even or odd) = P(even) + P(odd)
$= \frac{25}{50} + \frac{25}{50}$
$= \frac{50}{50}$
$= 1$

3.39 sample space (prime) = {2, 3, 5, 7, 11, 13 17, 19, 23, 29, 31, 37 41, 43, 47}

sample space (overlap) = {2}
P(even or prime) = P(even) + P(prime) – P(overlap)
$= \frac{25}{50} + \frac{15}{50} - \frac{1}{50}$
$= \frac{39}{50}$

3.40 sample space = {5, 10, 15, 20, 25, 30, 35, 40, 45, 50}

$P(\text{multiple of } 5) = \frac{10}{50} = \frac{1}{5}$

3.41 P(multiple of 5 or even)
= P(multiple of 5) + P(even) – P(overlap)

P(overlap) = {10, 20, 30, 40, 50}

P(multiple of 5 or even)
$= \frac{10}{50} + \frac{25}{50} - \frac{5}{50}$
$= \frac{30}{50} = \frac{3}{5}$

3.42 0

3.43 sample space = {11, 12, 13, 14, 15, 16, 17, 18, 19}

P(between 10 and 200 $= \frac{9}{50}$

3.44 sample space = {1, 4, 6, 8, 9, 10, 12, 14, 15, 16, 18, 20, 21, 22, 24, 25, 26, 27, 28, 30, 32 33, 34, 35, 36, 38, 39, 40 ,42, 44, 45, 46, 48, 49, 50}

$P(\text{not prime}) = \frac{35}{50} = \frac{7}{10}$

3.45 Let A = event of drawing a prime number and

Let B = event of drawing a multiple of 3 and

Let C = overlap of A and B.

sample space A = {2, 3, 5, 7, 11, 13, 17, 19, 23, 29, 31, 37, 41, 43, 47}

sample space B = {3, 6, 9, 12, 15, 18, 21, 24, 27, 30, 33, 36, 39, 42, 45, 48}

sample space C = {3}

$P(A \text{ or } B) = P(A) + P(B) - P(C)$

$= \frac{15}{50} + \frac{16}{50} - \frac{1}{50} = \frac{30}{50} = \frac{3}{5}$

3.46 $\frac{1}{5} \times \frac{1}{4} = \frac{1}{20}$

3.47 $\frac{1}{5} \times \frac{1}{4} = \frac{1}{20}$

3.48 $\frac{1}{5} \times \frac{1}{4} = \frac{1}{20}$

3.49 $\frac{1}{5} \times \frac{3}{4} = \frac{3}{20}$

3.50 $\frac{1}{5} \times \frac{3}{4} = \frac{3}{20}$

3.51 $\frac{1}{4} \times \frac{1}{2} = \frac{1}{8}$

3.52 $\frac{1}{4} \times \frac{1}{2} + \frac{1}{4} \times \frac{1}{2} = \frac{1}{8} + \frac{1}{8} = \frac{2}{8} = \frac{1}{4}$

3.53 $\frac{26}{36} = \frac{13}{18}$

3.54 $\frac{10}{36} = \frac{5}{18}$

3.55 $\frac{1}{36}$

3.56 $\frac{1}{36}$

3.57 $\frac{5}{36}$

3.58 $\frac{10}{36} \times \frac{26}{36} = \frac{5}{18} \times \frac{13}{18} = \frac{65}{324}$

3.59 $\frac{26}{36} \times \frac{26}{36} = \frac{13}{18} \times \frac{13}{18} = \frac{169}{324}$

3.60 $\frac{10}{36} \times \frac{10}{36} = \frac{5}{18} \times \frac{5}{18} = \frac{25}{324}$

3.61 $\frac{1}{36} \times \frac{1}{36} = \frac{1}{1{,}296}$

3.62 $\frac{1}{36} \times \frac{1}{36} = \frac{1}{1{,}296}$

3.63 $\frac{1}{36} \times \frac{26}{36} = \frac{1}{\cancel{36}_{18}} \times \frac{\cancel{26}^{13}}{36} = \frac{13}{648}$

3.64 $\frac{1}{36} \times \frac{1}{36} \times \frac{1}{36} = \frac{1}{46{,}656}$

3.65 $0.500 = \frac{500}{1{,}000} = \frac{1}{2}$

$0.300 = \frac{300}{1{,}000} = \frac{3}{10}$

$\frac{3}{10} \times \frac{1}{2} = \frac{3}{20}$

I. SECTION ONE

1.1	-5
1.2	-18
1.3	3
1.4	-40
1.5	22
1.6	1
1.7	-173
1.8	325
1.9	-9
1.10	15
1.11	true
1.12	true
1.13	false
1.14	true
1.15	false
1.16	4, 5, 6, 7, ...
1.17	-2, -1
1.18	-2, -1, 0
1.19	2, 4, 6, 8, ...
1.20	0, 2, 4, 6, ...
1.21	0
1.22	-1, -3, -5, ...
1.23	-1, -3, -5, ...
1.24	4, 2, 0, -2, ...
1.25	-3, -1, 1
1.26	<
1.27	>
1.28	>
1.29	<
1.30	<
1.31	<
1.32	<
1.33	>
1.34	<
1.35	<
1.36	>
1.37	>
1.38	<
1.39	>
1.40	>
1.41	-2 < 0 < 7 < 15 < 19
1.42	-1,000 < -500 < -75 < -5 < 10
1.43	-17 < -9 < 9 < 14 < 23
1.44	-321 < -312 < -231 < -213 < -132 < -123 < 123 < 132 < 213 < 231 < 312 < 321
1.45	teacher check
1.46	8
1.47	2
1.48	15
1.49	172
1.50	12
1.51	47
1.52	1

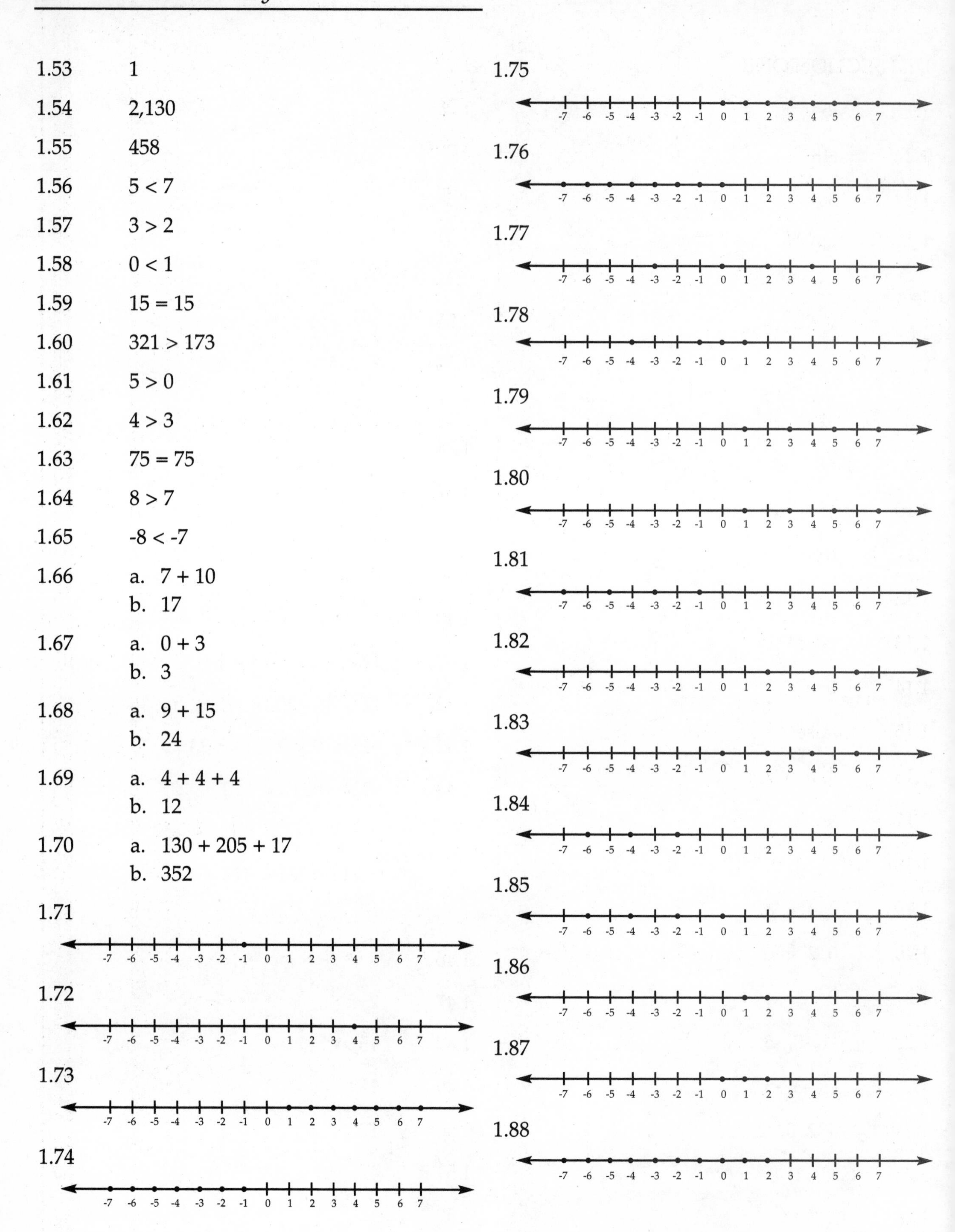

1.53 1

1.54 2,130

1.55 458

1.56 5 < 7

1.57 3 > 2

1.58 0 < 1

1.59 15 = 15

1.60 321 > 173

1.61 5 > 0

1.62 4 > 3

1.63 75 = 75

1.64 8 > 7

1.65 -8 < -7

1.66 a. 7 + 10
b. 17

1.67 a. 0 + 3
b. 3

1.68 a. 9 + 15
b. 24

1.69 a. 4 + 4 + 4
b. 12

1.70 a. 130 + 205 + 17
b. 352

1.71

1.72

1.73

1.74

1.75

1.76

1.77

1.78

1.79

1.80

1.81

1.82

1.83

1.84

1.85

1.86

1.87

1.88

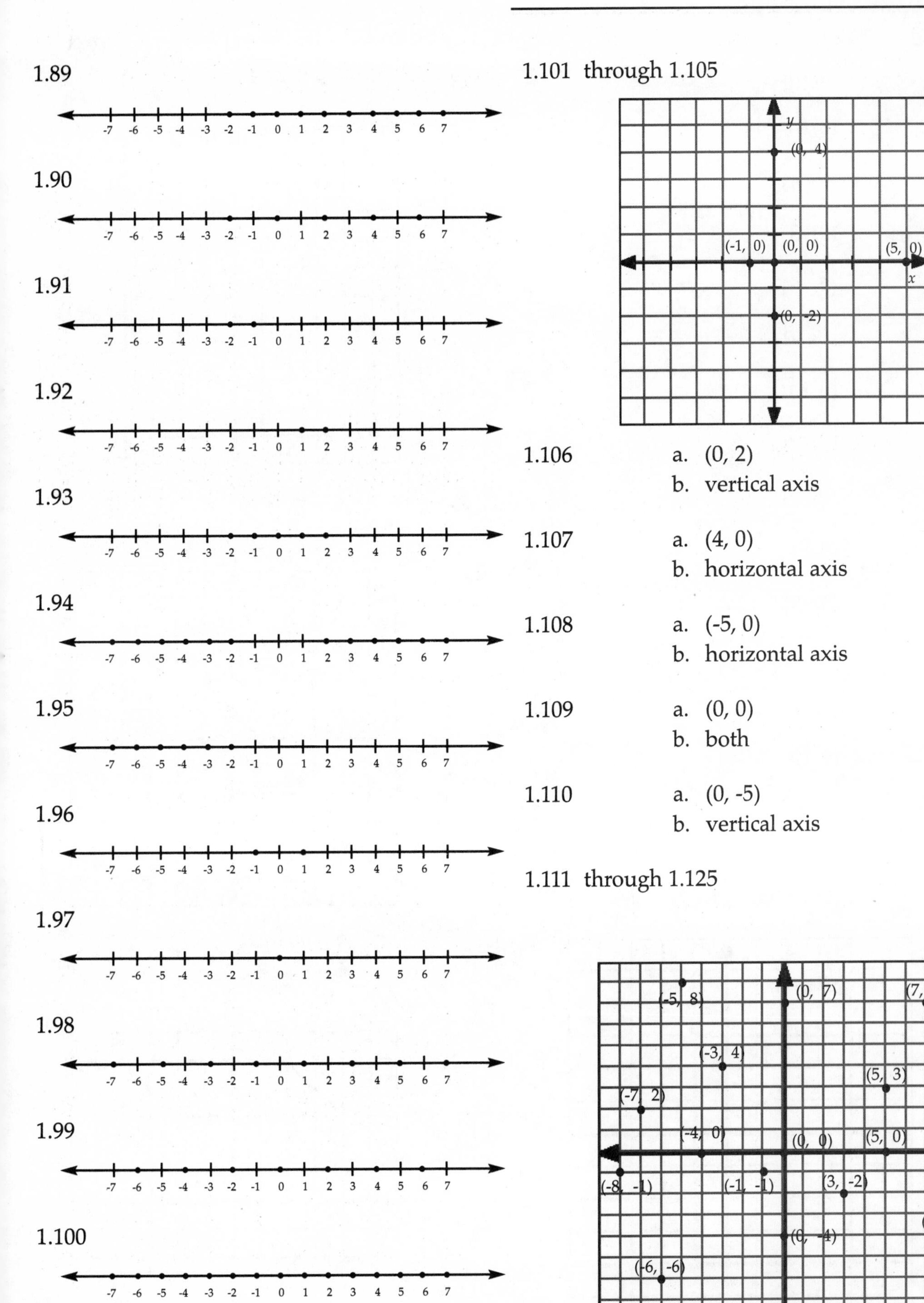

1.106 a. (0, 2)
b. vertical axis

1.107 a. (4, 0)
b. horizontal axis

1.108 a. (-5, 0)
b. horizontal axis

1.109 a. (0, 0)
b. both

1.110 a. (0, -5)
b. vertical axis

1.126 (3, 8)

1.127 (8, 0)

1.128 (-6, 4)

1.129 (0, -3)

1.130 (7, -6)

1.131 (-8, -2)

1.132 (0, 6)

1.133 (-2, -8)

1.134 (-5, 0)

1.135 (9, 9)

1.136 (3, -8)

1.137 (-4, 9)

1.138 (-2, 2)

1.139 (2, -2)

1.140 (0, 0)

1.141 $A = lw$
$A = 7 \cdot 5 = 35$

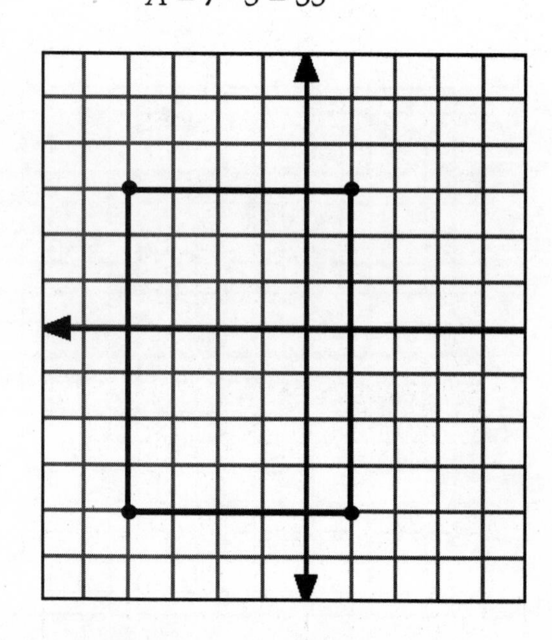

1.142 $A = s^2$
$= 4 \cdot 4 = 16$

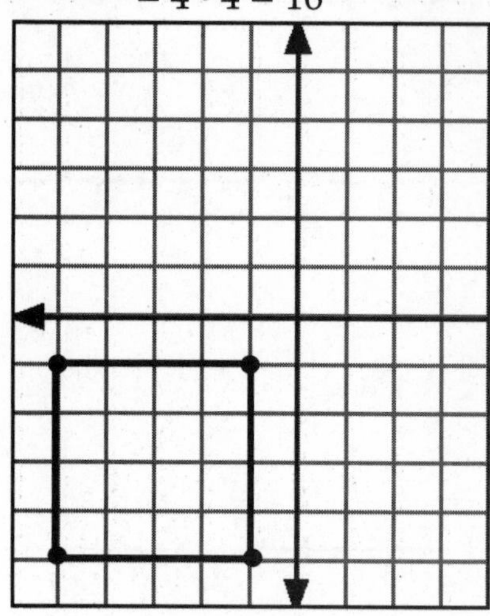

1.143 $A = \frac{bh}{2}$
$A = \frac{4 \cdot 4}{2} = \frac{16}{2} = 8$

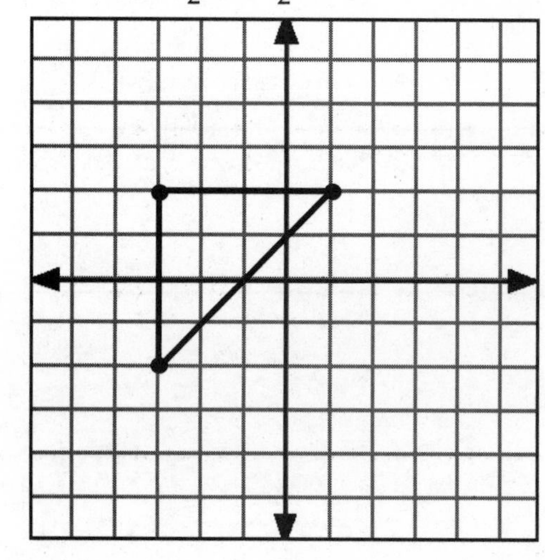

1.144 $A = bh$
$A = 6 \cdot 2 = 12$

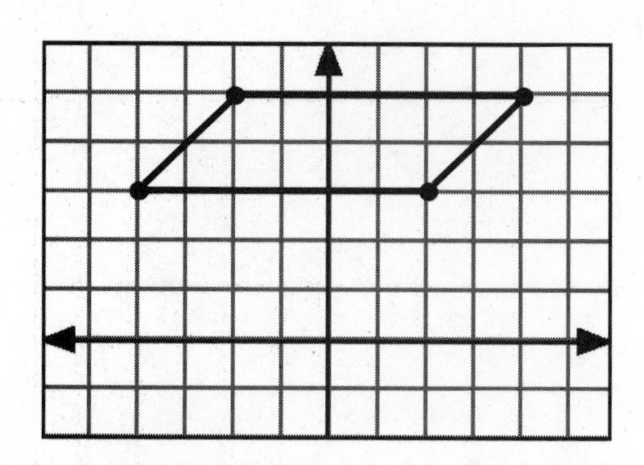

1.145 $A = \frac{(a+b)h}{2}$

$A = \frac{(4+10)3}{2} = \frac{14(3)}{2} = \frac{42}{2} = 21$

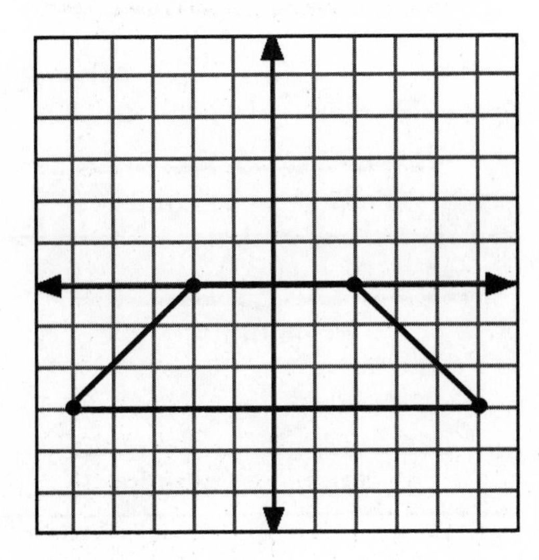

II. SECTION TWO

2.1 7, 6, 5, 4, 3, 2, 1, 0, -1, -2

2.2 10, 8, 6, 4, 2, 0, -2, -4, -6, -8

2.3 1, 0 -1, -2, -3, -4, -5, -6

2.4 -4, -3, -2, -1, 0, 1, 2, 3

2.5 -8, -6, -4, -2, 0, 2, 4, 6, 8

2.6 3, 3, 3, 3, 3, 3, 3, 3, 3

2.7 7

2.8 a. 2
b. 2

2.9 a. -1
b. (-5)
c. -1

2.10 a. 3
b. 3
c. 3

2.11 a. -2
b. -2 + 0 = -2

2.12 a. -6
b. -1 + (-5) = -6

2.13 a. 6
b. 11 + (-5) = 6

2.14 a. -3
b. -2 + (-1) = -3

2.15 a. -6
b. 4 + (-10) = -6

2.16

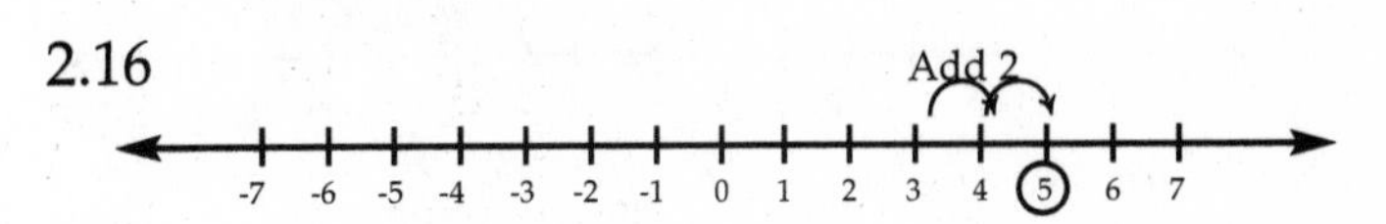

2.17

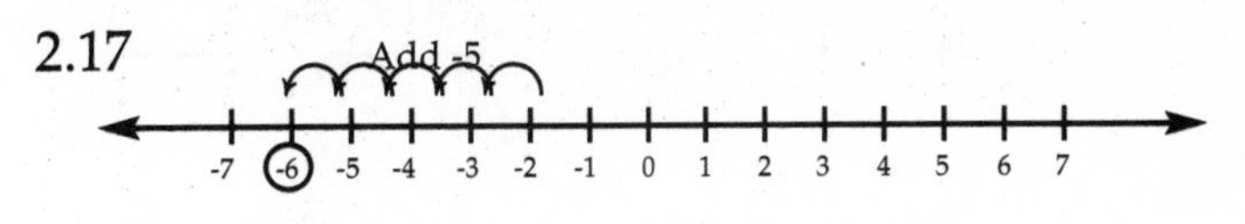

2.18

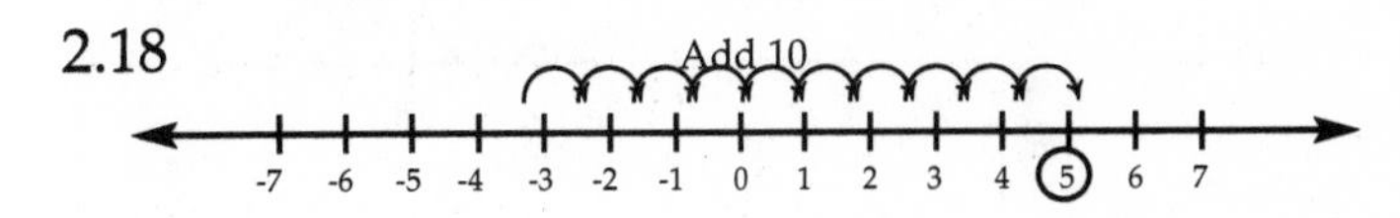

2.19

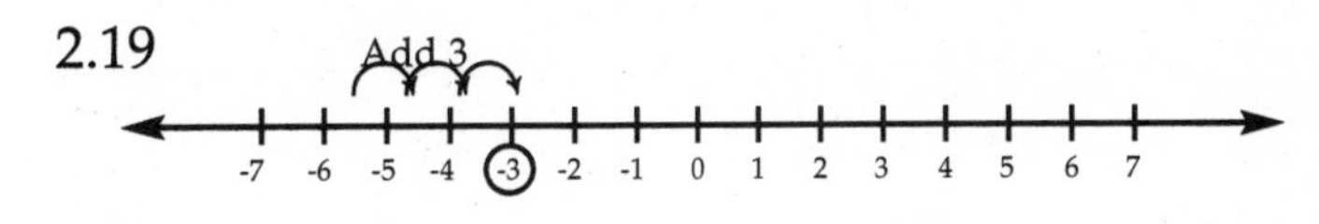

2.20

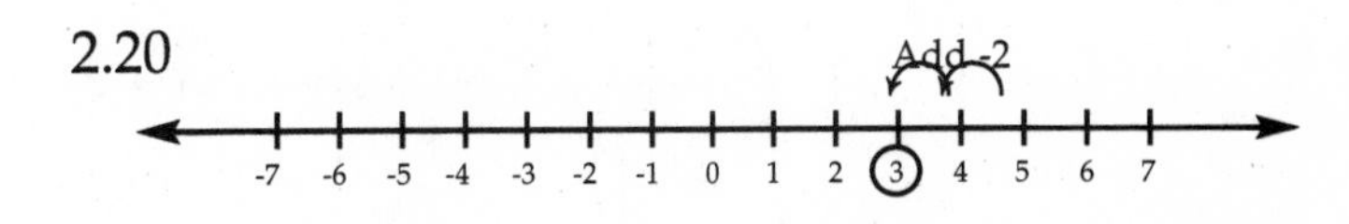

2.21

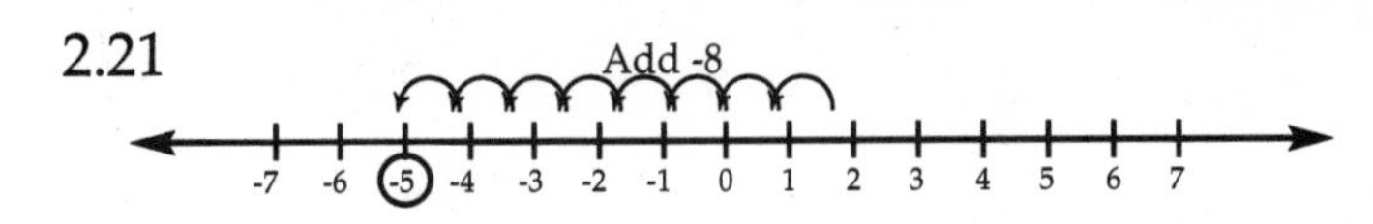

2.22

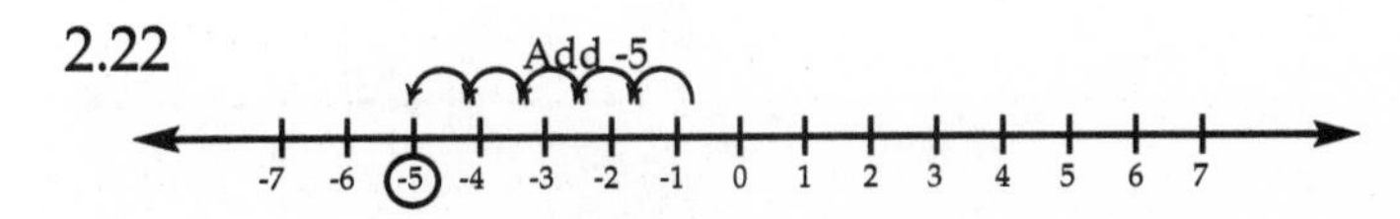

2.23

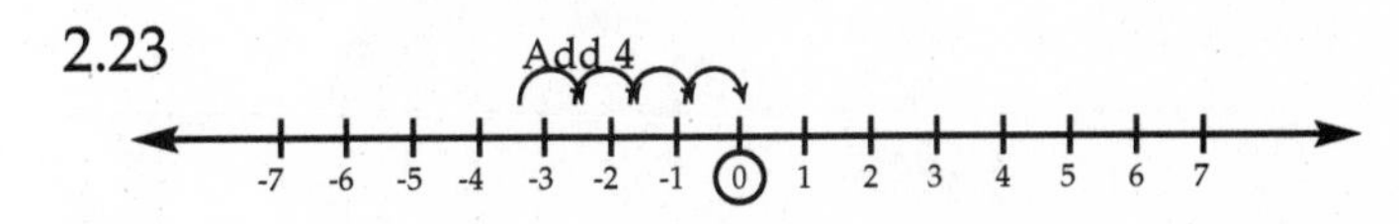

2.24

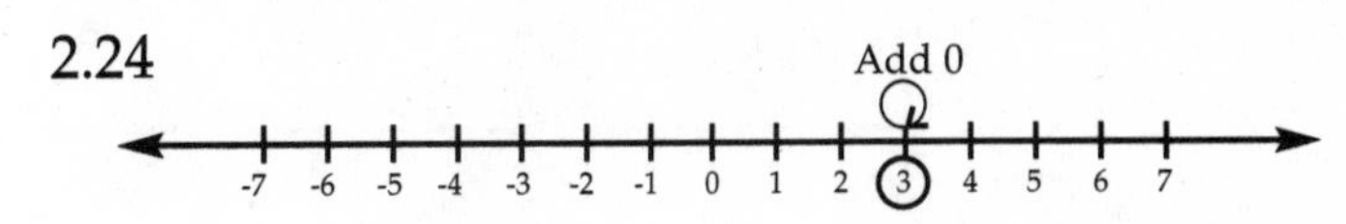

2.25

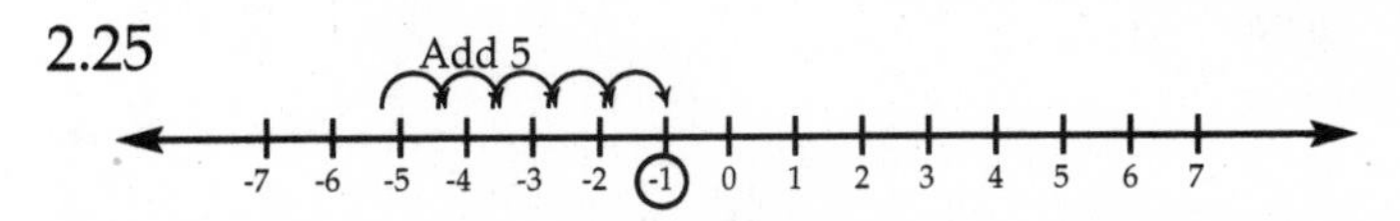

2.26

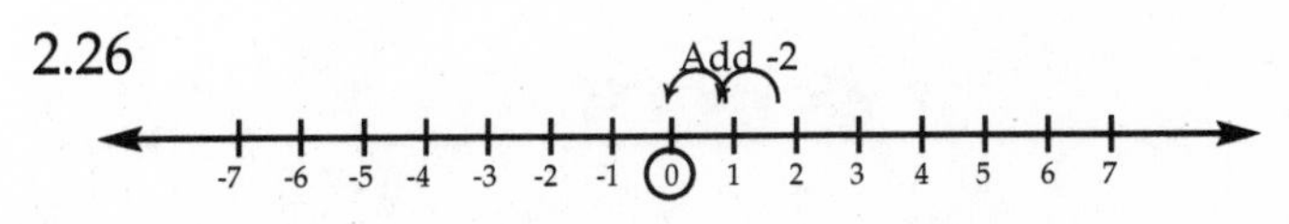

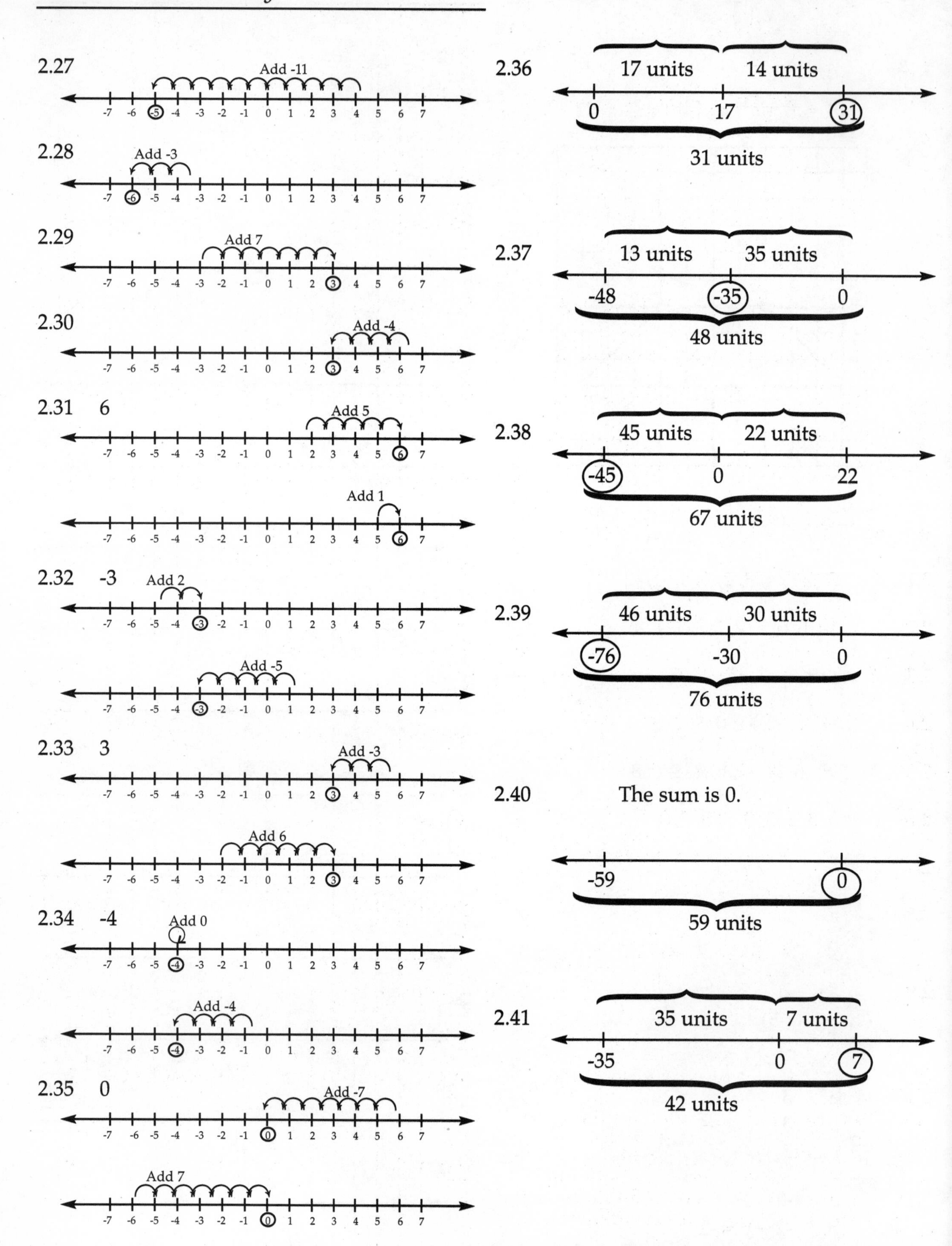

2.27
Add -11
2.28
Add -3
2.29
Add 7
2.30
Add -4
2.31 6
Add 5
Add 1
2.32 -3
Add 2
Add -5
2.33 3
Add -3
Add 6
2.34 -4
Add 0
Add -4
2.35 0
Add -7
Add 7
2.36
17 units
14 units
0
17
31
31 units
2.37
13 units
35 units
-48
-35
0
48 units
2.38
45 units
22 units
-45
0
22
67 units
2.39
46 units
30 units
-76
-30
0
76 units
2.40
The sum is 0.
-59
0
59 units
2.41
35 units
7 units
-35
0
7
42 units

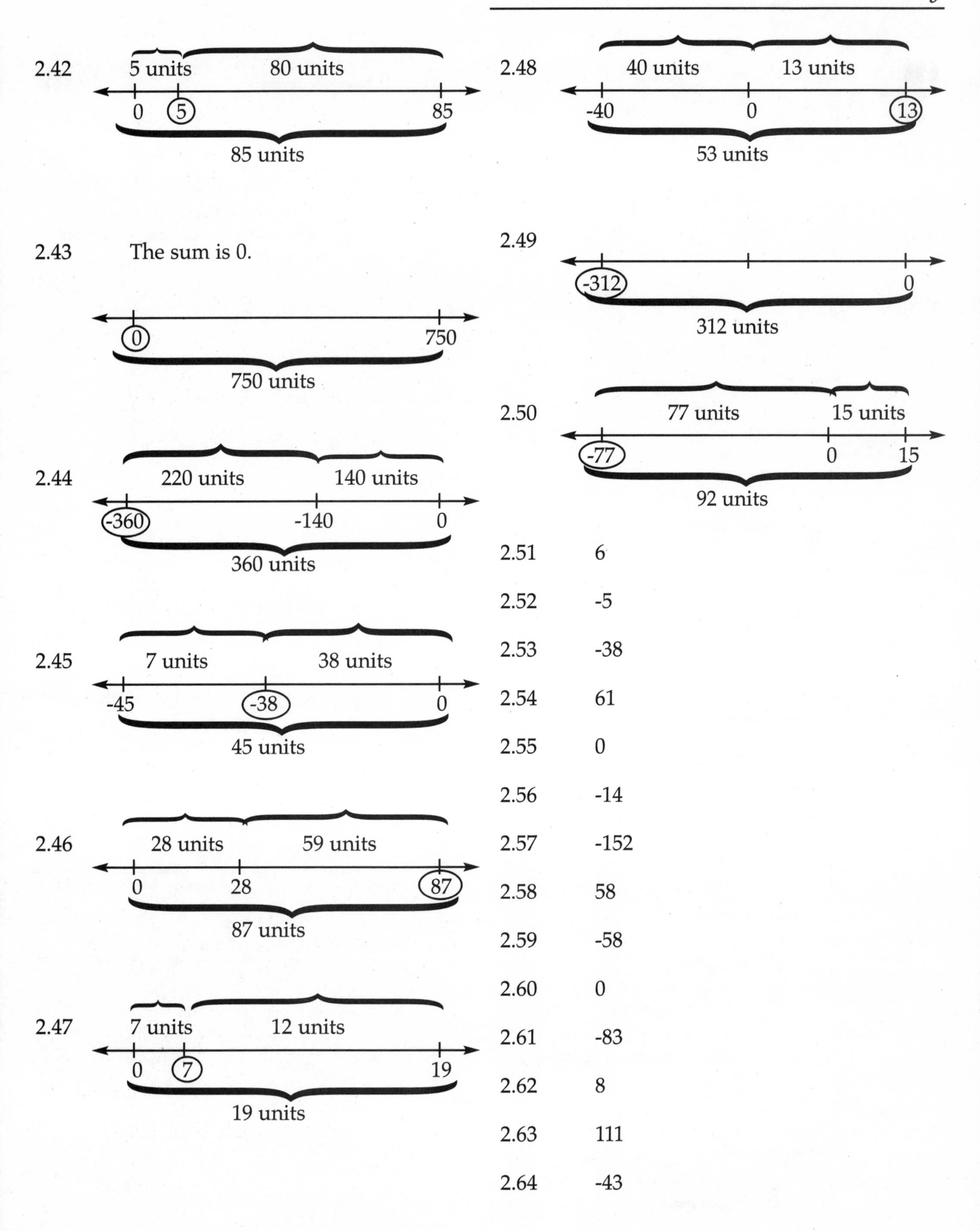

2.51 6

2.52 -5

2.53 -38

2.54 61

2.55 0

2.56 -14

2.57 -152

2.58 58

2.59 -58

2.60 0

2.61 -83

2.62 8

2.63 111

2.64 -43

2.65 119

2.66 -6

2.67 -82

2.68 5

2.69 -5

2.70 -23

2.71
$6 + 2 + (-5) + 3 + (-7)$
$= 8 + (-5) + 3 + (-7)$
$= 3 + 3 + (-7)$
$= 6 + (-7)$
$= -1$

2.72
$-8 + 9 + 3 + (-9) + 5$
$= 1 + 3 + (-9) + 5$
$= 4 + (-9) + 5$
$= -5 + 5$
$= 0$

2.73
$25 + (-11) + (-15) + 7 + (-8) + 17$
$= 14 + (-15) + 7 + (-8) + 17$
$= -1 + 7 + (-8) + 17$
$= 6 + (-8) + 17$
$= -2 + 17$
$= 15$

2.74
$-420 + 300 + (-250)$
$= -120 + (-250)$
$= -370$

2.75
$-119 + (-317) + 529 + 760 + (-237)$
$= -436 + 529 + 760 + (-237)$
$= 93 + 760 + (-237)$
$= 853 + (-237)$
$= 616$

2.76 3, 2, 1, 0, -1, ,-2 , -3

2.77 3, 4, 5, 6, 7, 8, 9

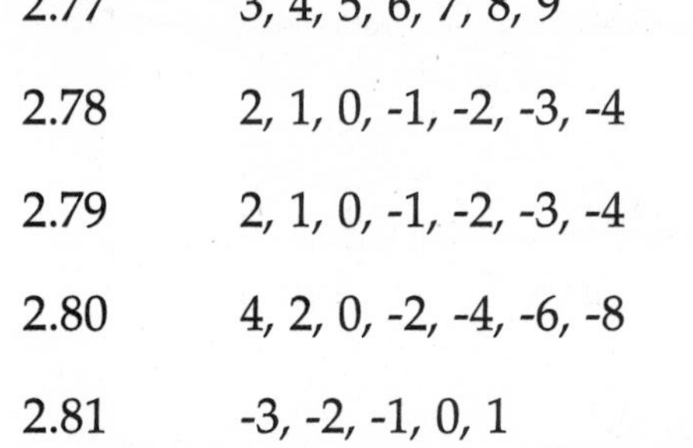

2.78 2, 1, 0, -1, -2, -3, -4

2.79 2, 1, 0, -1, -2, -3, -4

2.80 4, 2, 0, -2, -4, -6, -8

2.81 -3, -2, -1, 0, 1

2.82

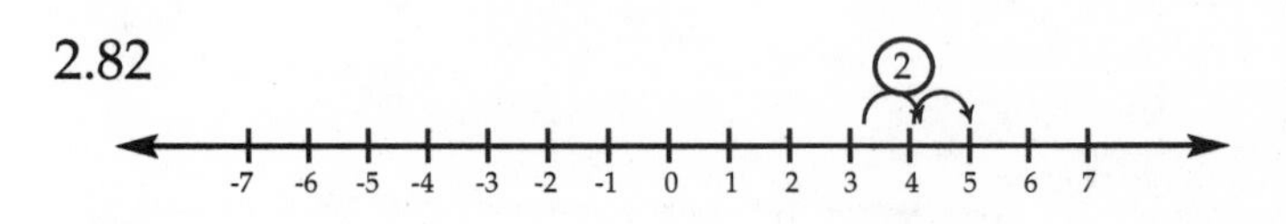

2.83

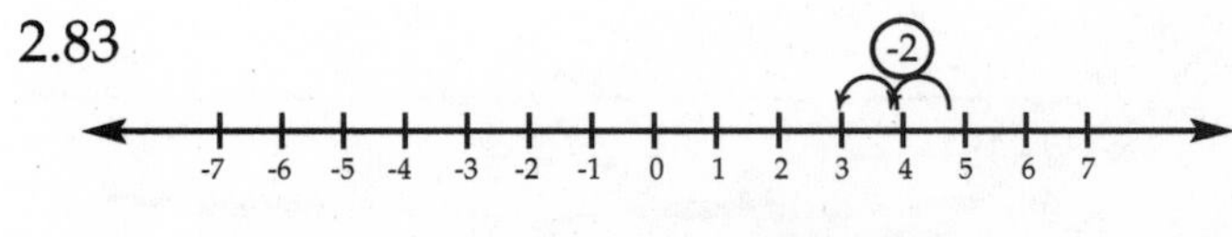

2.84

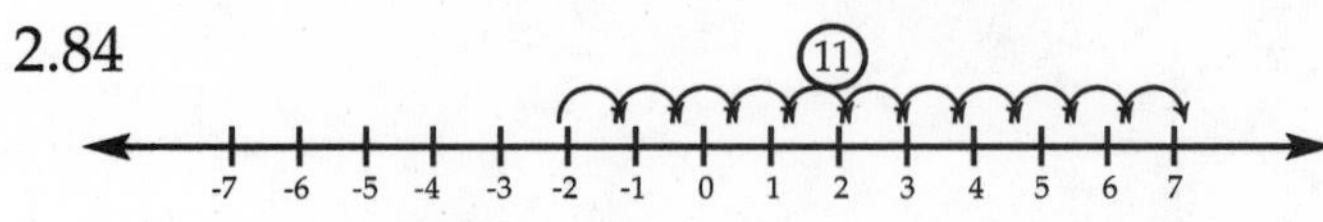

2.85

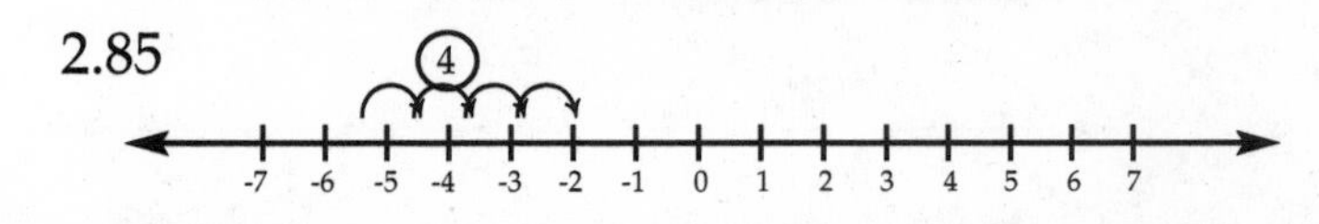

2.86

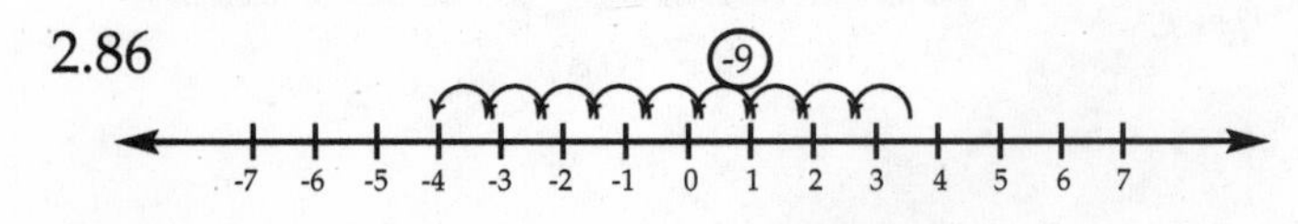

2.87

2.88

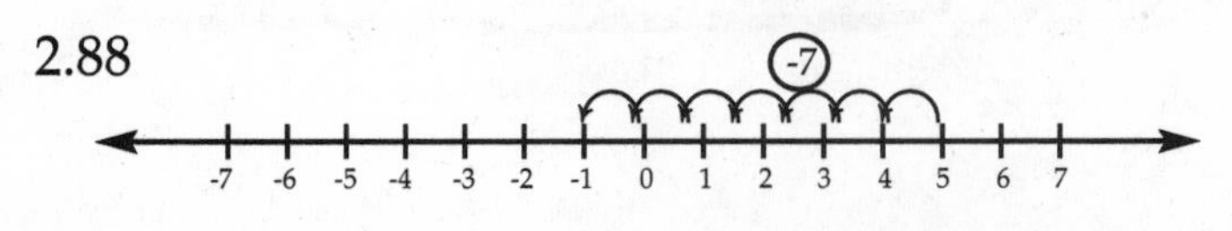

2.89

2.90

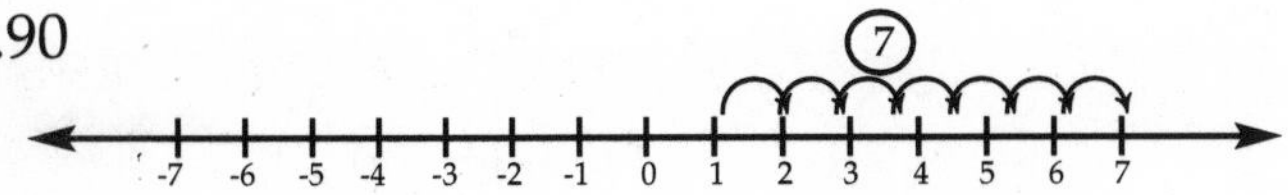

2.91

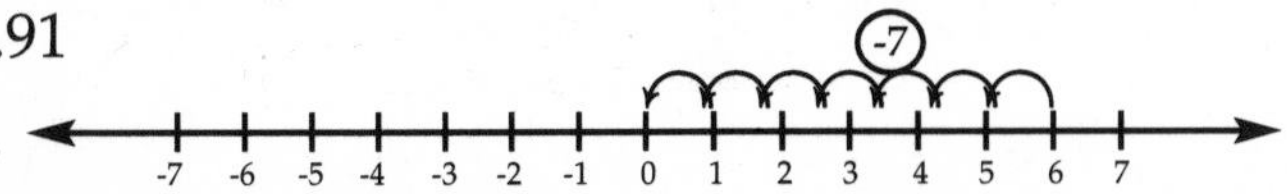

2.92

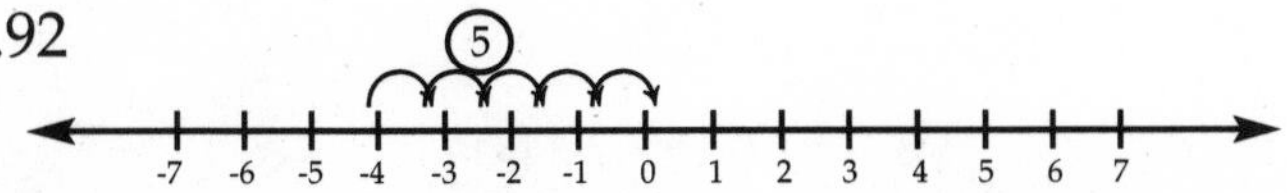

2.93

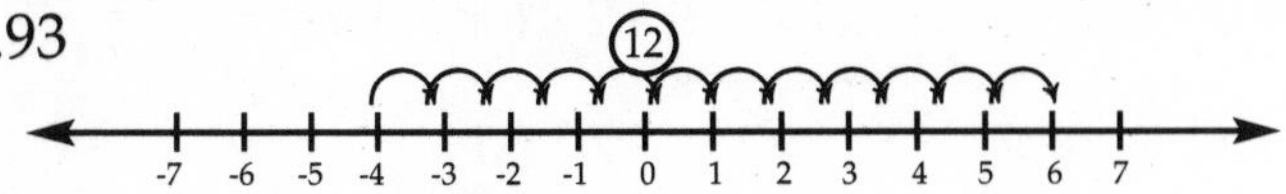

2.94

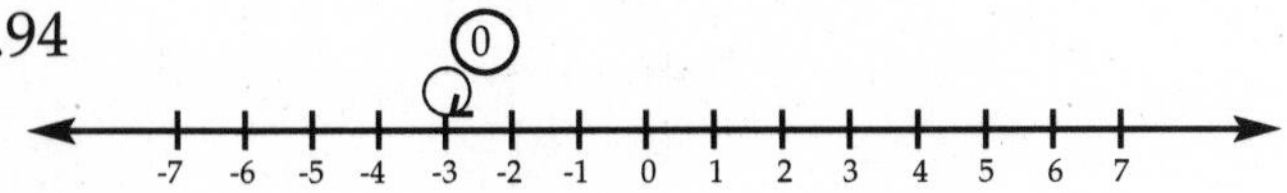

2.95

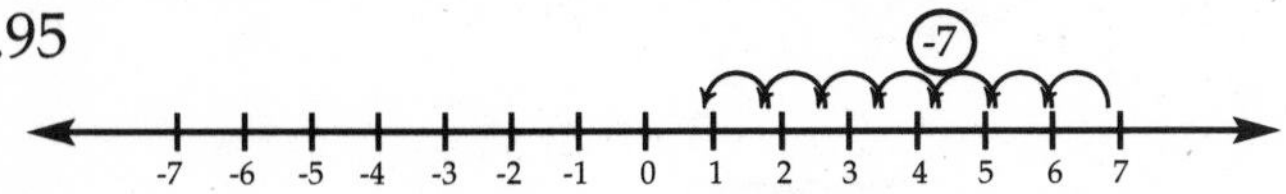

2.96 8

2.97 12

2.98 -9

2.99 -6

2.100 14

2.101 -38

2.102 0

2.103 70

2.104 -121

2.105 500

2.106 -6

2.107 17

2.108 14

2.109 0

2.110 -125

2.111 -125

2.112 $8 - 3 - (-6)$

$= \underbrace{8 + (-3)} + 6$

$= \underbrace{5 + 6}$

$= 11$

2.113 $-40 - (-9) - 7$

$= \underbrace{-40 + 9} + (-7)$

$= \underbrace{-31 + (-7)}$

$= -38$

2.114 $-10 - (-20) - (-30) - (-40)$

$= \underbrace{-10 + 20} + 30 + 40$

$= \underbrace{10 + 30} + 40$

$= \underbrace{40 + 40}$

$= 80$

2.115 $-10 - 20 - 30 - 40$

$= \underbrace{-10 + (-20)} + (-30) + (-40)$

$= \underbrace{-30 + (-30)} + (-40)$

$= \underbrace{-60 + (-40)}$

$= -100$

2.116 6, 4, 2, 0, -2, -4, -6

2.117 30, 20 , 10, 0, -10, -20, -30

2.118 -8, -4, 0, 4, 8

2.119 12, 6, 0, -6, -12

2.120 -3, 0, 3, 6, 9, 12

2.121 -12, -12, -12, -12, -12

2.122 6

2.123 a. -4
b. -4

2.124 a. -20
b. (-4)
c. -20

2.125 a. 0
b. 0
c. 0

2.126 a. 3
b. $-1 \cdot (-3) = 3$

2.127 a. -12
b. $-2 \cdot 6 = -12$

2.128 a. 0
b. $0 \cdot (-3) = 0$

2.129 a. 18
b. $-9 \cdot (-2) = 18$

2.130 a. -30
b. $5 \cdot (-6) = -30$

2.131 -48

2.132 40

2.133 -17

2.134 1,500

2.135 -108

2.136 105

2.137 0

2.138 -47

2.139 -56

2.140 196

2.141 $\underbrace{6 \cdot (-2)} \cdot (-1) \cdot 3$
$= \underbrace{-12 \cdot (-1)} \cdot 3$
$= \underbrace{12 \cdot 3}$
$= 36$

2.142 0

2.143 $-1 \cdot (-1) \cdot (-1) \cdot (-1) \cdot (-1)$
$= \underbrace{1 \cdot (-1)} \cdot (-1) \cdot (-1)$
$= \underbrace{-1 \cdot (-1)} \cdot (-1)$
$= \underbrace{1 \cdot (-1)}$
$= -1$

2.144 $-5 \cdot 4 \cdot (-3) \cdot 2 \cdot (-1)$
$= \underbrace{-20 \cdot (-3)} \cdot 2 \cdot (-1)$
$= \underbrace{60 \cdot 2} \cdot (-1)$
$= \underbrace{120 \cdot (-1)}$
$= -120$

2.145 $5 \cdot (-4) \cdot 3 \cdot (-2) \cdot 1$
$= \underbrace{-20 \cdot 3} \cdot (-2) \cdot 1$
$= \underbrace{-60 \cdot (-2)} \cdot 1$
$= \underbrace{120 \cdot 1}$
$= 120$

2.146 $4^3 = 4 \cdot 4 \cdot 4 = 64$

2.147 $(-3)^2 = (-3) \cdot (-3) = 9$

2.148 $(-7)^3 = (-7) \cdot (-7) \cdot (-7) = -343$

2.149 $15^2 = 15 \cdot 15 = 225$

2.150 $(-6)^4 = (-6) \cdot (-6) \cdot (-6) \cdot (-6) = 1{,}296$

2.151 $(-11)^3 = (-11) \cdot (-11) \cdot (-11) = -1{,}331$

2.152 $(-2)^5 = (-2) \cdot (-2) \cdot (-2) \cdot (-2) \cdot (-2)$
$= -32$

2.153 $9^3 = 9 \cdot 9 \cdot 9 = 729$

2.154 $(-1)^7 = (-1) \cdot (-1) \cdot (-1) \cdot (-1) \cdot$
$(-1) \cdot (-1) \cdot (-1)$
$= -1$

2.155 $(-1)^8 = (-1) \cdot (-1) \cdot (-1) \cdot (-1) \cdot$
$(-1) \cdot (-1) \cdot (-1) \cdot (-1)$
$= 1$

2.156 $(-10)^4 = (-10) \cdot (-10) \cdot (-10) \cdot (-10) = 10{,}000$

2.157 $(-10)^5 = (-10) \cdot (-10) \cdot (-10) \cdot (-10) \cdot (-10) = -100{,}000$

2.158 $2^{10} = 2 \cdot 2 \cdot 2 \cdot 2 \cdot 2 \cdot 2 \cdot 2 \cdot 2 \cdot 2 \cdot 2 = 1{,}024$

2.159 $8^2 = 8 \cdot 8 = 64$

2.160 $(-8)^2 = (-8) \cdot (-8) = 64$

2.161 $(-8)^3 = (-8) \cdot (-8) \cdot (-8) = -512$

2.162
$2 \cdot 4^2 + (-3) =$
$2 \cdot 4 \cdot 4 + (-3) =$
$32 + (-3) =$
29

2.163
$3^2 \cdot 2^3 - (-7)^2 =$
$\underbrace{3 \cdot 3} \cdot 2 \cdot 2 \cdot 2 - \underbrace{(-7)\,(-7)} =$
$\underbrace{9 \cdot 8} - 49 =$
$\underbrace{72 - 49} =$
23

2.164
$(-4)^4 + (-3)^3 + (-2)^2 =$

$(-4) \cdot (-4) \cdot (-4) \cdot (-4) +$
$(-3) \cdot (-3) \cdot (-3) +$
$(-2) \cdot (-2) =$
$\underbrace{256 + (-27)} + 4 =$
$\underbrace{229 + 4} =$
233

2.165
$-6 \cdot (-1)^5 - (-4)^3 =$
$-6 \cdot (-1) \cdot (-1) \cdot (-1) \cdot (-1) \cdot (-1) -$
$(-4) \cdot (-4) \cdot (-4) =$
$\underbrace{-6 \cdot (-1)} - (-64) =$
$\underbrace{6 + 64} =$
70

2.166 $81, 27, 9, 3, 1, \frac{1}{3}, \frac{1}{9}$

2.167 $5^2, 5^1, 5^0, 5^{-1}, 5^{-2}, 5^{-3}$

2.168 $-512, 64, -8, 1, -\frac{1}{8}, \frac{1}{64}, -\frac{1}{512}$

2.169 6^{-2}

2.170 10^{-5}

2.171 $(-17)^{-2}$

2.172 $(-3)^{-9}$

2.173 4^{-3}

2.174 $(-8)^{-5}$

2.175 $\frac{1}{(-3)^5} = -\frac{1}{243}$

2.176 $\frac{1}{2^6} = \frac{1}{64}$

2.177 $\frac{1}{10^4} = \frac{1}{10{,}000}$

2.178 $\frac{1}{(-1)^{10}} = \frac{1}{1} = 1$

2.179 $\frac{1}{(-2)^3} = -\frac{1}{8}$

2.180 $\frac{1}{4^1} = \frac{1}{4}$

2.181 8

2.182 -4

2.183 -2

2.184 17

2.185 0

2.186 -9

2.187 -1

2.188 undefined

2.189 0

2.190 1

2.191 -3

2.192 -276

2.193 13

2.194 -5

2.195 undefined

2.196 -59

2.197 4

2.198 -5

2.199 80

2.200 -1

III. SECTION THREE

3.1 $3 \cdot 2 = 6$

3.2 $3 \cdot (-5) = -15$

3.3 $3 \cdot 0 = 0$

3.4 $2 + (-5) = -3$

3.5 $2 - (-5) = 2 + 5 = 7$

3.6 $-5 \cdot (-3) = 15$

3.7 $-5 \cdot (-3) \cdot (-1) = -15$

3.8 $-5 \cdot (-3) \cdot 0 = 0$

3.9 $-3 + 3 = 0$

3.10 $2 \cdot (-3) + 3 = -6 + 3 = -3$

3.11 $\frac{12}{-3} = -4$

3.12 $\frac{12}{2} + \frac{12}{-1} - \frac{12}{-3} =$

$\underbrace{6 - 12} + \frac{12}{-3} =$

$\underbrace{-6 + 4} =$

-2

3.13 $\underbrace{2 \cdot (-1)} + 5 \cdot 0 =$

$\underbrace{-2 + 0} =$

-2

3.14 $\underbrace{5 \cdot (-1)} + 2 \cdot 0 =$

$\underbrace{-5 + 0} =$

-5

3.15 $\underbrace{2 + (-5)} + (-3) =$

$\underbrace{-3 + (-3)} =$

-6

3.16 $2 + (-5) - (-3) =$

$\underbrace{2 + (-5)} + 3 =$

$\underbrace{-3 + 3} =$

0

3.17 $2 - (-5) + (-3) =$

$\underbrace{2 + 5} + (-3) =$

$\underbrace{7 + (-3)} =$

4

3.18 $\underbrace{-2 + (-5)} + (-3) =$

$\underbrace{-7 + (-3)} =$

-10

3.19 $-2 - (-5) - (-3) =$

$\underbrace{-2 + 5} + 3 =$

$\underbrace{3 + 3} =$

6

3.20 $\frac{2 + (-5)}{3} = \frac{-3}{3} = -3 \div 3 = -1$

3.21 $\frac{-5 + (-3)}{2} = \frac{-8}{2} = -8 \div 2 = -4$

3.22 $\frac{2}{-1} = 2 \div (-1) = -2$

3.23 $\frac{-5}{-1} = -5 \div (-1) = 5$

3.24 $\frac{0}{-1} = 0$

3.25 $\frac{-1}{0}$ is undefined

3.26 $2 \cdot (-5) \cdot (-3) + (-1) \cdot 0 =$

$-10 \cdot (-3) + (-1) \cdot 0 =$

$30 + (-1) \cdot 0 =$

$30 + 0 =$

30

3.27 $2 \cdot (-5) \cdot (-3) - (-1) \cdot 0 =$

$-10 \cdot (-3) - (-1) \cdot 0 =$

$30 - (-1) \cdot 0 =$

$30 - 0 =$

30

3.28 $-1 + 0 + 1 =$

$-1 + 1 =$

0

3.29 $-3 \cdot 2 \cdot (-5) + (-3) \cdot (-1) =$

$-6 \cdot (-5) + (-3) \cdot (-1) =$

$30 + (-3) \cdot (-1) =$

$30 + 3 =$

33

3.30 $-3 \cdot 2 \cdot (-5) - (-3) \cdot (-1) =$

$-6 \cdot (-5) - (-3) \cdot (-1) =$

$30 - (-3) \cdot (-1) =$

$30 - 3 =$

27

3.31 $\frac{2 \cdot (-3)}{-5 + (-1)} = \frac{-6}{-5 - 1} = \frac{-6}{-6} =$

$-6 \div (-6) = 1$

3.32 $\frac{30}{-2} - \frac{30}{-(-5)} + \frac{30}{-(-3)} =$

$\frac{30}{-2} - \frac{30}{5} + \frac{30}{3} =$

$30 \div (-2) - 30 \div 5 + 30 \div 3 =$

$-15 - 6 + 10 =$

$-21 + 10 =$

-11

3.33 $2 + 2 \cdot (-5) + 3 \cdot (-3) =$

$2 + (-10) + (-9) =$

$-8 + (-9) =$

-17

3.34 $3 \cdot 2 + 2 \cdot (-5) + (-3) =$

$6 + (-10) + (-3) =$

$-4 + (-3) =$

-7

3.35 $\frac{10}{-5} - 2 \cdot (-1)$

$10 \div (-5) - 2 \cdot (-1) =$

$-2 - (-2) =$

$-2 + 2 =$

0

3.36 $2 \cdot (-1) + (-5) \cdot (-1) + (-3) \cdot (-1) =$

$-2 + 5 + 3 =$

$3 + 3 =$

6

3.37 $2 \cdot (-1) + (-5) \cdot 0 + (-3) =$
$-2 + 0 + (-3) =$
$-2 + (-3) =$
-5

3.38 $2 \cdot (-5) \cdot (-3) =$
$-10 \cdot (-3) =$
30

3.39 $2 \cdot (-5) \cdot (-3) \cdot (-1) =$
$-10 \cdot (-3) \cdot (-1) =$
$30 \cdot (-1) =$
-30

3.40 $2 \cdot (-5) \cdot (-3) \cdot (-1) \cdot 0 = 0$

3.41 $(-2)^2 = 4$

3.42 $5^2 = 25$

3.43 $0^2 = 0$

3.44 $(-7)^2 + 4 =$
$49 + 4 =$
53

3.45 $(-7)^2 + 4^2 =$
$49 + 16 =$
65

3.46 $-7 + 4^2 =$
$-7 + 16 =$
9

3.47 $2 \cdot 5^3 =$
$2 \cdot 125 =$
250

3.48 $3 \cdot 5^2 =$
$3 \cdot 25 =$
75

3.49 $(-2)^5 = -32$

3.50 $5^{-2} = \frac{1}{5^2} = \frac{1}{25}$

3.51 $0^4 = 0$

3.52 $4^0 = 1$

3.53 $-2 \cdot 5 = -10$

3.54 $(-2)^2 \cdot 5 =$
$4 \cdot 5 =$
20

3.55 $(-2)^2 \cdot 5^2 =$
$4 \cdot 25 =$
100

3.56 $-2 \cdot 5^2 =$
$-2 \cdot 25 =$
-50

3.57 $(-1)^5 = -1$

3.58 $(-1)^{-7} = \frac{1}{(-1)^7} = \frac{1}{-1} = -1$

3.59 $(-1)^0 = 1$

3.60 $(-1)^4 = 1$

3.61 $(-2)^3 + 5^2 - (-7) =$
$(-2)^3 + 5^2 + 7 =$
$-8 + 25 + 7 =$
$17 + 7 =$
24

3.62 $5 \cdot 5 + 4^3 =$
$25 + 64 =$
89

3.63 $-2 \cdot 5 \cdot (-7) \cdot 0 \cdot 4 = 0$

3.64 $-2 + 5 + (-7) + 0 + 4 =$
$3 + (-7) + 0 + 4 =$
$-4 + 0 + 4 =$
$-4 + 4 =$
0

3.65 $(-2)^2 \cdot 5^2 \cdot (-7)^2 \cdot 0^2 \cdot 4^2 = 0$

3.66 $(-2)^2 + 5^2 + (-7)^2 + 0^2 + 4^2 =$

$4 + 25 + 49 + 0 + 16 =$

$29 + 49 + 0 + 16 =$

$78 + 0 + 16 =$

$78 + 16 =$

94

3.67 $(-2)^4 + 4^{-2} =$

$16 + \frac{1}{4^2} =$

$16 + \frac{1}{16} =$

$16\frac{1}{16}$

3.68 $10^4 = 10{,}000$

3.69 $(-10)^4 = 10{,}000$

3.70 $10^{-2} = \frac{1}{10^2} = \frac{1}{100}$

3.71 $(-10)^{-2} = \frac{1}{(-10)^2} = \frac{1}{100}$

3.72 $10^0 = 1$

3.73 $10 \cdot (-7)^3 =$

$10\ (-343) =$

$-3{,}430$

3.74 $-10 \cdot (-7)^3 =$

$-10\ (-343) =$

$3{,}430$

3.75 $5 \cdot (-2)^3 + 8 \cdot 5 =$

$5 \cdot (-8) + 8 \cdot 5 =$

$-40 + 40 =$

0

3.76 $5\ (-2)^3 - 8 \cdot 5 =$

$5 \cdot (-8) - 8 \cdot 5 =$

$-40 - 40 =$

-80

3.77 $\frac{0^2}{(-7)^2} = 0^2 \div (-7)^2 = 0$

3.78 $\frac{(-2)^4}{4^2} = (-2)^4 \div 4^2 =$

$16 \div 16 = 1$

3.79 $-2 \cdot 5 \cdot (-7) - 0^2 \cdot 4 =$

$-2 \cdot 5 \cdot (-7) - 0 =$

$-10 \cdot (-7) - 0 =$

$70 - 0 =$

70

3.80 $0 \cdot 4^2 - (-2) \cdot 5 \cdot (-7) =$

$0 - (-2) \cdot 5 \cdot (-7) =$

$0 - (-10) \cdot (-7) =$

$0 - 70 =$

-70

3.81

a	1	2
$b = 3 - a$	$3 - 1$	$3 - 2$
b	2	1

b	3	4
$a = 3 - b$	$3 - 3$	$3 - 4$
a	0	-1

The number pairs are (1, 2), (2, 1), (0, 3), and (-1, 4).

3.82

a	4	0
$b = a$	4	0
b	4	0

b	2	-1
$a = b$	2	-1
a	2	-1

The number pairs are (4, 4), (2, 2), (0, 0), and (-1, -1).

3.83

a	5	0	-7	-10
$3a$	$3 \cdot 5$	$3 \cdot 0$	$3 \cdot (-7)$	$3 \cdot (-10)$
b	15	0	-21	-30

The number pairs are (5, 15), (0, 0), (-7, -21), and (-10, -30).

3.84

a	-2	-1	1	2
$-3a$	$-3 \cdot (-2)$	$-3 \cdot (-1)$	$-3 \cdot 1$	$-3 \cdot 2$
b	6	3	-3	-6

The number pairs are (-2, 6), (-1, 3), (1, -3), and (2, -6).

3.85

a	3	1	-1	-3
a^2	3^2	1^2	$(-1)^2$	$(-3)^2$
b	9	1	1	9

The number pairs are (3, 9), (1, 1), (-1, 1), and (-3, 9).

3.86

m	-2	0	4
$5m + 4$	$5 \cdot (-2) + 4$ $-10 + 4$	$5 \cdot 0 + 4$ $0 + 4$	$5 \cdot 4 + 4$ $20 + 4$
n	-6	4	24

The number pairs are (-2, -6), (0, 4), and (4, 24).

3.87

m	-5	-2	3
$-2m - 1$	$-2 \cdot (-5) - 1$ $10 - 1$	$-2 \cdot (-2) - 1$ $4 - 1$	$-2 \cdot 3 - 1$ $-6 - 1$
n	9	3	-7

The number pairs are (-5, 9), (-2, 3), and (3, -7).

3.88

m	-2	0	2
$m^2 + 3$	$(-2)^2 + 3$ $4 + 3$	$0^2 + 3$ $0 + 3$	$2^2 + 3$ $4 + 3$
n	7	3	7

The number pairs are (-2, 7), (0, 3), and (2, 7).

3.89

m	4	3	-1
$10 - m^2$	$10 - 4^2$ $10 - 16$	$10 - 3^2$ $10 - 9$	$10 - (-1)^2$ $10 - 1$
n	-6	1	9

The number pairs are (4, -6), (3, 1), and (-1, 9).

3.90

m	2	0	-2
$m^3 + m^2$	$2^3 + 2 \cdot 2^2$ $8 + 2 \cdot 4$ $8 + 8$	$0^3 + 2 \cdot 0^2$ $0 + 2 \cdot 0$ $0 + 0$	$(-2)^3 + 2 \cdot (-2)^2$ $-8 + 2 \cdot 4$ $-8 + 8$
n	16	0	0

The number pairs are (2 ,16), (0, 0), and (-2, 0).

3.91

a	-4	-3	-2
$b = 1 - a$	$1 - (-4)$ $1 + 4$	$1 - (-3)$ $1 + 3$	$1 - (-2)$ $1 + 2$
b	5	4	3

-1	0	1	2
$1 - (-1)$ $1 + 1$	$1 - 0$	$1 - 1$	$1 - 2$
2	1	0	-1

3	4	5
$1 - 3$	$1 - 4$	$1 - 5$
-2	-3	-4

The number pairs are (-4, 5), (-3, 4), (-2, 3), (-1, 2), (0, 1), (1, 0), (2, -1), (3 -2), (4, -3), and (5, -4).

3.91 cont.

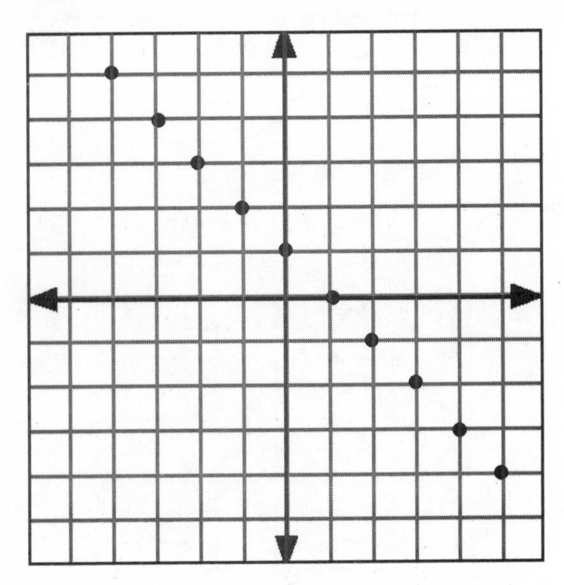

3.92

a	-5	-4	-3	-2	-1
$b = a$	-5	-4	-3	-2	-1
b	-5	-4	-3	-2	-1

0	1	2	3	4	5
0	1	2	3	4	5
0	1	2	3	4	5

The number pairs are (-5, -5), (-4, -4), (-3, -3), (-2, -2), (-1, -1), (0, 0), (1, 1), (2, 2), (3, 3), (4, 4), (5, 5).

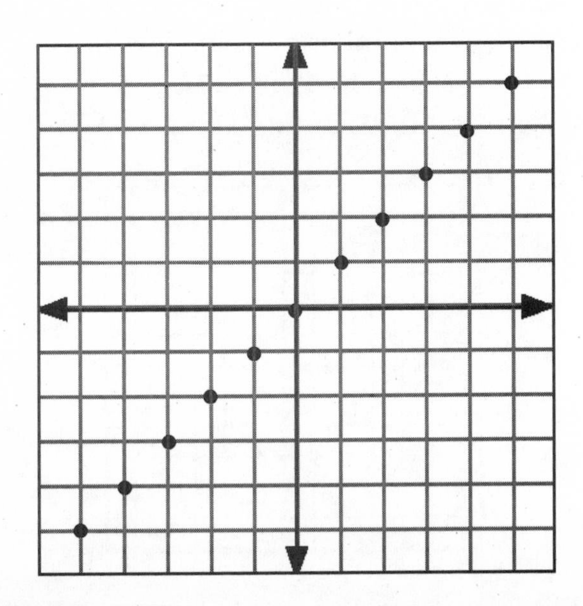

3.93

m	-2	-1	0
$2m$	$2 \cdot (-2)$	$2 \cdot (-1)$	$2 \cdot 0$
n	-4	-2	0

1	2
$2 \cdot 1$	$2 \cdot 2$
2	4

The number pairs are (-2, -4), (-1, -2), (0, 0), (1, 2), and (2, 4).

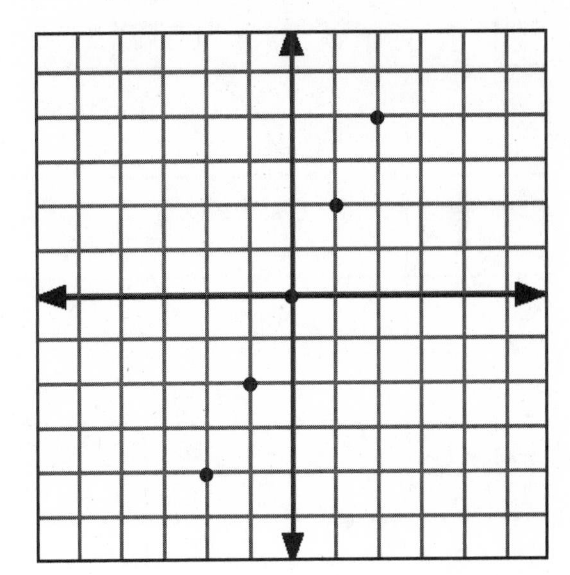

3.94

m	0	1	2
$2m - 4$	$2 \cdot 0 - 4$ $0 \ - 4$	$2 \cdot 1 - 4$ $2 \ - 4$	$2 \cdot 2 - 4$ $4 \ - 4$
n	-4	-2	0

3	4
$2 \cdot 3 - 4$ $6 \ - 4$	$2 \cdot 4 - 4$ $8 \ - 4$
2	4

The number pairs are (0, -4), (1, -2), (2, 0), (3, 2), and (4, 4).

3.94 cont.

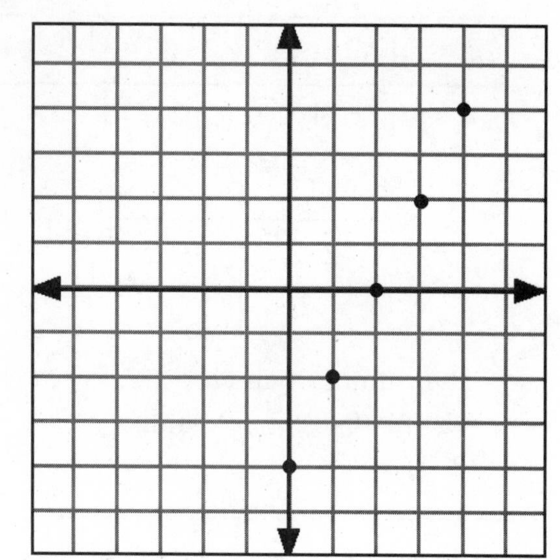

3.95

m	-4	-3
$-2m - 4$	$-2 \cdot (-4) - 4$ $8 - 4$	$-2 \cdot (-3) - 4$ $6 - 4$
n	4	2

-2	-1	0
$-2 \cdot (-2) - 4$ $4 - 4$	$-2 \cdot (-1) - 4$ $2 - 4$	$-2 \cdot (0) - 4$ $0 - 4$
0	-2	-4

The number pairs are (-4, 4), (-3, 2), (-2, 0), (-1, -2), and (0, -4).

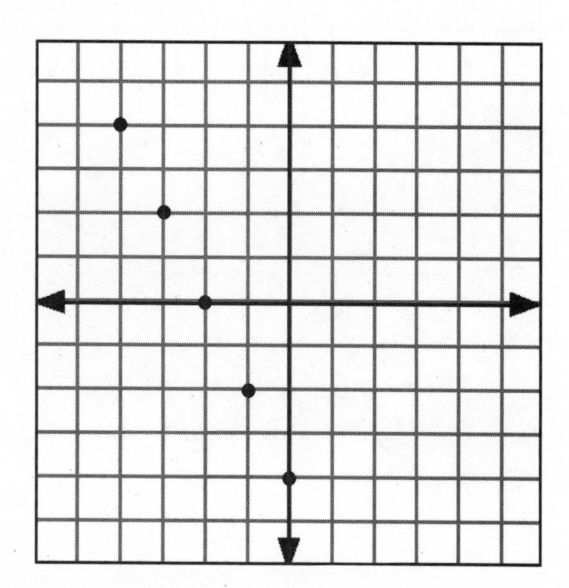

3.96

x	-2	-1	0	1	2
x^2	$(-2)^2$	$(-1)^2$	0^2	1^2	2^2
y	4	1	0	1	4

The number pairs are (-2, 4), (-1, 1), (0, 0), (1, 1), and (2, 4).

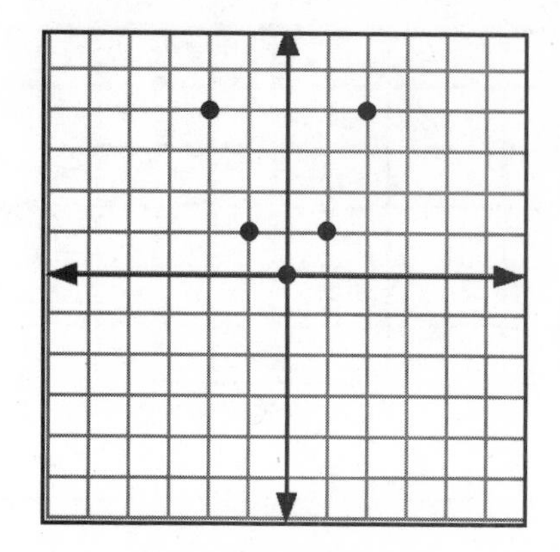

3.97

x	-3	-2	-1
$x^2 - 4$	$(-3)^2 - 4$ $9 - 4$	$(-2)^2 - 4$ $4 - 4$	$(-1)^2 - 4$ $1 - 4$
y	5	0	-3

0	1	2	3
$0^2 - 4$ $0 - 4$	$1^2 - 4$ $1 - 4$	$2^2 - 4$ $4 - 4$	$3^2 - 4$ $9 - 4$
-4	-3	0	5

The number pairs are (-3, 5), (-2, 0), (-1, -3), (0, -4), (1, -3), (2, 0), and (3, 5).

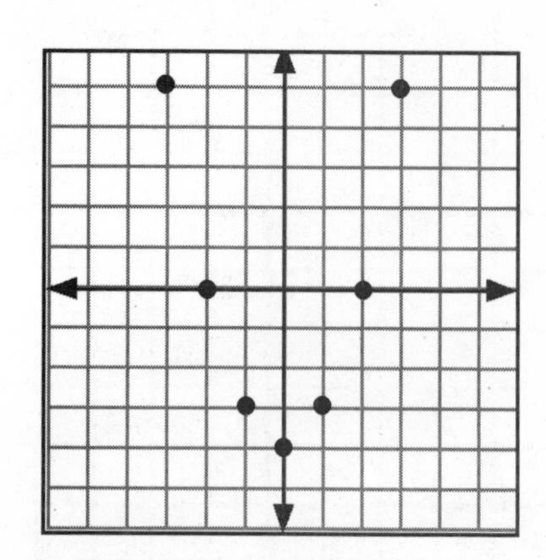

3.98

x	-3	-2	-1	0
$5 - x^2$	$5 - (-3)^2$ $5 - 9$	$5 - (-2)^2$ $5 - 4$	$5 - (-1)^2$ $5 - 1$	$5 - 0^2$ $5 - 0$
y	-4	1	4	5

1	2	3
$5 - 1^2$ $5 - 1$	$5 - 2^2$ $5 - 4$	$5 - 3^2$ $5 - 9$
4	1	-4

The number pairs are (-3, -4), (-2, 1), (-1, 4), (0, 5), (1, 4), (2, 1), and (3, -4).

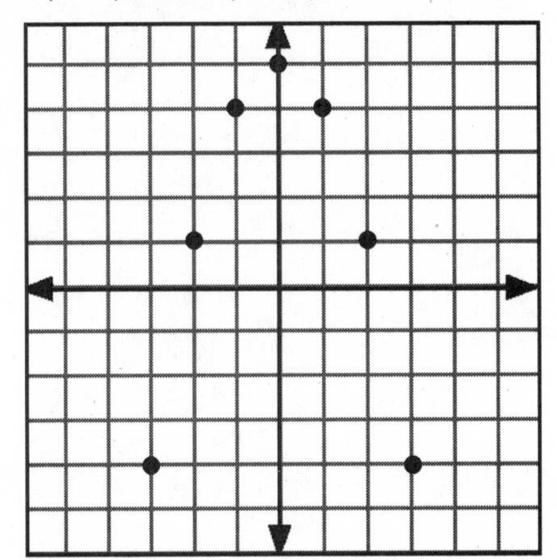

3.99

x	-2	-1
$-2x^2 + 5$	$-2 \cdot (-2)^2 + 5$ $-2 \cdot 4 \quad + 5$ $-8 \quad + 5$	$-2 \cdot (-1)^2 + 5$ $-2 \cdot 1 \quad + 5$ $-2 \quad + 5$
y	-3	3

0	1	2
$-2 \cdot 0^2 + 5$	$-2 \cdot 1^2 + 5$ $-2 \cdot 1 \quad +5$ $-2 \quad + 5$	$-2 \cdot 2^2 + 5$ $-2 \cdot 4 \quad + 5$ $-8 \quad + 5$
5	3	-3

The number pairs are (-2, -3), (-1, 3), (0, 5), (1, 3), and (2, -3).

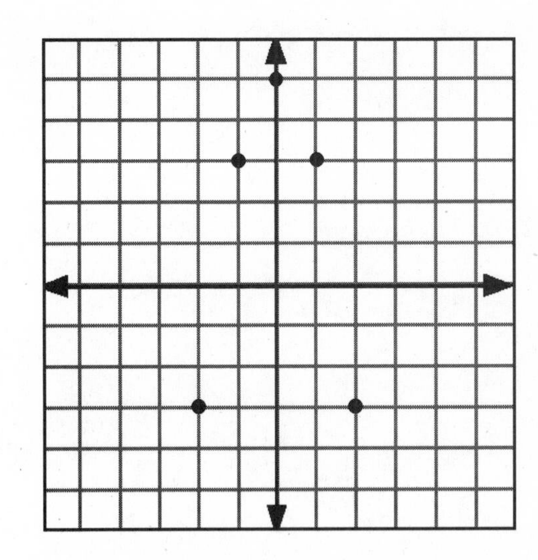

3.100

x	-1	0
$x^3 - x^2$	$(-1)^3 - (-1)^2$ $-1 \quad - \quad 1$	$0^3 - 0^2$ $0 - 0$
y	-2	0

1	2
$1^3 - 1^2$ $1 - 1$	$2^3 - 2^2$ $8 - 4$
0	4

The number pairs are (-1, -2), (0, 0), (1, 0), and (2, 4).

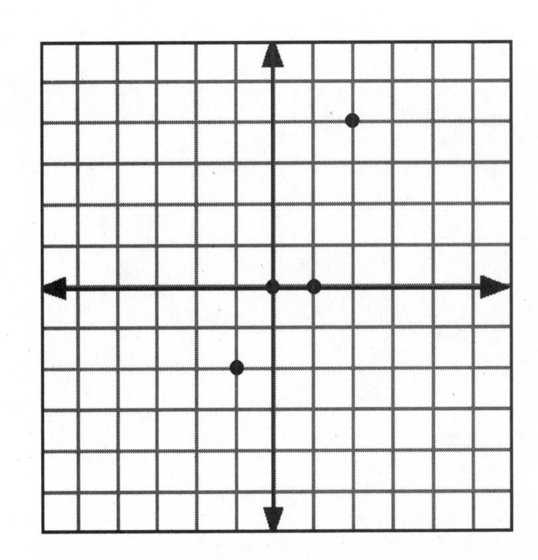

I. SECTION ONE

1.1
4. 1 2 → 4.1
$\sqrt{17.\ 00\ 00}$
16
8 1 | 1 00
81
82 2 | 19 00
16 44
2 56

1.2
1 3. 6 → 13.6
$\sqrt{1\ 84.\ 96}$
1
2 3 | 0 84
69
26 6 | 15 96
15 96
0

1.3 9

1.4 2

1.5 5

1.6 8

1.7 1

1.8 7

1.9 4

1.10 11

1.11 6

1.12 12

1.13 3

1.14 10

1.15 0.4

1.16
5. 8 3 → 5.8
$\sqrt{34.\ 00\ 00}$
25
10 8 | 9 00
8 64
116 3 | 36 00
34 89
1 11

1.17
1 0. 6 → 10.6
$\sqrt{1\ 12.\ 36}$
1
2 06 | 0 12 36
12 36
0

1.18
6. 6 → 6.6
$\sqrt{43.\ 56}$
36
12 6 | 7 56
7 56
0

1.19
2 0 → 20
$\sqrt{4\ 00}$
4
0 00

1.20
3. 6 0 → 3.6
$\sqrt{13.\ 00\ 00}$
9
6 6 | 4 00
3 96
72 0 | 4 00
00
4 00

1.21 A number is squared by multiplying it by itself; $4^2 = 4 \cdot 4 = 16$.

1.22 A square root is a number that produces a given number when multiplied by itself; $\sqrt{25} = 5$.

1.23 The decimal point in the answer is placed directly above the decimal in the number under the radical.

1.24 $5^2 = 25$ is too small.
$6^2 = 36$ is too large.
Guess 5.5.

$$5.5\overline{)31.360} = 5.70$$

```
       5.70
5.5)31.360
    27 5
     3 86
     3 85
       10
```

$\frac{5.7 + 5.5}{2} = \frac{11.2}{2} = 5.6$

1.25 $6^2 = 36$ is too small.
$7^2 = 49$ is too large.
Guess 6.8.

```
       6.65
6.8)45.250
    40 8
     4 45
     4 08
       370
       340
        30
```

$\frac{6.65 + 6.8}{2} = \frac{13.45}{2} = 6.725$

Round to 6.7.

1.26 $10^2 = 100$ is too small.
$11^2 = 121$ is too large.
Guess 10.5

```
         10.85
10.5)114.000
     105
       900
       840
        600
        525
         75
```

Stop. Do not round up.

$\frac{10.85 + 10.5}{2} = \frac{21.35}{2} = 10.675$

Round to 10.7.

1.27 $20^2 = 400$ is too small.
$21^2 = 441$ is too large.
Guess 20.4 since the number is a little less than half-way between.

1.27 cont.

```
         20.4
20.4)416.16
     408
       816
       816
         0
```

1.28 $C = 2 \cdot \pi \cdot r$
$= 2 \cdot 3.14 \cdot 3.2$
$= 6.28 \cdot 3.2$
$= 20.096$
$= 20.1$ cm

1.29 $r = \frac{1}{2} \cdot d = \frac{1}{2} \cdot 8 = 4$ ft.

$A = \pi \cdot r^2$
$= 3.14 \cdot 4^2$
$= 3.14 \cdot 16$
$= 50.24$ sq. ft.

1.30 $C = \pi \cdot d$
$= 3.14 \cdot 40$
$= 125.6$ ft.

1.31 $r = \frac{1}{2} \cdot d = \frac{1}{2} \cdot 40 = 20$ ft.
$A = \pi \cdot r^2$
$= 3.14 \cdot 20^2$
$= 3.14 \cdot 400$
$= 1{,}256$ ft.^{2} or sq. ft.

1.32 $d = 42 + 8 + 8 = 58''$

1.33 $C = 2 \cdot \pi \cdot r$
$18.84 = 2 \cdot 3.14 \cdot r$
$18.84 = 6.28 \cdot r$
$18.84 \div 6.28 = r$
$3 = r$
$r = 3$ cm

1.34 $C = 2 \cdot \pi \cdot r$
$= 2 \cdot 3.14 \cdot 3.5$
$= 6.28 \cdot 3.5$
$= 21.98$ ft.

1.35 a. $A = \pi \cdot r^2$
$= 3.14 \cdot 50^2$
$= 3.14 \cdot 2{,}500$
$= 7{,}850 \text{ mm}^2$

b. $A = \pi \cdot r^2$
$= 3.14 \cdot 40^2$
$= 3.14 \cdot 1{,}600$
$= 5{,}024 \text{ mm}^2$

Area of ring = 7,850 – 5,024 = 2,826 mm^2

1.36 $C = \pi \cdot d$
$= 3.14 \cdot 12$
$= 37.68$ ft.

1.37 The following answers will be true for a regulation basketball.
circumference = $29\frac{3}{4}$ inches
$C = 2 \cdot \pi \cdot r$
$29.75 = 2 \cdot 3.14 \cdot r$
$29.75 = 6.28 \cdot r$
$29.75 \div 6.28 = r$
$4.737 = r$
$r = 4.737$ inches
Answers will vary for nonregulation basketballs.

1.38 Area of red: $A = \pi \cdot r^2$
$= 3.14 \cdot 1^2$
$= 3.14 \text{ ft.}^2$

Area of white:
red + white: $A = \pi \cdot r^2$
$= 3.14 \cdot 4^2$
$= 3.14 \cdot 16$
$= 50.24 \text{ ft.}^2$
Subtract red: $= 50.24 - 3.14$
$= 47.10$
$= 47.1 \text{ ft.}^2$

Area of blue:
r + w + b: $A = \pi \cdot r^2$
$= 3.14 \cdot 6^2$
$= 3.14 \cdot 36$
$= 113.04 \text{ ft.}^2$
Subtract $r + w$: $= 113.04 - 50.24$
$= 62.80$
$= 62.8 \text{ ft.}^2$

1.39 $P = 9.4 + 6.8 + 3.6$
$= 19.8$ meters

1.40 a. $A = \frac{1}{2} \cdot b \cdot h$
$= \frac{1}{2} \cdot 27 \cdot 50$
$= 675 \text{ ft.}^2$

b. radius of circle = $\frac{1}{2} \cdot 58 = 29$ ft.
area of circle $= \pi \cdot r^2$
$= 3.14 \cdot 29^2$
$= 3.14 \cdot 841$
$= 2{,}640.74 \text{ ft.}^2$

area of circle:	2,640.74
area of triangle:	– 675.00
	1,965.74 ft.^2

1.41 a. 3.8 cm $\doteq 1\frac{1}{2}$ in.

b. 5.8 cm $\doteq 2\frac{5}{16}$ in.

c. 7.5 cm $\doteq 2\frac{15}{16}$ in.

d. 2.9 cm $\doteq 1\frac{1}{8}$ in.

e. $A = \frac{1}{2} \cdot b \cdot h$
$= \frac{1}{2} \cdot 7.5 \cdot 2.9$
$= 10.875$ sq. cm

$A = \frac{1}{2} \cdot b \cdot h$
$= \frac{1}{2} \cdot 2\frac{15}{16} \cdot 1\frac{1}{8}$
$= \frac{1}{2} \cdot \frac{47}{16} \cdot \frac{9}{8}$
$= \frac{423}{256}$
$= 1.65$ or $1\frac{5}{8}$ sq. inches

f. $P = 3.8 + 5.8 + 7.5$
$= 17.1$ cm

$P = 1\frac{1}{2} + 2\frac{5}{16} + 2\frac{15}{16}$
$= 1\frac{8}{16} + 2\frac{5}{16} + 2\frac{15}{16}$
$= 5\frac{28}{16}$
$= 6\frac{3}{4}$ inches

1.42 $P = 27 + 50 + 58$
$= 135$ ft.

1.43 Perimeter is 135 ft.

How many packages with 8-foot coverage are needed?

$135 \div 8 = 16.875$

Round up because cannot buy partial packages

17 packages · \$2.89 = \$49.13

1.44 $a = 6$
$b = 8$
$c =$ unknown

Unknown side is hypotenuse.

$6^2 + 8^2 = c^2$
$36 + 64 = c^2$
$\sqrt{100} = \sqrt{c^2}$
10 in. = c (perfect square solution)

1.45 $a = 9$
$b =$ unknown
$c = 15$
$9^2 + b^2 = 15^2$
$81 + b^2 = 225$
$b^2 = 225 - 81$
$b^2 = 144$
$\sqrt{b^2} = \sqrt{144}$
$b = 12$ meters (perfect square solution)

1.46 $a = 18$
$b =$ unknown
$c = 20$
$18^2 + b^2 = 20^2$
$324 + b^2 = 400$
$b^2 = 400 - 324$
$b^2 = 76$
$\sqrt{b^2} = \sqrt{76}$
$b \doteq 8.71 = 8.7$ yds.

1.47 $a = 40$
$b = 40$
$c =$ unknown
$40^2 + 40^2 = c^2$
$1{,}600 + 1{,}600 = c^2$
$3{,}200 = c2$
$\sqrt{c^2} = \sqrt{3{,}200}$
$c \doteq 56.56 \doteq 56.6$ ft.

1.48 $a = 5$
$b = 9$
$c =$ unknown
$5^2 + 9^2 = c^2$
$25 + 81 = c^2$
$106 = c^2$
$\sqrt{c^2} = \sqrt{106}$
$c \doteq 10.29563 \doteq 10.3$ ft.

1.49 $a = 4.5$ (half of 9)
$b = 5$
$c =$ unknown
$4.5^2 + 5^2 = c^2$
$20.25 + 25 = c^2$
$45.25 = c^2$
$\sqrt{c^2} = \sqrt{45.25}$
$c \doteq 6.72 \doteq 6.7$ ft.

1.50 $a = 4$
$b = 10$
$c =$ unknown
$4^2 + 10^2 = c^2$
$16 + 100 = c^2$
$116 = c^2$
$\sqrt{c^2} = \sqrt{116}$
$c \doteq 10.77 \doteq 10.8$ ft.

1.51 Examples:
To form a triangle so the gate will be stronger. *or*

To be more rigid. *or*

To keep the gate from sagging.

1.52
a. trapezoid: two sides parallel

b. $A = \frac{1}{2} \cdot h \cdot (b_1 + b_2)$

c. $A = \frac{1}{2} \cdot 2 \cdot (8 + 6)$

$= \frac{1}{2} \cdot 2 \cdot 14$

$= 1 \cdot 14$

$= 14$ ft.^{2} or sq ft.

d. $P = 2.5 + 6 + 2.5 + 8$

$= 19$ ft.

1.53
a. rhombus: all four sides equal and parallel
b. $A = b \cdot h$
c. $A = 40 \cdot 35$
$= 1{,}400$ cm^2
d. $P = 40 + 40 + 40 + 40$ or $4 \cdot 40$
$= 160$ cm

1.54
a. parallelogram: two pairs of parallel sides
b. $A = b \cdot h$
c. $A = 24 \cdot 10$
$= 240 \text{ in.}^2$
d. $P = 16 + 24 + 16 + 24$
$= 80 \text{ in.}$

1.55
a. trapezoid: two sides parallel
b. $A = \frac{1}{2} \cdot h \cdot (b_1 + b_2)$
c. $A = \frac{1}{2} \cdot 0.9 \cdot (3.7 + 1.9)$
$= \frac{1}{2} \cdot 0.9 \cdot 5.6$
$= 0.45 \cdot 5.6$
$= 2.52 \text{ cm}^2$
d. $P = 1.1 + 1.9 + 1.3 + 3.7$
$= 8.0 \text{ cm}$

1.56
a. $P = 4 \cdot s$
$= 4 \cdot 27$
$= 108 \text{ ft.}$
b. $A = s^2$
$= 27^2$
$= 729 \text{ ft.}^2$
c. $729 \text{ ft.}^2 \cdot \frac{1 \text{ yd.}^2}{9 \text{ ft.}^2} =$

$729 \cancel{\text{ft.}^2} \cdot \frac{1 \text{ yd.}^2}{9 \cancel{\text{ft.}^2}} =$

$\frac{729 \text{ yd.}^2}{9} = 81 \text{ yd.}^2$

1.57
a. $A = l \cdot w$
$= 300 \cdot 165$
$= 49{,}500 \text{ ft.}2$
b. $P = (2 \cdot l) + (2 \cdot w)$
$= (2 \cdot 300) + (2 \cdot 165)$
$= 600 + 330$
$= 930 \text{ ft.}$

1.58
$A = l \cdot w$
$= 360 \cdot 225$
$= 81{,}000 \text{ ft.}^2$

1.59
8 ft.
a
6 ft.
b
$2\frac{1}{2}$ ft.
8 ft. 4 ft.
12 ft.

$a = 6 - 2\frac{1}{2} = 3\frac{1}{2}$ ft.
$b = 12 - 8 \quad = 4$ ft.

Rectangle 6 ft. by 8 ft:
$A = l \cdot w$
$= 8 \cdot 6$
$= 48 \text{ ft.}^2$

Rectangle $2\frac{1}{2}$ ft. by 4 ft:
$A = l \cdot w$
$= 4 \cdot 2.5$
$= 10 \text{ ft.}^2$

Total area $= 48 \text{ ft.}^2 + 10 \text{ ft.}^2$
$= 58 \text{ ft.}^2$

1.60
a. approximately 8″ x $10\frac{11}{16}$″
b. $8 \times \frac{171}{16} = \frac{1{,}368}{16} = 85.5 \text{ in.}^2$

1.61
Rectangle:
$A = l \cdot w$
$= 13 \cdot 6$
$= 78 \text{ m}^2$

Two half circles (= one circle):
$r = \frac{1}{2} \cdot 6 = 3$ meters
$A = \pi \cdot r^2$
$= 3.14 \cdot 3^2$
$= 3.14 \cdot 9$
$= 28.26 \text{ m}^2$

Total area:
78.00 m^2
$+28.26 \text{ m}^2$
106.26 m^2

1.62
$A = s^2$
$= 11^2$
$= 121 \text{ m}^2$

1.63 a. $15 \div 3 = 5$ cookies on length

$10 \div 2 = 5$ cookies on width

$5 \cdot 5 = 25$ cookies altogether

b. $1 \cdot \frac{1}{2} = \frac{1}{2}$ cup raisins

$\frac{1}{4} \cdot \frac{1}{2} = \frac{1}{8}$ cup water

$\frac{1}{4} \cdot \frac{1}{2} = \frac{1}{8}$ tsp. baking soda

$\frac{3}{4} \cdot \frac{1}{2} = \frac{3}{8}$ cup flour

$1 \cdot \frac{1}{2} = \frac{1}{2}$ cup sugar

$1 \cdot \frac{1}{2} = \frac{1}{2}$ tsp. baking powder

$\frac{1}{4} \cdot \frac{1}{2} = \frac{1}{8}$ tsp. salt

$3 \cdot \frac{1}{2} = \frac{3}{2} = 1\frac{1}{2}$ eggs

c. $1 \cdot 2 = 2$ cups raisins

$\frac{1}{4} \cdot 2 = \frac{1}{2}$ cup water

$\frac{1}{4} \cdot 2 = \frac{1}{2}$ tsp. baking soda

$\frac{3}{4} \cdot 2 = \frac{3}{2} = 1\frac{1}{2}$ cups flour

$1 \cdot 2 = 2$ cups sugar

$1 \cdot 2 = 2$ tsp. baking powder

$\frac{1}{4} \cdot 2 = \frac{1}{2}$ tsp. salt

$3 \cdot 2 = 6$ eggs

1.64 a. $A = l \cdot w$
$= 16 \cdot 6$
$= 96$ ft.^{2}

$96 \text{ ft.}^2 \cdot \frac{1 \text{ yd.}^2}{9 \text{ ft.}^2} = 96 \cancel{\text{ft.}^2} \cdot \frac{1 \text{ yd.}^2}{9 \cancel{\text{ft.}^2}} =$

$\frac{96 \text{ yd.}^2}{9} = 10\frac{2}{3}$ yd.^{2}

Round to 11 yd.^{2}
$11 \text{ yd.}^2 \cdot \$5.50 = \$60.50$

1.64 (Cont.)
b. $A = l \cdot w$
$= 19\frac{1}{3} \cdot 12\frac{1}{3}$

$= \frac{58}{3} \cdot \frac{37}{3}$

$= \frac{2{,}146}{9}$

$= 238\frac{4}{9}$

$\doteq 238.44$ sq. ft.
Alternate solution with decimals:
$A = 19.3 \cdot 12.3$
$= 237.39$ sq. ft.

1.65 a. Large area $= 100 \cdot 30$
$= 3{,}000$ ft.^{2}

Small area $= 32 \cdot 30\frac{5}{12}$

$= \frac{\overset{8}{\cancel{32}}}{1} \cdot \frac{365}{\underset{3}{\cancel{12}}}$

$= \frac{2920}{3}$

$= 973\frac{1}{3}$

$\doteq 973.33$ ft.^{2}

b. $A = 100 \cdot 60$
$= 6{,}000$ ft.

c. Round 973.33 ft.^{2} to 973 ft.^{2}

$3{,}000 \text{ ft.}^2 + 973 \text{ ft.}^2 =$

$3{,}973$ ft.^{2}

$3{,}973 \text{ ft.}^2 \cdot \frac{1 \text{ yd.}^2}{9 \text{ ft.}^2} =$

$3{,}973 \cancel{\text{ft.}^2} \cdot \frac{1 \text{ yd.}^2}{9 \cancel{\text{ft.}^2}} =$

$\frac{3{,}973 \text{ yd.}^2}{9} =$

$441\frac{4}{9}$ yd.^{2}

441.4 yd.^{2}

$441.4 \cdot \$20 = \$8{,}828$

II. SECTION TWO

2.1 $20 \text{ yd.} \cdot \frac{3 \text{ ft.}}{1 \text{ yd.}} = 60 \text{ ft.}$

$10 \text{ yd.} \cdot \frac{3 \text{ ft.}}{1 \text{ yd.}} = 30 \text{ ft.}$

$V = l \cdot w \cdot h$
$= 60 \cdot 30 \cdot 5$
$= 1{,}800 \cdot 5$
$= 9{,}000 \text{ ft.}^3$

2.2 Area $A = 60 \text{ ft.} \cdot 30 \text{ ft.}$
$= 1{,}800 \text{ ft.}^2$
$B = 60 \text{ ft.} \cdot 5 \text{ ft.}$
$= 300 \text{ ft.}^2$
$C = 30 \text{ ft.} \cdot 5 \text{ ft.}$
$= 150 \text{ ft.}^2$
$D = 60 \text{ ft.} \cdot 5 \text{ ft.}$
$= 300 \text{ ft.}^2$
$E = 30 \text{ ft.} \cdot 5 \text{ ft.}$
$= 150 \text{ ft.}^2$
Total area $= 2{,}700 \text{ ft.}^2$

2.3 $2{,}700 \div 300 = 9$ gallons
at \$11 each $= 9 \cdot 11 = \$99$

2.4 $V = l \cdot w \cdot h$
$= 10 \cdot 8 \cdot 3.5$
$= 80 \cdot 3.5$
$= 280 \text{ ft.}^3$

2.5 $1 \text{ yd.}^3 = 3 \text{ ft.} \cdot 3 \text{ ft.} \cdot 3 \text{ ft.}$
$= 27 \text{ ft.}^3$

$V = 280 \text{ ft.}^3 \cdot \frac{1 \text{ yd.}^3}{27 \text{ ft.}^3}$

$= 280 \cancel{\text{ft.}^3} \cdot \frac{1 \text{ yd.}^3}{27 \cancel{\text{ft.}^3}}$

$= \frac{280 \text{ yd.}^3}{27}$

$= 10\frac{10}{27} \text{ yd.}^3$

$= 10.37 \text{ yd.}^3$

2.6 long center piece:
$V = (3 + 2 + 6) \cdot 2 \cdot 1\frac{1}{2}$
$= 11 \cdot {}^1\cancel{2} \cdot \frac{3}{\cancel{2}_1}$
$= 11 \cdot 3$
$= 33 \text{ ft.}^3$

short side piece:
$V = 3 \cdot 2 \cdot 1\frac{1}{2}$
$= 3 \cdot {}^1\cancel{2} \cdot \frac{3}{\cancel{2}_1}$
$= 3 \cdot 3$
$= 9 \text{ ft.}^3$

Total volume $= 33 \text{ ft.}^3 + 9 \text{ ft.}^3 + 9 \text{ ft.}^3$
$= 51 \text{ ft.}^3$

2.7 $V = l \cdot w \cdot h$
$= 132 \cdot 70 \cdot 40$
$= 9{,}240 \cdot 40$
$= 369{,}600 \text{ cm}^3$

$\frac{3}{{}_1\cancel{4}} \cdot \frac{\cancel{369{,}600}^{92{,}400}}{1} = 277{,}200$ cubic cm or cm^3

2.8 $V = l \cdot w \cdot h$
$= 60 \cdot 35 \cdot 9$
$= 2{,}100 \cdot 9$
$= 18{,}900 \text{ ft.}^3$

2.9 $18{,}900 \cancel{\text{ft.}^3} \cdot \frac{1 \text{ yd.}^3}{27 \cancel{\text{ft.}^3}} =$

$\frac{18{,}900 \text{ yd.}^3}{27} = 700 \text{ yd.}^3$

2.10 teacher check

2.11

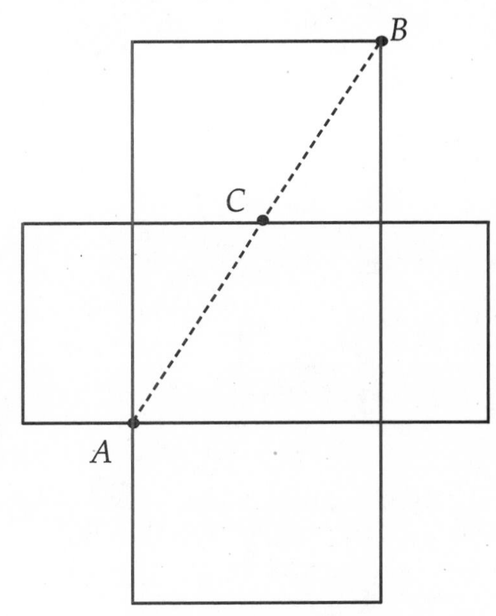

From A to B through C

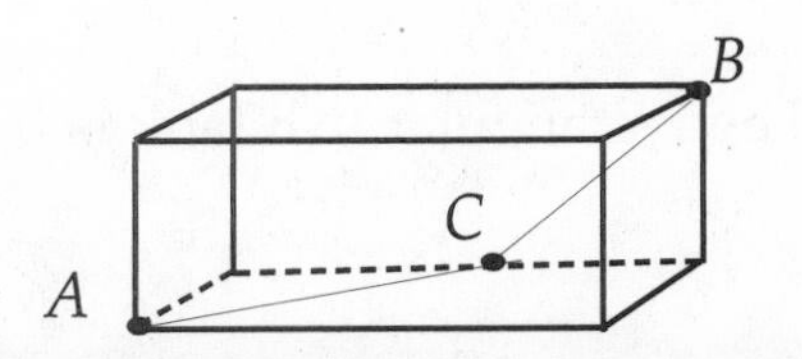

2.12 $V = \frac{1}{2} \cdot b_{\Delta} \cdot h_{\Delta} \cdot h_p$
$= \frac{1}{2} \cdot 3 \cdot 4 \cdot 8$
$= \frac{1}{2} \cdot 12 \cdot 8$
$= 6 \cdot 8$
$= 48 \text{ m}^3$

2.13 3 faces: $5 \cdot 8 = 40 \text{ m}^2$
$4 \cdot 8 = 32 \text{ m}^2$
$3 \cdot 8 = 24 \text{ m}^2$
2 triangle bases:
$2 \cdot \frac{1}{2} \cdot 3 \cdot 4 = 12 \text{ m}^2$

Total Surface Area =
$40 \text{ m}^2 + 32 \text{ m}^2 + 24 \text{ m}^2 + 12 \text{ m}^2 =$
108 m^2

2.14 $V = \frac{1}{2} \cdot b_{\Delta} \cdot h_{\Delta} \cdot h_p$
$= \frac{1}{2} \cdot 40 \cdot 22.4 \cdot 60$
$= 20 \cdot 22.4 \cdot 60$
$= 448 \cdot 60$
$= 26{,}880 \text{ ft.}^3$

2.15 Area of two faces $= 2 \cdot l \cdot w$
$= 2 \cdot 60 \cdot 30$
$= 120 \cdot 30$
$= 3{,}600 \text{ ft.}^2$

2.16 8 ft. = 96 in. $(8 \cdot 12 = 96)$
Prism:
$V = \frac{1}{2} \cdot b_{\Delta} \cdot h_{\Delta} \cdot h_p$
$= \frac{1}{2} \cdot 12 \cdot 6 \cdot 96$
$= 6 \cdot 6 \cdot 96$
$= 36 \cdot 96$
$= 3{,}456 \text{ in.}^3$
Rectangular solid:
3 ft. = 36 in. $(3 \cdot 12 = 36)$
2 ft. = 24 in. $(2 \cdot 12 = 24)$
$V = l \cdot w \cdot h$
$= 36 \cdot 9 \cdot 24$
$= 324 \cdot 24$
$= 7{,}776 \text{ in.}^3$
The rectangular fish tank is larger.

2.17 Prism has 12" by 8 ft. = 1 ft. by 8 ft. rectangular front viewing face.
$A = 1 \cdot 8 = 8 \text{ ft.}^2$
Rectangular solid has 2 ft. by 3 ft. rectangular front viewing face.
$A = 2 \cdot 3 = 6 \text{ ft.}^2$
The prism has more viewing area.

2.18 $V = \frac{1}{3} \cdot B \cdot h$
$\frac{1}{3} \cdot s^2 \cdot h$
$\frac{1}{3} \cdot 25^2 \cdot 15$
$\frac{1}{3} \cdot 625 \cdot 15$
$5 \cdot 625$
$3{,}125 \text{ cm}^3$

2.19 total area (TA) = $L.A.$ + area of base
$L.A. = 4 \cdot \frac{1}{2} \cdot b \cdot h$
$= 4 \cdot \frac{1}{2} \cdot 3 \cdot 2$
$= 2 \cdot 3 \cdot 2$
$= 6 \cdot 2$
$= 12 \text{ yd.}^2$
area of base $= s^2$
$= 3^2$
$= 9 \text{ yd.}^2$
$TA = 12 + 9 = 21 \text{ yd.}^2$

III. SECTION THREE

Note: When solving problems involving a fraction and π, do not clear the denominator by dividing it into π as an intermediate step toward the solution. If you were to do so, the resulting answer could vary from the one given in the answer key.

3.1 teacher check

3.2 Example:
A value for π was probably discovered by experiments such as Problem 3.1. Mathematicians probably measured the parts of circles and found that the circumference divided by the diameter was always very close to 3.14.

3.3 a. $r = \frac{1}{2} \cdot d = \frac{1}{2} \cdot 7 = 3.5$
$TA = 2 \cdot \pi \cdot r \cdot (r + h)$
$= 2 \cdot 3.14 \cdot 3.5 \cdot (3.5 + 7.5)$
$= 2 \cdot 3.14 \cdot 3.5 \cdot 11$
$= 6.28 \cdot 3.5 \cdot 11$
$= 21.98 \cdot 11$
$= 241.78$ in.^{2}
b. $V = \pi \cdot r^2 \cdot h$
$= 3.14 \cdot 3.5 \cdot 7.5$
$= 3.14 \cdot 12.25 \cdot 7.5$
$= 38.465 \cdot 7.5$
$= 288.4875$ in.^{3}

3.4 $V = \pi \cdot r^2 \cdot h$
$= 3.14 \cdot 10^2 \cdot 12$
$= 3.14 \cdot 100 \cdot 12$
$= 3.14 \cdot 12$
$= 3{,}768$ m^3 or cubic meters

3.5 $r = \frac{1}{2} \cdot d = \frac{1}{2} \cdot 3 = \frac{3}{2} = 1.5$ in.
$V = \pi \cdot r^2 \cdot h$
$= 3.14 \cdot 1.5^2 \cdot 3$
$= 3.14 \cdot 2.25 \cdot 3$
$= 7.065 \cdot 3$
$= 21.195$ in.^{3}

3.6 area of rectangle:
$A = 2 \cdot \pi \cdot r \cdot h$
$= 2 \cdot 3.14 \cdot 1.5 \cdot 3$
$= 6.28 \cdot 1.5 \cdot 3$
$= 9.42 \cdot 3$
$A = 28.26$ in^2
area of bottom:
$A = \pi \cdot r^2$
$= 3.14 \cdot 1.5$
$= 3.14 \cdot 2.25$
$= 7.065$ in.^{2}
total area: 28.26 in.^{2}
+ 7.065 in.^{2}
35.325 in.^{2}

3.7 a. $LA = \pi \cdot r \cdot l$
$= 3.14 \cdot 5 \cdot 8$
$= 15.7 \cdot 8$
$= 125.6$ in.^{2}
b. $B = \pi \cdot r^2$
$= 3.14 \cdot 5^2$
$= 3.14 \cdot 25$
$= 78.5$ in.^{2}
c. $TA = LA + B$
$= 125.6 + 78.5$
$= 204.1$ in.^{2}

3.8 $V = \frac{1}{3} \cdot \pi \cdot r^2 \cdot h$
$= \frac{1}{3} \cdot 3.14 \cdot 4^2 \cdot 6$
$= \frac{1}{3} \cdot 3.14 \cdot 16 \cdot 6$
$= 3.14 \cdot 16 \cdot 2$
$= 50.24 \cdot 2$
$= 100.48$ ft.^{3}

3.9 $V = \frac{1}{3} \cdot \pi \cdot r^2 \cdot h$
$= \frac{1}{3} \cdot 3.14 \cdot 8^2 \cdot 9$
$= \frac{1}{3} \cdot 3.14 \cdot 64 \cdot 9$
$= 3.14 \cdot 64 \cdot 3$
$= 200.96 \cdot 3$
$= 602.88$ in.^{3}

3.10 $S = 4 \cdot \pi \cdot r^2$
$= 4 \cdot 3.14 \cdot 10^2$
$= 4 \cdot 3.14 \cdot 100$
$= 12.56 \cdot 100$
$= 1{,}256$ ft.^{2}

3.11 $V = \frac{4}{3} \cdot \pi \cdot r^3$
$= \frac{4}{3} \cdot 3.14 \cdot 10^3$
$= \frac{4}{3} \cdot 3.14 \cdot 1{,}000$
$= \frac{4}{3} \cdot 3{,}140$
$= \frac{12{,}560}{3}$
$= 4{,}186.\overline{6}$ cu. ft. or ft.^{3}

3.12 a. Area = 1,256 ft.^{2}
1,256 ÷ 250 = 5.024 gal.
Since the answer is greater than 5, to do a good job, you will need 6 gallons.

3.12 cont. $6 \cdot \$13.45 = \$80.70$

b. $4\% = 0.04$

$\$80.70 \cdot 0.04 = 3.228 = \$3.23$

$\$80.70 + \$3.23 = \$83.93$

3.13 $r = \frac{1}{2} \cdot d = \frac{1}{2} \cdot 8 = 4$ ft.

volume of hemisphere =

$\frac{1}{2}$ volume of sphere =

$\frac{1}{2} \cdot \frac{4}{3}\pi \cdot r^3 =$

$\frac{1}{2} \cdot \frac{4}{3} \cdot 3.14 \cdot 4^3 =$

$\frac{1}{2} \cdot \frac{4}{3} \cdot 3.14 \cdot 64 =$

$\frac{2}{3} \cdot 3.14 \cdot 64 =$

$\frac{2}{3} \cdot 200.96 =$

$\frac{401.92}{3} =$

$133.97\overline{3}$ ft.^{3}

3.14 Example:
Measure circumference with tape measure; then use the circumference formula to find radius.

$C = 2 \cdot \pi \cdot r$

$C = 2 \cdot 3.14 \cdot r$

$C = 6.28 \cdot r$

$\frac{C}{6.28} = \frac{6.28 \cdot r}{6.28}$

$\frac{C}{6.28} = r$

Now the volume formula can be used.

3.15 $A = l \cdot w$

$= 400 \cdot 300$

$= 120{,}000$ m^2

3.16 $A = b \cdot h$

$= 49 \cdot 38$

$= 1{,}862$ sq. ft. or ft.^{2}

3.17 $A = \frac{1}{2} \cdot h \cdot (b_1 + b_2)$

$= \frac{1}{2} \cdot 64 \cdot (209 + 145)$

$= \frac{1}{2} \cdot 64 \cdot 354$

$= 32 \cdot 354$

$= 11{,}328$ cm^2

3.18 $a = 490$

$b = 380$

c = unknown (hypotenuse)

$490^2 + 380^2 = c^2$

$240{,}100 + 144{,}400 = c^2$

$384{,}500 = c^2$

$\sqrt{c^2} = \sqrt{384{,}500}$

$c \doteq 620.0806$

$c \doteq 620.1$

3.19 $V = \frac{4}{3} \cdot \pi \cdot r^3$

$= \frac{4}{3} \cdot 3.14 \cdot 21^3$

$= \frac{4}{3} \cdot 3.14 \cdot 9{,}261$

$= \frac{4}{3} \cdot 29{,}079.54$

$= \frac{116{,}318.16}{3}$

$= 38{,}772.72$ cu. in.

3.20 $V = \frac{1}{3} \cdot \pi \cdot r^2 \cdot h$

$= \frac{1}{3} \cdot 3.14 \cdot 13^2 \cdot 22$

$= \frac{1}{3} \cdot 3.14 \cdot 169 \cdot 22$

$= \frac{1}{3} \cdot 530.66 \cdot 22$

$= \frac{1}{3} \cdot 11{,}674.52$

$= 3{,}891.5067$ cu. in.

3.21 $r = \frac{1}{2} \cdot d = \frac{1}{2} \cdot 46 = 23$ ft.

$TA = 2 \cdot \pi \cdot r \cdot (r + h)$

$= 2 \cdot 3.14 \cdot 23 \cdot (23 + 142)$

$= 2 \cdot 3.14 \cdot 23 \cdot 165$

$= 6.28 \cdot 23 \cdot 165$

$= 144.44 \cdot 165$

$= 23{,}832.6$ ft.^{2}

3.22 a. Lateral area is area of the four faces.

$L.A. = 4 \cdot \frac{1}{2} \cdot b \cdot l$

$= 4 \cdot \frac{1}{2} \cdot 29 \cdot 18$

$= 2 \cdot 29 \cdot 18$

$= 58 \cdot 18$

$= 1{,}044$ ft.^{2}

3.22 cont. b. $A = s^2$
$= 29^2$
$= 841 \text{ ft.}^2$

c. total area = *L.A.* + area of base
$= 1{,}044 + 841$
$= 1{,}885 \text{ ft.}^2$

3.23 $V = \frac{1}{2} \cdot b_{\Delta} \cdot h_{\Delta} \cdot h_p$
$= \frac{1}{2} \cdot 20 \cdot 8 \cdot 39$
$= 10 \cdot 8 \cdot 39$
$= 80 \cdot 39$
$= 3{,}120 \text{ ft.}^3$

I. SECTION ONE

1.1 $5 + 4 = 9$ and $4 + 5 = 9$, so $5 + 4 = 4 + 5$ is true.

1.2 $10 + 1 = 11$ and $1 + 10 = 11$, so $10 + 1 = 1 + 10$ is true.

1.3 $18 + 16 = 34$ and $16 + 18 = 34$, so $18 + 16 = 16 + 18$ is true.

1.4 $47 + 19 = 66$ and $19 + 47 = 66$, so $47 + 19 = 19 + 47$ is true.

1.5 $146 + 218 = 364$ and $218 + 146 = 364$, so $146 + 218 = 218 + 146$ is true.

1.6 $814 + 961 = 1{,}775$ and $961 + 814 = 1{,}775$, so $814 + 961 = 961 + 814$ is true.

1.7 $2{,}847 + 1{,}896 = 4{,}743$ and $1{,}896 + 2{,}847 = 4{,}743$ so $2{,}847 + 1{,}896 = 1{,}896 + 2{,}847$ is true.

1.8 $37{,}416 + 51{,}914 = 89{,}330$ and $51{,}914 + 37{,}416 = 89{,}330$, so $37{,}416 + 51{,}914 = 51{,}914 + 37{,}416$ is true.

1.9 9

1.10 4

1.11 10

1.12 y

1.13 a

1.14 p

1.15 97

1.16 ab

1.17 $\odot$

1.18 Tuesday

1.19 No, because $14 - 8 = 6$ and $8 - 14 = -6$; $6 \neq -6$

1.20 No, because $16 - 14 = 2$ and $14 - 16 = -2$; $2 \neq -2$

1.21 Yes, because $8 + 6 = 14$ and $6 + 8 = 14$; $14 = 14$

1.22 Yes, because $100 + 141 = 241$ and $141 + 100 = 241$; $241 = 241$

1.23 no

1.24 no

1.25 Yes; they can be done in either order.

1.26 Yes; they can be done in either order

1.27 Yes, because $(14 - 2) + (8 - 6) = 12 + 2 = 14$ and $(8 - 6) + (14 - 2) = 2 + 12 = 14$; $14 = 14$

1.28 Yes, because $xyz + abc = abc + xyz$

1.29 $5 \cdot 8 = 40$ and $8 \cdot 5 = 40$ so $5 \cdot 8 = 8 \cdot 5$ is true.

1.30 $9 \cdot 7 = 63$ and $7 \cdot 9 = 63$, so $9 \cdot 7 = 7 \cdot 9$ is true.

1.31 $10 \cdot 6 = 60$ and $6 \cdot 10 = 60$, so $10 \cdot 6 = 6 \cdot 10$ is true.

1.32 $14 \cdot 8 = 112$ and $8 \cdot 14 = 112$, so $8 \cdot 14 = 14 \cdot 8$ is true.

1.33 $51 \cdot 10 = 510$ and $10 \cdot 51 = 510$, so $51 \cdot 10 = 10 \cdot 51$ is true.

1.34 $47 \cdot 26 = 1{,}222$ and $26 \cdot 47 = 1{,}222$, so $47 \cdot 26 = 26 \cdot 47$ is true.

1.35 $106 \cdot 16 = 1{,}696$ and $16 \cdot 106 = 1{,}696$ so $106 \cdot 16 = 16 \cdot 106$ is true.

1.36 $314 \cdot 267 = 83{,}838$ and $267 \cdot 314 = 83{,}838$, so $314 \cdot 267 = 267 \cdot 314$ is true

1.37 9

1.38 14

1.39 b

1.40 f

1.41 6

1.42 16

1.43 15

1.44 $(a - c)$

1.45 15

1.46 xyz

1.47 $(6 + 8) + 10 = 14 + 10 = 24$ and $6 + (8 + 10) = 6 + 18 = 24$, so $(6 + 8) + 10 = 6 + (8 + 10)$ is true.

1.48 $(14 + 6) + 9 = 20 + 9 = 29$ and $14 + (6 + 9) = 14 + 15 = 29$, so $(14 + 6) + 9 = 14 + (6 + 9)$ is true.

1.49 $116 + (1 + 59) = 116 + 60 = 176$ and $(116 + 1) + 59 = 117 + 59 = 176$, so $116 + (1 + 59) = (116 + 1) + 59$ is true.

1.50 $(191 + 216) + 4 = 407 + 4 = 411$ and $191 + (216 + 4) = 191 + 220 = 411$, so $(191 + 216) + 4 = 191 + (216 + 4)$ is true.

1.51 $(2a + 3a) + 4a = 5a + 4a = 9a$ and $2a + (3a + 4a) = 2a + 7a = 9a$, so $(2a + 3a) + 4a = 2a + (3a + 4a)$ is true.

1.52 (5 eggs + 6 eggs) + 9 eggs = 11 eggs + 9 eggs = 20 eggs and 5 eggs + (6 eggs + 9 eggs) = 5 eggs + 15 eggs = 20 eggs, so (5 eggs + 6 eggs) + 9 eggs = 5 eggs + (6 eggs + 9 eggs) is true.

1.53 $5 + (1 + 8)$

1.54 $16 + (9 + 14)$

1.55 $19 + (4 + 96)$

1.56 $(100 + 8) + 27$

1.57 $e + (f + g)$

1.58 $x + (y + z)$

1.59 9

1.60 19

1.61 a, b

1.62 e, f, g

1.63 o, q, p

1.64 A, E, G

1.65 $(5 \cdot 2) \cdot 3 = 10 \cdot 3 = 30$ and $5 \cdot (2 \cdot 3) = 5 \cdot 6 = 30$, so $(5 \cdot 2) \cdot 3 = 5 \cdot (2 \cdot 3)$ is true.

1.66 $8 \cdot (3 \cdot 5) = 8 \cdot 15 = 120$ and $(8 \cdot 3) \cdot 5 = 24 \cdot 5 = 120$, so $8 \cdot (3 \cdot 5) = (8 \cdot 3) \cdot 5$ is true

1.67 $(10 \cdot 4)\ 7 = 40 \cdot 7 = 280$ and $10 \cdot (4 \cdot 7) = 10 \cdot 28 = 280$, so $(10 \cdot 4) \cdot 7 = 10 \cdot (4 \cdot 7)$ is true.

1.68 $12 \cdot (20 \cdot 3) = 12 \cdot 60 = 720$ and $(12 \cdot 20)\ 3 = 240 \cdot 3 = 720$, so $12 \cdot (20 \cdot 3) = (12 \cdot 20) \cdot 3$ is true.

1.69 $(100 \cdot 10) \cdot 1 = 1{,}000 \cdot 1 = 1{,}000$ and $100 \cdot (10 \cdot 1) = 100 \cdot 10 = 1{,}000$, so $(100 \cdot 10) \cdot 1 = 100 \cdot (10 \cdot 1)$ is true.

1.70 $(\frac{1}{2} \cdot \frac{1}{4}) \cdot \frac{1}{3} = \frac{1}{8} \cdot \frac{1}{3} = \frac{1}{24}$ and $\frac{1}{2} \cdot (\frac{1}{4} \cdot \frac{1}{3}) = \frac{1}{2} \cdot \frac{1}{12} = \frac{1}{24}$, so $(\frac{1}{2} \cdot \frac{1}{4})\ \frac{1}{3} = \frac{1}{2} \cdot (\frac{1}{4} \cdot \frac{1}{3})$ is true.

1.71 $5 \cdot (6 \cdot 7)$

1.72 $(8 \cdot 2) \cdot 9$

1.73 $(14 \cdot 26) \cdot 71$

1.74 $27 \cdot (161 \cdot 87)$

1.75 $(a \cdot b) \cdot c$

1.76 $(x \cdot y) \cdot z$

1.77 3

1.78 7

1.79 7, 9

1.80 63, 101, 41

1.81 b, a, c

1.82 E, F, G

1.83 $A = s^2$
$A = 7^2$
$A = 49$

1.84 $b = \frac{A}{h}$
$b = \frac{27}{3}$
$b = 9$

1.85 $i = prt$
$i = 100 \cdot 0.05 \cdot 3$
$i = 15$

1.86 $A = \frac{1}{2}bh$
$A = \frac{1}{2} \cdot 6 \cdot 14$
$A = 42$

1.87 $C = \frac{5}{9} \cdot 90 - 32$
$C = 5 \cdot 10 - 32$
$C = 50 - 32$
$C = 18°$

1.88 $F = \frac{9}{5}C + 32$
$F = \frac{9}{5} \cdot 100 + 32$
$F = 180 + 32$
$F = 212°$

1.89 $B = 2S + 2D$
$B = 2 \cdot 16 + 2 \cdot 10$
$B = 32 + 20$
$B = 52$

1.90 $A = \pi r^2$
$A = 3.14 \cdot 2^2$
$A = 3.14 \cdot 4$
$A = 12.56$

1.91 $V = LWH$
$V = 9 \cdot 12 \cdot 3$
$V = 324$

1.92 $A = \frac{1}{2}h(a + b)$
$A = \frac{1}{2} \cdot 4(3 + 5)$
$A = 2(8)$
$A = 16$

1.93 $r = \frac{d}{t}$
$r = \frac{200}{4}$
$r = 50$

1.94 $s = kt^2$
$s = 32 \cdot 6^2$
$s = 32 \cdot 36$
$s = 1{,}152$

1.95 $A = b \cdot h$

1.96 $d = r \cdot t$

1.97 $V = B \cdot h$

1.98 $P = 4 \cdot s$

1.99 $W = r \cdot h$

1.100 $C = \pi \cdot d$

1.101 -2, -3, -4, -5

1.102 1, 2, 3, 4, 5

1.103 -3, -2, -1, 0, 1, 2

1.104 Example:
$-\frac{1}{2}, -\frac{1}{3}, -\frac{1}{4}$, and so on.

1.105 Example:
1.1, 1.2, 1.3, 1.4, and so on

1.106 zero (0)

1.107 5 spaces

-8 -7 -6 -5 -4 -3 -2 -1 0 1 2 3

The number is -3.

1.108 3 spaces

-5 -4 -3 -2 -1 0 1 2 3 4 5 6

The number is -3

1.109 7 spaces

-5 -4 -3 -2 -1 0 1 2 3 4 5 6

The number is - 3.

1.110 6 spaces

-5 -4 -3 -2 -1 0 1 2 3 4 5 6

The number is 4.

1.111 3 spaces 3 spaces

-12 -11 -10 -9 -8 -7 -6 -5 -4 -3 -2 -1

a. -10

b. -4

1.112 6 spaces 6 spaces

-2 -1 0 1 2 3 4 5 6 7 8 9 10

a. -2

b. 10

1.113 On the number line the distance between -4 and +5 is 9.

1.114 15

1.115 19

1.116 18

1.117 44

1.118 98

1.119 A double negative makes a positive number when multiplied together.

1.120 $2b + 3b - 7b + 9b = 5b - 7b + 9b = -2b + 9b = 7b$

1.121 $4t - 6t + 9t - 2t = -2t + 9t - 2t = 7t - 2t = 5t$

1.122 $16a - 14a + 7a + 6a = 2a + 7a + 6a = 9a + 6a = 15a$

1.123 $8xy + 2xy - 7xy - 9xy = 10xy - 7xy - 9xy = 3xy - 9xy = -6xy$

1.124 $14c + 6c - 20c + c = 20c - 20c + c = 0 + c = 0 + c = c$

1.125 $56pq + 14pq - 100pq = 70pq - 100pq = -30pq$

1.126 $8r + 9s - 4r - 7s = 8r - 4r + 9s - 7s = 4r + 2s$

1.127 $2a - 3a^2 + 4a + 4a^2 = 2a + 4a - 3a^2 + 4a^2 = 6a + a^2$

1.128 $x + y + z - 3x - 4y - 2z = x$

$3x + y - 4y + z - 2z = -2x - 3y - z$

1.129 $9 + 4b + c - 9b - 1 = 9 - 1 + 4b - 9b + c = 8 - 5b + c$

1.130 $4a - 3ab - 4ab + 3b - 6a = 4a - 6a - 3ab - 4ab + 3b = -2a - 7ab + 3b$

1.131 $7w - 8w - v + 8v - 2w = 7w - 8w - 2w - v + 8v = -3w + 7v$

1.132 $10q - 14p + 17q + 14p = 10q + 17q - 14p + 14p = 27q$

1.133 $0.16a - 0.41b - 0.32a + 0.52b = 0.16a - 0.32a - 0.41b + 0.52b = -0.16a + 0.11b$

1.134 $\frac{1}{2}x + \frac{1}{4}y - \frac{1}{4}x + \frac{1}{2}y = \frac{1}{2}x - \frac{1}{4}x + \frac{1}{4}y + \frac{1}{2}y = \frac{2}{4}x - \frac{1}{4}x + \frac{1}{4}y + \frac{2}{4}y = \frac{1}{4}x + \frac{3}{4}y$

1.135 $0.5m - 0.6n + \frac{1}{2}m - \frac{1}{4}n = 0.5m + \frac{1}{2}m - 0.6n - \frac{1}{4}n = \frac{5}{10}m + \frac{5}{10}m - \frac{60}{100}n - \frac{25}{100}n = m - \frac{85}{100}n$

or $m - 0.85n$

1.136 2, 3, 4, 5

1.137 17, 4, 8.6, 9

1.138 15, $\frac{1}{2}$, 450, 0.01

1.139 1, 1, 14, 97

1.140 14%, 8.4, 1, 0.5

1.141 $\frac{1}{5}$, 14, $8, 90

1.142 $3(a + 6) = 3 \cdot a + 3 \cdot 6 = 3a + 18$

1.143 $5(2b - 1) = 5 \cdot 2b - 5 \cdot 1 = 10b - 5$

1.144 $7(x + 10) = 7 \cdot x + 7 \cdot 10 = 7x + 70$

1.145 $8(3c - 2d) = 8 \cdot 3c - 8 \cdot 2d = 24c - 16d$

1.146 $p(4 + x) = p \cdot 4 + p \cdot x = 4p + px$

1.147 $y(a - 6) = y \cdot a - y \cdot 6 = ya - 6y$

1.148 $4(x - 2y + 6z) = 4 \cdot x - 4 \cdot 2y + 4 \cdot 6z =$ $4x - 8y + 24z$

1.149 $6(2p - 3q - 8r) = 6 \cdot 2p - 6 \cdot 3q - 6 \cdot 8r =$ $12p - 18q - 48r$

1.150 $w(3 + 5a - 6b) = w \cdot 3 + w \cdot 5a - w \cdot 6b =$ $3w + 5aw - 6bw$

1.151 $2x(x + x^2 - x^3) = 2x \cdot x + 2x \cdot x^2 -$ $2x \cdot x^3 = 2x^2 + 2x^3 - 2x^4$

1.152 $4x - 8 = 4 \cdot x - 4 \cdot 2 = 4(x - 2)$

1.153 $16a + 32 = 16a + 16 \cdot 2 = 16(a + 2)$

1.154 $3x + ax = 3 \cdot x + a \cdot x = x(3 + a)$

1.155 $12a + ab = 12 \cdot a + a \cdot b = a(12 + b)$

1.156 $3p - 6q + 12r = 3 \cdot p - 3 \cdot 2q +$ $3 \cdot 4r = 3(p - 2q + 4r)$

1.157 $5a + 10b - 25c = 5 \cdot a + 5 \cdot 2b - 5 \cdot 5c =$ $5(a + 2b - 5c)$

1.158 $16cx + 12cy - 8cz = 4c \cdot 4x + 4c \cdot 3y -$ $4c \cdot 2z = 4c(4x + 3y - 2z)$

1.159 $100a^2 - 50a = 50a \cdot 2a - 50a \cdot 1 = 50a$ $(2a - 1)$

1.160 $t^2 - ts + 3t = t \cdot t - t \cdot s + 3 \cdot t =$ $t(t - s + 3)$

1.161 $2g^3 + 4g^2 - 6g = 2g \cdot g^2 + 2g \cdot 2g -$ $2g \cdot 3 = 2g(g^2 + 2g - 3)$

1.162 $4(a - b) = 4 \cdot a - 4 \cdot b = 4a - 4b$ The missing terms are a, 4, $4a$, and $4b$.

1.163 $5(s + t) = 5 \cdot s + 5 \cdot t = 5s + 5t$ The missing terms are 5, t, $5s$, and $5t$.

1.164 $2(x + y + z) = 2 \cdot x + 2 \cdot y + 2 \cdot z = 2x +$ $2y + 2z$
The missing terms are $2y$ and $2z$.

1.165 $6c - 6d = 6 \cdot c - 6 \cdot d = 6(c - d)$ The missing terms are c and d.

1.166 $t(y - 5) = t \cdot y - t \cdot 5 = ty - 5t$
The missing term is ty.

1.167 $7m = 7 \cdot m$
$7(m + 7) = 7 \cdot m + 7 \cdot 7 = 7m + 49$
The missing terms are 7 and 49.

1.168 $9h = 9 \cdot h$
$r = r$
$9k = 9 \cdot k$
$9(h + r - k) = 9h + 9r - 9k$
The missing terms are h, k, and $9r$.

1.169 $4a - 8b + 16c = 4 \cdot a - 4 \cdot 2b + 4 \cdot 4c =$ $4(a - 2b + 4c)$
The missing terms are 4, a, $2b$, and $4c$.

1.170 $A(B - 2C + 5D) = A \cdot B - A \cdot 2C +$ $A \cdot 5D = AB - 2AC + 5AD$
The missing terms are AB, $2AC$, and $5AD$.

1.171 $xy(3 + 2c + 4g) = xy \cdot 3 + xy \cdot 2c +$ $xy \cdot 4g = 3xy + 2cxy + 4gxy$
The missing terms are $2cxy$ and gxy.

1.172 $(x + 2)(y + 3) = xy + 3x + 2y + 6$

1.173 $(g + 3)(h + 5) = gh + 5g + 3h + 15$

1.174 $(3s + 1)(4t + 3) = 12st + 9s + 4t + 3$

1.175 $(2w + 3)(3v + 4) = 6wv + 8w + 9v + 12$

1.176 $(x + 4)(x - 3) = x^2 - 3x + 4x - 12 = x^2$ $+ x - 12$

1.177 $(y - 2)(2y + 3) = 2y^2 + 3y - 4y - 6 =$ $2y^2 - y - 6$

1.178 $(x + y)(x - y) = x^2 - xy + xy - y^2 = x^2 - y^2$

1.179 $(t + 4)(t - 4) = t^2 - 4^t + 4t - 16 = t^2 - 16$

1.180 $(g - 3)(g - 3) = g^2 - 3g - 3g + 9 = g^2 - 6g + 9$

1.181 $(3a + 4b)(3a + 4b) = 9a^2 + 12ab + 12ab + 16b^2 = 9a^2 + 24ab + 16b^2$

II. SECTION TWO

2.1
$x - 2 = 6$
$2 = 2$
$x = 8$
Check:
$8 - 2 = 6$
$6 = 6$

2.2
$x - 5 = 7$
$5 = 5$
$x = 12$
Check:
$12 - 5 = 7$
$7 = 7$

2.3
$x - 10 = 2$
$10 = 10$
$x = 12$
Check:
$12 - 10 = 2$
$2 = 2$

2.4
$x - 14 = 1$
$14 = 14$
$x = 15$
Check:
$15 - 14 = 1$
$1 = 1$

2.5
$a - 1 = 0$
$1 = 1$
$a = 1$
Check:
$1 - 1 = 0$
$0 = 0$

2.6
$d - 8 = -3$
$8 = 8$
$d = 5$
Check:
$5 - 8 = 3$
$-3 = -3$

2.7
$c - 1 = -9$
$1 = 1$
$c = -8$
Check:
$-8 - 1 = -9$
$9 = 9$

2.8
$x - 20 = -1$
$20 = 20$
$x = 19$
Check:
$19 - 20 = 1$
$-1 = -1$

2.9
$b - 12 = 12$
$12 = 12$
$b = 24$
Check:
$24 - 12 = 12$
$12 = 12$

2.10
$t - 6 = -6$
$6 = 6$
$t = 0$
Check:
$0 - 6 = -6$
$-6 = -6$

2.11
$m - 1 = 6$
$1 = 1$
$m = 7$
Check:
$7 - 1 = 6$
$6 = 6$

2.12
$x - 4 = -9$
$4 = 4$
$x = -5$
Check:
$-5 - 4 = -9$
$-9 = -9$

2.13
$x - 0.5 = 1.5$
$0.5 = 0.5$
$x = 2.0$
Check :
$2 - 0.5 = 1.5$
$1.5 = 1.5$

2.14
$y - \frac{1}{2} = 1\frac{1}{2}$
$\frac{1}{2} = \frac{1}{2}$
$y = 2$

Check :

$2 - \frac{1}{2} = 1\frac{1}{2}$

$1\frac{1}{2} = 1\frac{1}{2}$

2.15 $6 = x - 2$

$2 = \quad 2$

$8 = x$

Check:

$6 = 8 - 2$

$6 = 6$

2.16 $7 = w - 5$

$5 = \quad 5$

$12 = w$

Check:

$7 = 12 - 5$

$7 = 7$

2.17 $15 = -2 + z$

$2 = 2$

$17 = \quad z$

Check:

$15 = -2 + 17$

$15 = 15$

2.18 $8 = -5 + q$

$5 = 5$

$13 = \quad q$

Check:

$8 = -5 + 13$

$8 = 8$

2.19 $x - 14.6 = 14.6$

$14.6 = 14.6$

$x = 29.2$

Check :

$29.2 - 14.6 = 14.6$

$14.6 = 14.6$

2.20 $y - 400 = 0$

$400 = 400$

$y = 400$

Check:

$400 - 400 = 0$

$0 = 0$

2.21 $x + 1 = 2$

$1 = -1$

$x = 1$

Check:

$1 + 1 = 2$

$2 = 2$

2.22 $y + 2 = 5$

$2 = -2$

$y = 3$

Check:

$3 + 2 = 5$

$5 = 5$

2.23 $t + 7 = 4$

$-7 = -7$

$t = -3$

Check:

$-3 + 7 = 4$

$4 = 4$

2.24 $x + 13 = 15$

$-13 = -13$

$x = 2$

Check:

$2 + 13 = 15$

$15 = 15$

2.25 $y + 11 = 11$

$-11 = -11$

$y = 0$

Check:

$0 + 11 = 11$

$11 = 11$

2.26 $a + 1 = 12$

$-1 = -1$

$a = 11$

Check:

$11 + 1 = 12$

$12 = 12$

2.27 $x + 1 = 8$

$-1 = -1$

$x = 7$

Check:

$7 + 1 = 8$

$8 = 8$

2.28
$$m + 6 = 1$$
$$-6 = -6$$
$$m = -5$$
Check:
$-5 + 6 = 1$
$1 = 1$

2.29
$$c + 0.5 = 3.0$$
$$-0.5 = -0.5$$
$$c = 2.5$$
Check:
$2.5 + 0.5 = 3.0$
$3.0 = 3.0$

2.30
$$d + 1.6 = 3.2$$
$$-1.6 = 1.6$$
$$d = 1.6$$
Check:
$1.6 + 1.6 = 3.2$
$3.2 = 3.2$

2.31
$$5 = x + 4$$
$$-4 = -4$$
$$1 = x$$
Check:
$5 = 1 + 4$
$5 = 5$

2.32
$$7 = y + 2$$
$$-2 = -2$$
$$5 = y$$
Check:
$7 = 5 + 2$
$7 = 7$

2.33
$$8 = r + 9$$
$$-9 = -9$$
$$-1 = r$$
Check:
$8 = -1 + 9$
$8 = 8$

2.34
$$10 = 1 + q$$
$$-1 = -1$$
$$9 = q$$
Check:
$10 = 1 + 9$
$10 = 10$

2.35
$$45 = w + 50$$
$$-50 = -50$$
$$-5 = w$$
Check:
$45 = -5 + 50$
$45 = 45$

2.36
$$x + 1 = 14$$
$$-1 = -1$$
$$x = 13$$
Check:
$13 + 1 = 14$
$14 = 14$

2.37
$$x + 1 = 8.4$$
$$-1 = -1.0$$
$$x = 7.4$$
Check:
$7.4 + 1 = 8.4$
$8.4 = 8.4$

2.38
$$14 = v + 18.5$$
$$-18.5 = -18.5$$
$$-4.5 = v$$
Check:
$14 = -4.5 + 18.5$
$14 = 14$

2.39
$$0 = 8 + p$$
$$-8 = -8$$
$$-8 = p$$
Check:
$0 = 8 - 8$
$0 = 0$

2.40
$$y + \frac{1}{2} = 4\frac{1}{2}$$
$$-\frac{1}{2} = -\frac{1}{2}$$
$$y = 4$$
Check:
$4 + \frac{1}{2} = 4\frac{1}{2}$
$4\frac{1}{2} = 4\frac{1}{2}$

2.41

$$\begin{aligned} t + 1 &= 1.1 \\ -1 &= 1.0 \\ \hline t \quad &= 0.1 \end{aligned}$$

Check:

$0.1 + 1 = 1.1$

$1.1 = 1.1$

2.42

$$\begin{aligned} c + 8 &= -8 \\ -8 &= -8 \\ \hline c \quad &= -16 \end{aligned}$$

Check:

$-16 + 8 = -8$

$-8 = -8$

2.43

$$\begin{aligned} x + 3 - 6 &= 4 + 7 \\ x \quad - 3 &= 11 \\ 3 &= 3 \\ \hline x \quad &= 14 \end{aligned}$$

Check:

$14 + 3 - 6 = 4 + 7$

$11 = 11$

2.44

$$\begin{aligned} y - 6 + 4 &= 9 - 1 \\ y - 2 \quad &= 8 \\ 2 &= 2 \\ \hline y \quad &= 10 \end{aligned}$$

Check:

$10 - 6 + 4 = 9 - 1$

$8 = 8$

2.45

$\frac{x}{3} = 5$

$3 \cdot \frac{x}{3} = 3 \cdot 5$

$x = 15$

Check :

$\frac{15}{3} = 5$

$5 = 5$

2.46

$\frac{d}{4} = 8$

$4 \cdot \frac{d}{4} = 4 \cdot 8$

$d = 32$

Check:

$\frac{32}{4} = 8$

$8 = 8$

2.47

$\frac{c}{5} = 2$

$5 \cdot \frac{c}{5} = 5 \cdot 2$

$c = 10$

Check :

$\frac{10}{5} = 2$

$2 = 2$

2.48

$\frac{1}{6}a = 1$

$6 \cdot \frac{1}{6}a = 6 \cdot 1$

$a = 6$

Check:

$\frac{1}{6} \cdot 6 = 1$

$1 = 1$

2.49

$\frac{b}{7} = 3$

$7 \cdot \frac{b}{7} = 7 \cdot 3$

$b = 21$

Check :

$\frac{21}{7} = 3$

$3 = 3$

2.50 $\frac{1}{8}p = 4$

$8 \cdot \frac{1}{8}p = 8 \cdot 4$

$p = 32$

Check:

$\frac{1}{8} \cdot 32 = 4$

$4 = 4$

2.51 $\frac{3}{5}y = 3$

$\frac{5}{3} \cdot \frac{3}{5}y = \frac{5}{3} \cdot 3$

$y = 5$

Check:

$\frac{3}{5} \cdot 5 = 3$

$3 = 3$

2.52 $\frac{4}{7}h = 8$

$\frac{7}{4} \cdot \frac{4}{7}h = \frac{7}{4} \cdot 8h = 14$

Check:

$\frac{4}{7} \cdot 14 = 8$

$8 = 8$

2.53 $\frac{5}{6}k = 10$

$\frac{6}{5} \cdot \frac{5}{6}k = \frac{6}{5} \cdot 10$

$k = 12$

Check:

$\frac{5}{6} \cdot 12 = 10$

$10 = 10$

2.54 $\frac{7}{10}a = 21$

$\frac{10}{7} \cdot \frac{7}{10}a = \frac{10}{7} \cdot 21$

$a = 30$

Check:

$\frac{7}{10} \cdot 30 = 21$

$21 = 21$

2.55 $\frac{4x}{5} = 12$

$\frac{5}{4} \cdot \frac{4x}{5} = \frac{5}{4} \cdot 12$

$x = 15$

Check:

$4 \cdot \frac{15}{5} = 12$

$12 = 12$

2.56 $\frac{5y}{8} = 15$

$\frac{8}{5} \cdot \frac{5y}{8} = \frac{8}{5} \cdot 15$

$y = 24$

Check:

$5 \cdot \frac{24}{8} = 15$

$15 = 15$

2.57 $\frac{1}{2}x + 3 = 5$

$-3 = -3$

$\frac{1}{2}x = 2$

$2 \cdot \frac{1}{2}x = 2 \cdot 2$

$x = 4$

Check:

$\frac{1}{2} \cdot 4 + 3 = 5$

$2 + 3 = 5$

$5 = 5$

2.58 $\frac{a}{3} - 1 = 4$

$1 = 1$

$\frac{a}{3} = 5$

$3 \cdot \frac{a}{3} = 3 \cdot 5$

$a = 15$

Check:

$\frac{15}{3} - 1 = 4$

$5 - 1 = 4$

$4 = 4$

2.59

$$\begin{aligned} \tfrac{b}{5} + 2 &= 6 \\ -2 &= -2 \\ \hline \tfrac{b}{5} &= 4 \\ 5 \cdot \tfrac{b}{5} &= 5 \cdot 4 \\ b &= 20 \end{aligned}$$

Check:

$\frac{20}{5} + 2 = 6$

$4 + 2 = 6$

$6 = 6$

2.60

$$\begin{aligned} \tfrac{1}{8}c + 2 &= 5 \\ -2 &= -2 \\ \hline \tfrac{1}{8}c &= 3 \\ 8 \cdot \tfrac{1}{8}c &= 8 \cdot 3 \\ c &= 24 \end{aligned}$$

Check:

$\frac{1}{8} \cdot 24 + 2 = 5$

$3 + 2 = 5$

$5 = 5$

2.61

$$\begin{aligned} 2x - 3 &= x + 2 \\ 3 &= 3 \\ \hline 2x &= x + 5 \\ -x &= -x \\ \hline x &= 5 \end{aligned}$$

Check:

$2 \cdot 5 - 3 = 5 + 2$

$10 - 3 = 7$

$7 = 7$

2.62

$$\begin{aligned} 4y + 1 &= 2y - 7 \\ -1 &= -1 \\ \hline 4y &= 2y - 8 \\ -2y &= -2y \\ \hline 2y &= -8 \\ \tfrac{1}{2}(2y) &= \tfrac{1}{2}(-8) \\ y &= -4 \end{aligned}$$

Check:

$4(-4) + 1 = 2(-4) - 7$

$-16 + 1 = -8 - 7$

$-15 = -15$

2.63

$$\begin{aligned} 5x + 1 &= 4x + 6 \\ -1 &= -1 \\ \hline 5x &= 4x + 5 \\ -4x &= -4x \\ \hline x &= 5 \end{aligned}$$

Check:

$5 \cdot 5 + 1 = 4 \cdot 5 + 6$

$25 + 1 = 20 + 6$

$26 = 26$

2.64

$$\begin{aligned} 6a + 3 &= 6 + 3a \\ -3 &= -3 \\ \hline 6a &= 3 + 3a \\ -3a &= -3a \\ \hline 3a &= 3 \\ \tfrac{1}{3} \cdot 3a &= \tfrac{1}{3} \cdot 3 \\ a &= 1 \end{aligned}$$

Check:

$6 \cdot 1 + 3 = 6 + 3 \cdot 1$

$6 + 3 = 6 + 3$

$9 = 9$

2.65

$$\begin{aligned} 2b - 2 &= -b + 1 \\ 2 &= 2 \\ \hline 2b &= -b + 3 \\ b &= b \\ \hline 3b &= 3 \\ \tfrac{1}{3} \cdot 3b &= \tfrac{1}{3} \cdot 3 \\ b &= 1 \end{aligned}$$

Check:
$2 \cdot 1 - 2 = -1 + 1$
$2 - 2 = -1 + 1$
$0 = 0$

2.66
$4c + 8 = 18 - c$
$-8 = -8$
$4c = 10 - c$
$c = c$
$5c = 10$
$\frac{1}{5} \cdot 5c = \frac{1}{5} \cdot 10$
$c = 2$

Check:
$4 \cdot 2 + 8 = 18 - 2$
$8 + 8 = 16$
$16 = 16$

2.67
$\frac{3}{4}x - 1 = \frac{1}{4}x + 2$
$1 = 1$
$\frac{3}{4}x = \frac{1}{4}x + 3$
$\frac{1}{4}x = -\frac{1}{4}$
$\frac{2}{4}x = 3$
$\frac{4}{2} \cdot \frac{2}{4}x = \frac{4}{2} \cdot 3$
$x = 6$

Check:
$\frac{3}{4} \cdot 6 - 1 = \frac{1}{4} \cdot 6 + 2$
$\frac{18}{4} - 1 = \frac{6}{4} + 2$
$4\frac{1}{2} - 1 = 1\frac{1}{2} + 2$
$3\frac{1}{2} = 3\frac{1}{2}$

2.68
$\frac{4}{5}y + 4 = 5 - \frac{1}{5}y$
$-4 = -4$
$\frac{4}{5}y = 1 - \frac{1}{5}y$
$\frac{1}{5}y = \frac{1}{5}y$
$y = 1$

Check:
$\frac{4}{5} \cdot 1 + 4 = 5 - \frac{1}{5} \cdot 1$
$4\frac{4}{5} = 4\frac{4}{5}$

2.69
$2x = 6$
$\frac{2x}{2} = \frac{6}{2}$
$x = 3$

Check:
$2 \cdot 3 = 6$
$6 = 6$

2.70
$4y = 12$
$\frac{4y}{4} = \frac{12}{4}$
$y = 3$

Check
$4 \bullet 3 = 12$
$12 = 12$

2.71
$5c = 20$
$\frac{5c}{5} = \frac{20}{5}$
$c = 4$

Check:
$5 \cdot 4 = 20$
$20 = 20$

2.72
$6b = -12$
$\frac{6b}{6} = -\frac{12}{6}$
$b = -2$

Check:
$6(-2) = -12$
$-12 = -12$

2.73
$3x = 36$
$\frac{3x}{3} = \frac{36}{3}$
$x = 12$

Check:
$3 \cdot 12 = 36$
$36 = 36$

2.74

$3a = 3$

$\frac{3a}{3} = \frac{3}{3}$

$a = 1$

Check:

$3 \cdot 1 = 3$

$3 = 3$

2.75

$14t = 42$

$\frac{14t}{14} = \frac{42}{14}$

$t = 3$

Check:

$14 \cdot 3 = 42$

$42 = 42$

2.76

$5q = 100$

$\frac{5q}{5} = \frac{100}{5}$

$q = 20$

Check:

$5 \cdot 20 = 100$

$100 = 100$

2.77

$10p = 5$

$\frac{10p}{10} = \frac{5}{10}$

$p = \frac{1}{2}$

Check:

$10 \cdot \frac{1}{2} = 5$

$5 = 5$

2.78

$8x = 6$

$\frac{8x}{8} = \frac{6}{8}$

$x = \frac{3}{4}$

Check:

$8 \cdot \frac{3}{4} = 6$

$6 = 6$

2.79

$$\begin{aligned} 3a + 2 &= 8 \\ -2 &= -2 \\ \hline 3a &= 6 \\ \frac{3a}{3} &= \frac{6}{3} \\ a &= 2 \end{aligned}$$

Check:

$$\begin{aligned} 3 \cdot 2 + 2 &= 8 \\ 6 + 2 &= 8 \\ 8 &= 8 \end{aligned}$$

2.80

$$\begin{aligned} 4w + 3 &= 15 \\ -3 &= -3 \\ \hline 4w &= 12 \\ \frac{4w}{4} &= \frac{12}{4} \\ w &= 3 \end{aligned}$$

Check:

$$\begin{aligned} 4 \cdot 3 + 3 &= 15 \\ 12 + 3 &= 15 \\ 15 &= 15 \end{aligned}$$

2.81

$$\begin{aligned} 7x + 8 &= 29 \\ -8 &= -8 \\ \hline 7x &= 21 \\ \frac{7x}{7} &= \frac{21}{7} \\ x &= 3 \end{aligned}$$

Check:

$$\begin{aligned} 7 \cdot 3 + 8 &= 29 \\ 21 + 8 &= 29 \\ 29 &= 29 \end{aligned}$$

2.82

$$\begin{aligned} y - 7 &= 1 \\ 7 &= 7 \\ \hline y &= 8 \end{aligned}$$

Check:

$$\begin{aligned} 8 - 7 &= 1 \\ 1 &= 1 \end{aligned}$$

2.83

$$\begin{aligned} 3a - 4 &= 20 \\ 4 &= 4 \\ \hline 3a &= 24 \\ \frac{3a}{3} &= \frac{24}{3} \\ a &= 8 \end{aligned}$$

Check:
$3 \cdot 8 - 4 = 20$
$24 - 4 = 20$
$20 = 20$

2.84
$6 = 2w$
$\frac{6}{2} = \frac{2w}{2}$
$3 = w$

Check:
$6 = 2 \cdot 3$
$6 = 6$

2.85
$12 = 3y + 6$
$-6 = \quad -6$
$6 = 3y$
$\frac{6}{3} = \frac{3y}{3}$
$2 = y$

Check:
$12 = 3 \cdot 2 + 6$
$12 = 6 + 6$
$12 = 12$

2.86
$4t - 1 = 19$
$1 = 1$
$4t = 20$
$\frac{4t}{4} = \frac{20}{4}$
$t = 5$

Check:
$4 \cdot 5 - 1 = 19$
$20 - 1 = 19$
$19 = 19$

2.87
$11x - 1 = 21$
$1 = 1$
$11x = 22$
$\frac{11x}{11} = \frac{22}{11}$
$x = 2$

Check:
$11 \cdot 2 - 1 = 21$
$22 - 1 = 21$
$21 = 21$

2.88
$18x + 11 = 29$
$-11 = -11$
$18x = 18$
$\frac{18x}{18} = \frac{18}{18}$
$x = 1$

Check:
$18 \cdot 1 + 11 = 29$
$29 = 29$

2.89
$2x + x + x = 6$
$4x = 6$
$\frac{4x}{4} = \frac{6}{4}$
$x = \frac{3}{2}$ or $1\frac{1}{2}$

Check:
$2 \cdot \frac{3}{2} + \frac{3}{2} + \frac{3}{2} = 6$
$3 + 1\frac{1}{2} + 1\frac{1}{2} = 6$
$6 = 6$

2.90
$4y - 6 = y + 2y + 2$
$4y - 6 = 3y + 2$
$6 = 6$
$4y = 3y + 8$
$-3y = -3y$
$y = 8$

Check:
$4 \cdot 8 - 6 = 8 + 2 \cdot 8 + 2$
$32 - 6 = 8 + 16 + 2$
$26 = 26$

2.91
$5c + 6 - c = 2c + 9 - c$
$4c + 6 = c + 9$
$-6 = -6$
$4c = c + 3$
$-c = -c$
$3c = 3$
$\frac{3c}{3} = \frac{3}{3}$
$c = 1$

Check:
$5 \cdot 1 + 6 - 1 = 2 \cdot 1 + 9 - 1$
$5 + 6 - 1 = 2 + 9 - 1$
$10 = 10$

2.92

$$\begin{aligned} 6d + 1 - 2d &= 3d - 3 \\ 4d + 1 &= 3d - 3 \\ -1 &= -1 \\ \hline 4d &= 3d - 4 \\ -3d &= -3d \\ \hline d &= -4 \end{aligned}$$

Check:

$$\begin{aligned} 6(-4) + 1 - 2(-4) &= 3(-4) - 3 \\ -24 + 1 + 8 &= -12 - 3 \\ -24 + 9 &= -12 - 3 \\ -15 &= -15 \end{aligned}$$

2.93

$$\begin{aligned} 4d + 3d + d &= 3d - 15 \\ 8d &= 3d - 15 \\ -3d &= -3d \\ \hline 5d &= -15 \\ \tfrac{5d}{5} &= -\tfrac{15}{5} \\ d &= -3 \end{aligned}$$

Check:

$$\begin{aligned} 4(-3) + 3(-3) + (-3) &= 3(-3) - 15 \\ -12 - 9 - 3 &= -9 - 15 \\ -24 &= -24 \end{aligned}$$

2.94

$$\begin{aligned} 15B - 6 &= 8B - 3B + 14 \\ 15B - 6 &= 5B + 14 \\ 6 &= 6 \\ \hline 15B &= 5B + 20 \\ -5B &= -5B \\ \hline 10B &= 20 \\ \tfrac{10B}{10} &= \tfrac{20}{10} \\ B &= 2 \end{aligned}$$

Check:

$$\begin{aligned} 15 \cdot 2 - 6 &= 8 \cdot 2 - 3 \cdot 2 + 14 \\ 30 - 6 &= 16 - 6 + 14 \\ 24 &= 24 \end{aligned}$$

2.95

$$\begin{aligned} 7h + h - 6 &= 10 + 4h \\ 8h - 6 &= 10 + 4h \\ 6 &= 6 \\ \hline 8h &= 16 + 4h \\ -4h &= -4h \\ \hline -4h &= 16 \\ \tfrac{4h}{4} &= \tfrac{16}{4} \\ h &= 4 \end{aligned}$$

Check:

$$\begin{aligned} 7 \cdot 4 + 4 - 6 &= 10 + 4 \cdot 4 \\ 28 + 4 - 6 &= 10 + 16 \\ 26 &= 26 \end{aligned}$$

2.96

$$\begin{aligned} k + 6k &= 8k - 4 \\ 7k &= 8k - 4 \\ -8k &= -8k - 4 \\ \hline -k &= -4 \\ k &= 4 \end{aligned}$$

Check:

$$\begin{aligned} 4 + 6 \cdot 4 &= 8 \cdot 4 - 4 \\ 4 + 24 &= 32 - 4 \\ 28 &= 28 \end{aligned}$$

2.97

$$\begin{aligned} 10p - 14 - p &= 4p - 9 \\ 9p - 14 &= 4p - 9 \\ 14 &= 14 \\ \hline 9p &= 4p + 5 \\ -4p &= -4p \\ \hline 5p &= 5 \\ \tfrac{5p}{5} &= \tfrac{5}{5} \\ p &= 1 \end{aligned}$$

Check:

$$\begin{aligned} 10 \cdot 1 - 14 - 1 &= 4 \cdot 1 - 9 \\ 10 - 14 - 1 &= 4 - 9 \\ -5 &= -5 \end{aligned}$$

2.98

$$\begin{aligned} x - 8x + 7 &= 2x - 11 \\ -7x + 7 &= 2x - 11 \\ -7 &= -7 \\ \hline -7x &= 2x - 18 \\ -2x &= -2x \\ \hline -9x &= -18 \\ \tfrac{-9x}{-9} &= \tfrac{-18}{-9} \\ x &= 2 \end{aligned}$$

Check:

$$\begin{aligned} 2 - 8 \cdot 2 + 7 &= 2 \cdot 2 - 11 \\ 2 - 16 + 7 &= 4 - 11 \\ -7 &= -7 \end{aligned}$$

2.99

$\frac{1}{4}x + \frac{2}{3} = x + \frac{3}{4}$

$12(\frac{1}{4}x + \frac{2}{3} = x + \frac{3}{4})$

$3x + 8 = 12x + 9$

$-8 = -8$

$3x = 12x + 1$

$-12x = -12x$

$-9x = 1$

$\frac{-9x}{-9} = \frac{1}{-9}$

$x = \frac{-1}{9}$

Check:

$\frac{1}{4}(-\frac{1}{9}) + \frac{2}{3} = -\frac{1}{9} = \frac{3}{4}$

$-\frac{1}{36} + \frac{2}{3} = -\frac{1}{9} = \frac{3}{4}$

$-\frac{1}{36} + \frac{24}{36} = -\frac{4}{36} + \frac{27}{36}$

$\frac{23}{36} = \frac{23}{36}$

2.100

$\frac{1}{2}y - \frac{2}{3} = \frac{2}{3}y + \frac{1}{2}$

$6(\frac{1}{2}y - \frac{2}{3} = \frac{2}{3}y + \frac{1}{2})$

$3y - 4 = 4y + 3$

$4 = 4$

$3y = 4y + 7$

$-4y = -4y$

$-y = 7$

$y = -7$

Check:

$\frac{1}{2}(-7) - \frac{2}{3} = \frac{2}{3}(-7) + \frac{1}{2}$

$-\frac{7}{2} - \frac{2}{3} = \frac{14}{3} + \frac{1}{2}$

$-\frac{21}{6} - \frac{4}{6} = -\frac{28}{6} + \frac{3}{6}$

$-\frac{25}{6} = -\frac{25}{6}$

2.101

$\frac{2h}{5} - \frac{1}{5} = \frac{3h}{5} + \frac{4}{5} - \frac{2h}{5}$

$5(\frac{2h}{5} - \frac{1}{5} = \frac{3h}{5} + \frac{4}{5} - \frac{2h}{5})$

$2h - 1 = 3h + 4 - 2h$

$1 = 1$

$2h = 3h + 5 - 2h$

$2h = h + 5$

$-h = -h$

$h = 5$

Check:

$\frac{2 \cdot 5}{5} - \frac{1}{5} = \frac{3 \cdot 5}{5} + \frac{4}{5} - \frac{2 \cdot 5}{5}$

$\frac{10}{5} - \frac{1}{5} = \frac{15}{5} + \frac{4}{5} - \frac{10}{5}$

$\frac{9}{5} = \frac{9}{5}$

2.102

$\frac{3c}{4} - \frac{c}{4} = \frac{c}{4} + 2$

$4(\frac{3c}{4} - \frac{c}{4} = \frac{c}{4} + 2)$

$3c - c = c + 8$

$2c = c + 8$

$-c = -c$

$c = 8$

Check:

$3 \cdot \frac{8}{4} - \frac{8}{4} = \frac{8}{4} + 2$

$6 - 2 = 2 + 2$

$4 = 4$

2.103

$\frac{4}{7}p + \frac{2}{7}p - 1 = \frac{2}{7}p + \frac{1}{2}$

$14(\frac{4}{7}p + \frac{2}{7}p - 1 = \frac{2}{7}p + \frac{1}{2})$

$8p + 4p - 14 = 4p + 7$

$12p - 14 = 4p + 7$

$14 = 14$

$12p = 4p + 21$

$-4p = -4p$

$8p = 21$

$\frac{8p}{8} = \frac{21}{8}$

$p = \frac{21}{8}$ or $2\frac{5}{8}$

Check:

$\frac{4}{7} \cdot \frac{21}{8} + \frac{2}{7} \cdot \frac{21}{8} - 1 = \frac{2}{7} \cdot \frac{21}{8} + \frac{1}{2}$

$\frac{84}{56} + \frac{42}{56} - 1 = \frac{42}{56} + \frac{1}{2}$

$\frac{126}{56} - \frac{56}{56} = \frac{42}{56} + \frac{28}{56}$

$\frac{70}{56} = \frac{70}{56}$

2.104

$$\begin{aligned} 3(x+1) &= 6 \\ 3x + 3 &= 6 \\ -3 &= -3 \\ \hline 3x &= 3 \\ \frac{3x}{3} &= \frac{3}{3} \\ x &= 1 \end{aligned}$$

Check:

$$\begin{aligned} 3(1+1) &= 6 \\ 3(2) &= 6 \\ 6 &= 6 \end{aligned}$$

2.105

$$\begin{aligned} 4(d-2) &= 2d + 4 \\ 4d - 8 &= 2d + 4 \\ 8 &= 8 \\ \hline 4d &= 2d + 12 \\ -2d &= -2d \\ \hline 2d &= 12 \\ \frac{2d}{2} &= \frac{12}{2} \\ d &= 6 \end{aligned}$$

Check:

$$\begin{aligned} 4(6-2) &= 2 \cdot 6 + 4 \\ 4 \cdot 4 &= 12 + 4 \\ 16 &= 16 \end{aligned}$$

2.106

$$\begin{aligned} 5(2y+3) &= 3y + 1 \\ 10y + 15 &= 3y + 1 \\ -15 &= -15 \\ \hline 10y &= 3y - 14 \\ -3y &= -3y \\ \hline \frac{7y}{7} &= \frac{-14}{7} \\ y &= -2 \end{aligned}$$

Check:

$$\begin{aligned} 5[2(-2) + 3] &= 3(-2) + 1 \\ 5(-4+3) &= -6 + 1 \\ 5(-1) &= -6 + 1 \\ -5 &= -5 \end{aligned}$$

2.107 $\frac{3}{4}(x+3) = \frac{3}{4}(2x-6)$

Divide the equation by $\frac{3}{4}$:

$$\begin{aligned} x + 3 &= 2x - 6 \\ -3 &= -3 \\ \hline x &= 2x - 9 \\ -2x &= -2x \\ \hline -x &= -9 \\ \hline x &= 9 \end{aligned}$$

Check:

$$\begin{aligned} \frac{3}{4}(9+3) &= \frac{3}{4}(2 \cdot 9 - 6) \\ \frac{3}{4}(12) &= \frac{3}{4}(12) \\ 9 &= 9 \end{aligned}$$

2.108 $\frac{2x}{5} - 1 = \frac{x}{5} + \frac{2}{3}$

$$\begin{aligned} 15\left(\frac{2x}{5} - 1 \right. &= \left. \frac{x}{5} + \frac{2}{3}\right) \\ 6x - 15 &= 3x + 10 \\ 15 &= 15 \\ \hline 6x &= 3x + 25 \\ -3x &= -3x \\ \hline 3x &= 25 \\ \frac{3x}{3} &= \frac{25}{3} \\ x &= \frac{25}{3} \text{ or } 8\frac{1}{3} \end{aligned}$$

Check:

$$\begin{aligned} \frac{2 \cdot \frac{25}{3}}{5} - 1 &= \frac{\frac{25}{3}}{5} + \frac{2}{3} \\ \frac{50}{3} \cdot \frac{1}{5} - 1 &= \frac{25}{3} \cdot \frac{1}{5} + \frac{2}{3} \\ \frac{10}{3} - \frac{3}{3} &= \frac{5}{3} + \frac{2}{3} \\ \frac{7}{3} &= \frac{7}{3} \end{aligned}$$

2.109

$$3(x-1)+4(2x-2)=x-1$$
$$3x-3+8x-8=x-1$$
$$11x-11=x-1$$
$$11 = \quad 11$$
$$11x \quad = x+10$$
$$-x \quad = -x$$
$$10x=10$$
$$\frac{10x}{10}=\frac{10}{10}$$
$$x=1$$

Check:

$$3(1-1)+4(2\cdot1-2)=1-1$$
$$3(0)+4(0)=1-1$$
$$0+0=0$$
$$0=0$$

2.110

$$7(3x-1)+6=2x+18$$
$$21x-7+6=2x+18$$
$$21x-1=2x+18$$
$$1 = \quad 1$$
$$21x \quad = 2x+19$$
$$-2x \quad = -2x$$
$$19x=19$$
$$\frac{19x}{19}=\frac{19}{19}$$
$$x=1$$

Check:

$$7(3\cdot1-1)+6=2\cdot1+18$$
$$7\cdot2+6=2+18$$
$$20=20$$

2.111

$$2(x+1)+3(2x-2)=4(x+3)+9$$
$$2x+2+6x-6=4x+12+9$$
$$8x-4=4x+21$$
$$4 = \quad 4$$
$$8x \quad = 4x+25$$
$$-4x \quad = -4x$$
$$4x=25$$
$$\frac{4x}{4}=\frac{25}{4}$$
$$x=\frac{25}{4} \text{ or } 6\frac{1}{4}$$

2.111 (cont.)

Check:

$$2(\tfrac{25}{4}+1)+3(2\cdot\tfrac{25}{4}-2)=4(\tfrac{25}{4}+3)+9$$
$$2(\tfrac{25}{4}+\tfrac{4}{4})+3(\tfrac{50}{4}-\tfrac{8}{4})=4(\tfrac{25}{4}+\tfrac{12}{4})+9$$
$$2(\tfrac{29}{4})+3(\tfrac{42}{4})=4(\tfrac{37}{4})+\tfrac{36}{4}$$
$$\frac{58}{4}+\frac{126}{4}=\frac{148}{4}+\frac{36}{4}$$
$$\frac{184}{4}=\frac{184}{4}$$

2.112

$$0.1x+0.6=0.06x+0.2$$
$$100(0.1x+0.6=0.06x+0.2)$$
$$10x+60=6x+20$$
$$-60= \quad -60$$
$$10x \quad = 6x-40$$
$$-6x \quad = -6x$$
$$4x=-40$$
$$\frac{4x}{4}=-\frac{40}{4}$$
$$x=-10$$

Check:

$$0.1(-10)+0.6=0.06(-10)+0.2$$
$$-1+0.6=-0.6+0.2$$
$$-0.4=-0.4$$

III. SECTION THREE

3.1 $n+4$

3.2 $2c+1$

3.3 $x-2$

3.4 $2k-5$

3.5 $10y-5$

3.6 $a+2a$

3.7 $8+n$

3. 8 $\frac{x}{2}$ or $\frac{1}{2}x$

3.9 $x+4$

3.10 xy

3.11 $6 - n$

3.12 $\frac{1}{2}x$ or $\frac{x}{2}$

3.13 $3t - 3$

3.14 $n + 8$

3.15 $2c - 7$

3.16 $x + 10$

3.17 $7 - n$

3.18 $\frac{m}{9}$

3.19 $w \cdot 8$

3.20 $b + 11$

3.21 $5 + 8n$

3.22 $0.08x$ or $8\%x$

3.23 $9 - x$

3.24 $4a + 2$

3.25 a number minus four *or* four less than a number

3.26 a number increased by three *or* three more than a number

3.27 twice a number

3.28 three times a number plus one *or* one more than three times a number

3.29 six plus a number

3.30 four times a number, decreased by eight

3.31 a number divided by 9

3.32 eleven less than six times a number

3.33 one-fourth of a number

3.34 the difference of a and b

3.35 eight minus twice a number

3.36 three times a number decreased by fourteen

3.37 a number divided by another number

3.38 twice a number minus twice another number

3.39 six more than five times a number

3.40 the sum of a number and two minus twice the number

3.41 eleven more than twice a number

3.42 fourteen divided by three times a number

3.43 twenty-six plus four times a number

3.44 a number divided by six, minus fifteen

3.45 seventeen more than twice a number divided by five

3.46 the sum of twice a number and five times the same number

3.47 four divided by a number, plus six

3.48 the difference of three times a number and five divided by the same number

3.49 $n + 2 = 13$

3.50 $2x - 7 = 41$

3.51 $8n = 56$

3.52 $a + 8 = 18$

3.53 $x + 5 = 21$

3.54 $\frac{c}{2} = 13$

3.55 $6y = 24$

3.56 $10 = d + 6$

3.57 $9 = x + 1$

3.58 $14 = 5 + n$

3.59 $p \cdot 8 = 72$

3.60 $\frac{7}{x} = 7$

3.61 $n - 10 = 15$

3.62 $9y = 36$

3.63 $7 = 2x - 2$

3.64 $\frac{3n}{6} = 4$

3.65 $4x - 10 = x + 3$

3.66 $\frac{n}{3} = 2n - 8$

3.67 $\frac{2}{3}t = t + 5$

3.68 $100 - 2x = x - 4$

3.69 $4a = 3a + 6$

3.70 $13 + 5n = 100$

3.71 $6g - 11 = 2g + 8$

3.72 $\frac{n}{3} + 6 = n - 10$

3.73 Three more than a number is six.

3.74 Six less than a number is eleven.

3.75 Three times a number is nine.

3.76 A number divided by four is five.

3.77 One less than twice a number is five.

3.78 Five times a number plus four is six.

3.79 Nine is two minus twice a number.

3.80 Three less than a number is two more than twice the number.

3.81 One-third of a number decreased by one is seven.

3.82 Ten less twice a number is three times the number.

3.83 The sum of five times a number and twice the same number is seventy.

3.84 Fourteen more than six times a number is twenty-one.

3.85 One more than a number divided by four minus a number is seven.

3.86 Twenty-seven minus three times a number is four times the number.

3.87 Three more than half a number is two-thirds the number.

3.88 Fourteen divided by a number plus seven is twice the number.

3.89 Eight plus two-fifths of a number is two minus the number.

3.90 The sum of seven times a number and the number is ten.

3.91 Fourteen plus a number is six minus the number *or* a number more than fourteen is six less the number.

3.92 The quotient of six divided by three times a number is negative ten.

3.93 Three plus three times a number is the number divided by three, minus one.

3.94 One-third less than twenty-six times a number is twenty- seven divided by three times the number.

3.95 Four-fifths is one less than the quotient of a number divided by three.

3.96 Five times a quantity minus sixteen is eight minus two times the quantity.

3.97
$$n - 10 = 2$$
$$10 = 10$$
$$n \quad = 12$$

3.98
$$3n = 18$$
$$\frac{3n}{3} = \frac{18}{3}$$
$$n = 6$$

3.99
$$n + 6 = 10$$
$$-6 = -6$$
$$n \quad = 4$$

3.100
$$\frac{n}{4} = 2$$
$$4 \cdot \frac{n}{4} = 4 \cdot 2$$
$$n = 8$$

3.101
$$2x - 6 = 12$$
$$6 = 6$$
$$\frac{2x}{2} = \frac{18}{2}$$
$$x = 9$$

3.102
$$15 = d + 3$$
$$-3 = \quad -3$$
$$12 = d$$
$$d = 12$$

3.103
$$5x - 5 = 2x + 4$$
$$5 = \quad 5$$
$$5x \quad = 2x + 9$$
$$-2x \quad = -2x$$
$$3x = 9$$
$$\frac{3x}{3} = \frac{9}{3}$$
$$x = 3$$

3.014
$$\frac{1}{3}n = 5$$
$$\frac{1}{3}n = 3 \cdot 5$$
$$n = 15$$

3.105
$$a + 7 = 20$$
$$-7 = -7$$
$$a \quad = 13$$

3.106
$$8n = 2n + 12$$
$$-2n = -2n$$
$$6n = 12$$
$$\frac{6n}{6} = \frac{12}{6}$$
$$n = 2$$

3.107
$$\frac{x}{3} = 25$$
$$3 \cdot \frac{x}{3} = 3 \cdot 25$$
$$x = 75$$

3.108
$$13 + n = 29$$
$$-13 \quad = -13$$
$$n = 16$$

3.109 Let n = the number
$$\frac{1}{5}n = 15$$
$$5 \cdot \frac{1}{5}n = 5 \cdot 15$$
$$n = 75$$

3.110 Let x = one number

$2x$ = another number

$x + 2x = 39$

$3x = 39$

$\frac{3x}{3} = \frac{39}{3}$

$x = 13$

$2x = 26$

The numbers are 13 and 26.

3.111 Let n = the number

$5n = 115$

$\frac{5n}{5} = \frac{115}{5}$

$n = 23$

3.112 Let d = one number

$3d$ = another number

$3d - d = 6$

$\frac{2d}{2} = \frac{6}{2}$

$d = 3$

$3d = 9$

The numbers are 3 and 9.

3.113 Let n = the number

$17 + n = 89$

$-17 \quad = -17$

$n = 72$

3.114 Let c = cost of one pound

$3c$ = \$7.98 or 798¢

$\frac{3c}{3} = \frac{798}{3}$

c = 266¢ or \$2.66

3.115 Let c = cost of each candy bar

$7c$ = \$1.33 or 133¢

$\frac{7c}{7} = \frac{133}{7}$

c = 19¢ or \$0.19

3.116 Let w = Trudy's weight

$\frac{5w}{5} = 225$

$\frac{5w}{5} = \frac{225}{5}$

w = 45 lb.

3.117 Let x = gross sales in a week

$0.12x = 100$

$\frac{0.12x}{0.12} = \frac{100}{0.12}$

x = \$833.33

3.118 Let c = commission

$4{,}000c = 200$

$\frac{4{,}000c}{4{,}000} = \frac{200}{4{,}000}$

$c = \frac{2}{40} = \frac{1}{20}$

$= 0.05 = 5\%$

3.119 Let w = the width

$2w$ = the length

$w + 2w + w + 2w = 66$

$6w = 66$

$\frac{6w}{6} = \frac{66}{6}$

$w = 11$

$2w = 22$

3.120 Let a = the largest angle

$a - 75 = 40$

$75 = 75$

$a \quad = 115°$

3.121 Let w = the width

$84 = 12w$

$\frac{82}{12} = \frac{12w}{12}$

$7 = w$

w = 7 ft.

3.122 Let s = the smallest side

$7s$ = the longest side

$4s$ = the other side

$s + 7s + 4s = 240$

$\frac{12s}{12} = \frac{240}{12}$
$s = 20\ mm$
$7s = 140\ mm$
$4s = 80\ mm$

3.123 Let p = one piece
$2p$ = other piece
$p + 2p = 24$
$3p = 24$
$\frac{3p}{3} = \frac{24}{3}$
$p = 8$ ft.
$2p = 16$ ft.

3.124 Let p = last year's population
$p + 26 = 658$
$-26 = -26$
$p \quad = 632$

3.125 Let p = this year's price
$p - 12{,}000 = 58{,}000$
$12{,}000 = 12{,}000$
$p \quad = \$70{,}000$

3.126 Let w = words on spelling test
$\frac{1}{6}w = 15$
$6 \cdot \frac{1}{6}w = 6 \cdot 15$
total words (w) = 90
words wrong = 15
words right = 75

3.127 Let e = total earnings
$\frac{1}{8}e = 4{,}200$
$8 \cdot \frac{1}{8}e = 8 \cdot 4{,}200$
$e = \$33{,}600$

3.128 Let w = Bob's weight
$w - 50 = 30$
$50 = 50$
$w \quad = 80$ lbs

3.129 Let n = one integer
$n + 1$ = next consecutive integer
$n + n + 1 = 11$
$2n + 1 = 11$
$-1 = -1$
$2n = 10$
$\frac{2n}{2} = \frac{10}{2}$
$n = 5$
$n + 1 = 6$

3.130 Let n = one integer
$n + 1$ = next consecutive integer
$n + n + 1 = 29$
$2n + 1 = 29$
$-1 = -1$
$2n = 28$
$\frac{2n}{2} = \frac{28}{2}$
$n = 14$
$n + 1 = 15$

3.131 Let n = first integer
$n + 1$ = 2nd consecutive integer
$n + 2$ = 3rd consecutive integer
$n + n + 1 + n + 2 = 12$
$3n + 3 = 12$
$-3 = -3$
$3n = 9$
$\frac{3n}{3} = \frac{9}{3}$
$n = 3$
$n + 1 = 4$
$n + 2 = 5$

3.132 Let n = first integer
$n + 1$ = 2nd consecutive integer
$n + 2$ = 3rd consecutive integer
$n + n + 1 + n + 2 = 54$
$3n + 3 = 54$
$-3 = -3$
$3n = 51$
$\frac{3n}{3} = \frac{51}{3}$
$n = 17$

$n + 1 = 18$
$n + 2 = 19$

3.133 Let n = first even ingeter
$n + 2$ = 2nd consecutive even integer
$n + n + 2 = 26$
$2n + 2 = 26$
$-2 = -2$
$2n = 24$
$\frac{2n}{2} = \frac{24}{2}$
$n = 12$
$n + 2 = 14$

3.134 Let n = first even integer
$n + 2$ = 2nd consecutive even integer
$n + 4$ = 3rd consecutive even integer
$n + n + 2 + n + 4 = 36$
$3n + 6 = 36$
$-6 = -6$
$3n = 30$
$\frac{3n}{3} = \frac{30}{3}$
$n = 10$
$n + 2 = 12$
$n + 4 = 14$

3.135 Let n = first even integer
$n + 2$ = 2nd consecutive even integer
$n + n + 2 = 58$
$2n + 2 = 58$
$-2 = -2$
$2n = 56$
$\frac{2n}{2} = \frac{56}{2}$
$n = 28$
$n + 2 = 30$

3.136 Let n = first even integer
$n + 2$ = 2nd consecutive even integer
$n + 4$ = 3rd consecutive even integer
$n + n + 2 + n + 4 = 54$
$3n + 6 = 54$
$-6 = -6$
$3n = 48$
$\frac{3n}{3} = \frac{48}{3}$

$n = 16$
$n + 2 = 18$
$n + 4 = 20$

3.137 Let n = first odd integer
$n + 2$ = 2nd consecutive odd integer
$n + n + 2 = 8$
$2n + 2 = 8$
$-2 = -2$
$2n = 6$
$\frac{2n}{2} = \frac{6}{2}$
$n = 3$
$n + 2 = 5$

3.138 Let n = first odd integer
$n + 2$ = 2nd consecutive odd integer
$n + n + 2 = 44$
$2n + 2 = 44$
$-2 = -2$
$2n = 42$
$\frac{2n}{2} = \frac{42}{2}$
$n = 21$
$n + 2 = 23$

3.139 Let n = first odd integer
$n + 2$ = 2nd consecutive odd integer
$n + 4$ = 3rd consecutive odd integer
$n + n + 2 + n + 4 = 39$
$3n + 6 = 39$
$-6 = -6$
$3n = 33$
$\frac{3n}{3} = \frac{33}{3}$
$n = 11$
$n + 2 = 13$
$n + 4 = 15$

3.140 Let n = first odd integer
$n + 2$ = 2nd consecutive odd integer
$n + 4$ = 3rd consecutive odd integer
$n + 6$ = 4th consecutive odd integer
$n + n + 2 + \text{n} + 4 + \text{n} + 6 = 24$
$4n + 12 = 24$

$$-12 = -12$$
$$4n = 12$$
$$\frac{4n}{4} = \frac{12}{4}$$
$$n = 3$$
$$n + 2 = 5$$
$$n + 4 = 7$$
$$n + 6 = 9$$

3.141 Let a = Cindy's age

$$5a = 35$$
$$\frac{5a}{5} = \frac{35}{5}$$
$$a = 7$$

3.142 Let a = Bill's age

$$a + 10 = 27$$
$$-10 = -10$$
$$a = 17 \text{ years}$$

3.143 Let a = Mac's age

$3a$ = Jim's age

$$a + 3a = 44$$
$$4a = 44$$
$$\frac{4a}{4} = \frac{44}{4}$$
$$a = 11 \text{ years}$$
$$3a = 33 \text{ years}$$

a. Mac is 11 years old.

b. Jim is 33 years old.

3.144 Let a = Joan's age now

$7a$ = Ann's age now

$a + 6$ = Joan's age in 6 years

$7a + 6$ = Ann's age in 6 years

$$7a + 6 = 5(a + 6)$$
$$7a + 6 = 5a + 30$$
$$-6 = -6$$
$$7a = 5a + 24$$
$$-5a = -5a$$
$$2a = 24$$
$$\frac{2a}{2} = \frac{24}{2}$$
$$a = 12 \text{ years}$$
$$7a = 84 \text{ years}$$

a. Joan is 12 yrs. old

b. Ann is 84 yrs. old

3.145 Let a = John's brother's age

$2a$ = John's age

$a - 4$ = John's brother's age four years ago

$2a - 4$ = John's age four years ago

$$4(a - 4) = 2a - 4$$
$$4a - 16 = 2a - 4$$
$$16 = 16$$
$$4a = 2a + 12$$
$$-2a = -2a$$
$$2a = 12$$
$$\frac{2a}{2} = \frac{12}{2}$$
$$a = 6 \text{ years}$$
$$2a = 12 \text{ years}$$

a. John is 12 yrs. old.

b. John's brother is 6 yrs old.

3.146 Let n = Robert's age

$n + 6$ = William's age

$n + 2$ = Robert's age in 2 years

$n + 8$ = William's age in 2 years

$$2(n + 2) = n + 8$$
$$2n + 4 = n + 8$$
$$-4 = -4$$
$$2n = n + 4$$
$$-n = -n$$
$$n = 4 \text{ years}$$
$$n + 6 = 10 \text{ years}$$

a. William is 10 yrs. old.

b. Robert is 4 yrs. old.

3.147 Let a = Jill's sister's age

$a - 2$ = Jill's sister's age 2 years ago

$3(a - 2)$ = Jill's age

$$3(a - 2) = 18$$
$$3a - 6 = 18$$
$$6 = 6$$
$$3a = 24$$
$$\frac{3a}{3} = \frac{24}{3}$$
$$a = 8 \text{ years}$$

Jill's sister is 8 yrs. old.

3.148 Let a = Bud's age

$\frac{1}{7}a = 8$

$7 \cdot \frac{1}{7}a = 7 \cdot 8$

$a = 56$ years

Bud's uncle is 56 yrs. old.

3.149 Let a = Irene's age

$2a$ = Don's age

$a + 16$ = Ed's age

$a + 2a + a + 16 = 60$

$4a + 16 = 60$

$-16 = -16$

$4a = 44$

$\frac{4a}{4} = \frac{44}{4}$

$a = 11$ years

$2a = 22$ years

$a + 16 = 27$ years

a. Irene is 11 yrs. old

b. Don is 22 yrs. old.

c. Ed is 27 yrs. old.

3.150 Let x = Anna's age

$2x - 10$ = Alice's age

$x + 2x - 10 = 26$

$3x - 10 = 26$

$10 = 10$

$3x = 36$

$\frac{3x}{3} = \frac{36}{3}$

$x = 12$ years

$2x - 10 = 24 - 10 = 14$ years

a. Anna is 12 yrs. old.

b. Alice is 14 yrs. old

3.151 Let d = distance needed

$\frac{120}{2} = \frac{d}{9}$

$2 \cdot d = 120 \cdot 9$

$2d = 1{,}080$

$2d = \frac{1{,}080}{2}$

$d = 540$ miles

3.152 Let t = time to read the entire book

$\frac{14}{30} = \frac{450}{t}$

$14 \cdot t = 30 \cdot 450$

$14t = 13{,}500$

$\frac{14t}{14} = \frac{13{,}500}{14}$

$t = 964.3$ min.

3. 153 Let v = winner's vote

$\frac{4}{3} = \frac{v}{5{,}760}$

$3 \cdot v = 4 \cdot 5{,}760$

$3v = 23{,}040$

$\frac{3v}{3} = \frac{23{,}040}{3}$

$v = 7{,}680$ votes

3.154 Let r = revolutions in 12 minutes

$\frac{400}{1} = \frac{r}{12}$

$1 \cdot r = 400 \cdot 12$

$r = 4{,}800$ revolutions

3.155 Let d = the distance traveled in 14 hours

$\frac{250}{4} = \frac{d}{14}$

$4 \cdot d = 250 \cdot 14$

$4d = 3{,}500$

$\frac{4d}{4} = \frac{3{,}500}{4}$

$d = 875$ miles

3.156 Let s = amount of sand

$\frac{2}{5} = \frac{17}{s}$

$2 \cdot s = 5 \cdot 17$

$2s = 85$

$\frac{2s}{2} = \frac{85}{2}$

$s = 42\frac{1}{2}$ bags

3.157 Let w = person's weight on Neptune

$\frac{5}{7} = \frac{125}{w}$

$5 \cdot w = 7 \cdot 125$

$5w = 875$

$\frac{5w}{5} = \frac{875}{5}$

$w = 175$ lb.

3.158 Let c = the needed charge of ore

$\frac{3}{2} = \frac{c}{86}$

$2 \cdot c = 3 \cdot 86$

$2c = 258$

$\frac{2c}{2} = \frac{258}{2}$

$c = 129$ tons

3.159 Let s = the necessary sulphur

$\frac{16}{49} = \frac{s}{40}$

$49 \cdot s = 16 \cdot 40$

$49s = 640$

$\frac{49s}{49} = \frac{640}{49}$

$s = 13.1$ kg

3.160 Let l = the needed lye

$\frac{18}{20} = \frac{950}{l}$

$18 \cdot l = 20 \cdot 950$

$18l = 19,000$

$\frac{18l}{18} = \frac{19,000}{18}$

$l = 1,055.6$ kg

3.161 Let b = the cost of 21 blenders

$\frac{9}{200} = \frac{21}{b}$

$9 \cdot b = 200 \cdot 21$

$9b = 4,200$

$\frac{9b}{9} = \frac{4,200}{9}$

$b = \$466.67$

3.162 Let s = the necessary sidewalk

$\frac{200}{20} = \frac{500}{s}$

$200 \cdot s = 20 \cdot 500$

$200s = 10,000$

$\frac{200s}{200} = \frac{10,000}{200}$

$s = 50$ feet

I. SECTION ONE

1.1	twenty-two
1.2	thirty-six
1.3	one hundred eighteen
1.4	three hundred nine
1.5	one thousand four hundred eighty-six
1.6	two thousand fifteen
1.7	three thousand eight hundred eighty-eight
1.8	ten thousand six hundred fourteen
1.9	sixty-three thousand four hundred seventy-one
1.10	two hundred thirteen thousand four hundred sixteen
1.11	56
1.12	85
1.13	168
1.14	304
1.15	948
1.16	1,437
1.17	8,500
1.18	78,000
1.19	100,396
1.20	516,555
1.21	300 + 40 + 6
1.22	1 = 1,000, 4 = 400, 6 = 60, 2 = 2
1.23	7,489
1.24	98,641
1.25	4,000 + 500 + 10 + 6
1.26	7,000,000,000; seven billion
1.27	27,000,000,000
1.28	8,000,000,000,000,000
1.29	ninety-six trillion
1.30	1,000,000 or one million
1.31	60
1.32	100
1.33	3,000
1.34	4,500
1.35	20,000
1.36	5, 1, 2
1.37	1, 2, 3, 5
1.38	141, 1, 6, 82
1.39	56, 146, 1,289
1.40	67
1.41	135
1.42	36
1.43	105
1.44	418
1.45	5,343
1.46	13,687
1.47	20,120,079
1.48	5,533,559
1.49	212
1.50	595

1.51 782

1.52 1,889

1.53 48,289

1.54 6,905

1.55 629

1.56 919

1.57 3,391

1.58 89,897

1.59 8,813

1.60 873

1.61 448

1.62 1,008

1.63 4,055

1.64 861

1.65 6,732

1.66 30,127

1.67 8,352

1.68 32,918

1.69 265,136

1.70 641,817

1.71 6,798,554

1.72 50,134,098

1.73 33,025,125

1.74 585,378,882

1.75 5

1.76 11

1.77 17

1.78 20

1.79 $127\frac{3}{4}$

1.80 $52\frac{1}{3}$

1.81 15

1.82 46

1.83 698

1.84 20

1.85 89

1.86 $810\frac{12}{17}$

1.87 56

1.88 125

1.89 413

1.90 \$54 + \$40 + \$35 + \$50 + \$47 = \$226

1.91 100 + 86 + 91 + 77 + 89 + 94 = 537

1.92 \$50.40 + \$6.47 + \$41.95 = \$98.82
\$460.54 – 98.82 = \$361.72 balance

1.93 62,314 – 51,266 = 11,048

1.94 0.25 • 24 = \$6.00

1.95 \$227 • 32 = \$7,264

1.96 84 ÷ 4 = 21 gals.

1.97 \$15.12 ÷ 36 = \$0.42 or 42¢ each

1.98 $p = 4 \cdot 4 = 16$ yds.
$p = 4 + 4 + 4 + 4 = 16$ yds.

1.99 12 • 8 = 96 in.

1.100 93 ÷ 3 = 31 in.

1.101 4 • 100 = 400 feet

1.102 $p = 5 + 6 + 8 + 10 + 13 = 42$ mm

1.103 $A = b \cdot h$
$A = 14 \cdot 6$
$A = 84$ sq. cm

1.104 6 = 6
8 – 2 = 6
2 • 3 = 6
1 + 2 + 3 = 6
9 – 3 = 6
4 + 6 = 10
VI = 6
4 + 6 does not represent 6

1.105 3 is the larger numeral; 9 is the larger number.

1.106 Examples: 2 • 9, XVIII, 9 + 9, 20 – 2

1.107 Yes; 4 + 7 = 11 and 13 – 2 = 11.

1.108 13

1.109 3

1.110 4

1.111 9

1.112 45¢

1.113 2

1.114 100,000 + 20,000 + 1,000 + 300 + 10 + 3 = 121,313

1.115 2,000,000 + 3,000 + 100 + 30 + 1 = 2,003,131

1.116 200 + 60 + 4 = 264

1.117 20,000 + 2,000 + 100 + 5 = 22,105

1.118 3,000 + 500 + 100 + 10 + 2 = 3,612

1.119 2,000 + 500 – 100 + 20 + 3 =
2,000 + 400 + 20 + 3 = 2,423

1.120 3,314 = 3,000 + 300 + 10 + 4 =

1.121 100,541 = 100,000 + 500 + 40 + 1 =

1.122 8,000,111 = 8,000,000 + 100 + 10 + 1 =

1.123 1,111,222 = 1,000,000 + 100,000 + 10,000 + 1,000 + 200 + 20 + 2 =

1.124 65 = 60 + 5 = ξ'ε'

1.125 143 = 100 + 40 + 3 = ρ'∠'∠'∠'∠'γ'

1.126 10,213 = 10,000 + 200 + 10 + 3 = ∠ρ'ρ'∠'γ'

1.127 20,369 = 20,000 + 300 + 60 + 9
= ∠∠ρ'ρ'ρ'ξ'9'

1.128 46 = 40 + 6 = XLVI

1.129 2,614 = 2,000 + 600 + 10 + 4
= MMDCXIV

1.130 1,979 = 1,000 + 900 + 70 + 9
= MCMLXXI X

1.131 2,444 = 2,000 + 400 + 40 + 4
= MMCDXLI V

1.132
$1 = 1_4$
$2 = 2_4$
$3 = 3_4$
$4 = 10_4$
$5 = 11_4$
$6 = 12_4$
$7 = 13_4$
$8 = 20_4$

1.133
$1 = 1_2$
$2 = 10_2$
$3 = 11_2$
$4 = 100_2$
$5 = 101_2$

1.134
$1 = 1_5$
$2 = 2_5$
$3 = 3_5$
$4 = 4_5$
$5 = 10_5$
$6 = 11_5$
$7 = 12_5$
$8 = 13_5$
$9 = 14_5$
$10 = 20_5$

1.135 $2,000 + 100 + 40 + 6$

1.136 $100_5 + 20_5 + 3_5$

1.137 $1000_2 + 1_2$

1.138 $2000_4 + 300_4 + 10_4$

1.139 6^4

1.140 8^7

1.141 $12_{10} = 2 \cdot 5 + 2 = 22_5$

1.142 $100_2 = 2^2 = 4$

1.143 $22_5 = 2 \cdot 5 + 2 = 12$

1.144 $8_{10} = 2^3 = 1000_2$

1.145 $5 \cdot 10^3 + 4 \cdot 10^2 + 6 \cdot 10^1 + 7 = 5,000 + 400 + 60 + 7 = 5,467$

1.146 231_5

1.147 1101_2

1.148 base 4

1.149 base 8

1.150 $3_5 + 3_5 = 11_5$

1.151 $10_2 + 11_2 + 100_2 = 1001_2$

1.152 {a, b, c, d, e, f, g, h, i, j}

1.153 {9, 10, 11, 12, 13, 14, 15, 16, 17, 18, 19, 20}

1.154 {the last three letters of the alphabet}

1.155 {the whole numbers between one and ten inclusive}

1.156 infinite

1.157 finite

1.158 infinite

1.159 $6 + 5 = 11$ and $5 + 6 = 11$, so $6 + 5 = 5 + 6$.

1.160 $17 \cdot 18 = 306$ and $18 \cdot 17 = 306$, so $17 \cdot 18 = 18 \cdot 17$.

1.161 $(9 + 8) + 7 = 17 + 7 = 24$ and $9 + (8 + 7) = 9 + 15 = 24$, so $(9 + 8) + 7 = 9 + (8 + 7)$.

1.162 $1 + 7 = 8$ and $7 + 1 = 8$, so $1 + 7 = 7 + 1$.

1.163 $(8 \cdot 7) \cdot 2 = 56 \cdot 2 = 112$ and $8 \cdot (7 \cdot 2) = 8 \cdot 14 = 112$, so $(8 \cdot 7) \cdot 2 = 8 \cdot (7 \cdot 2)$.

1.164 $A \cap B = \{b, c\}$

1.165 $X \cap Y = \{a, 1\}$

1.166 $P \cap Q = \{1, 2, 3, 6, 8\}$

1.167 $R \cap S = \{a, b, c, 1, 2, 3\}$

1.168

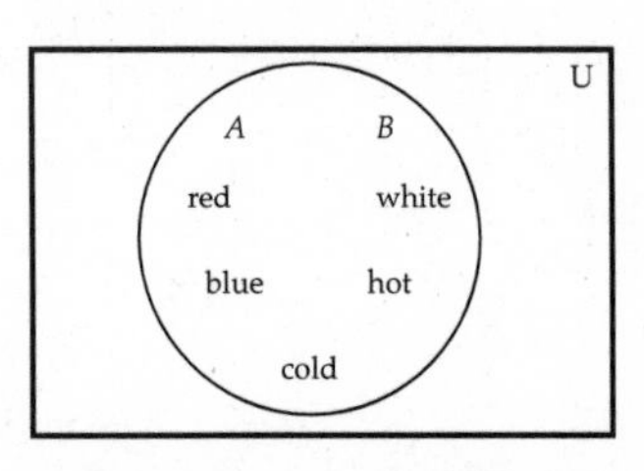

1.169

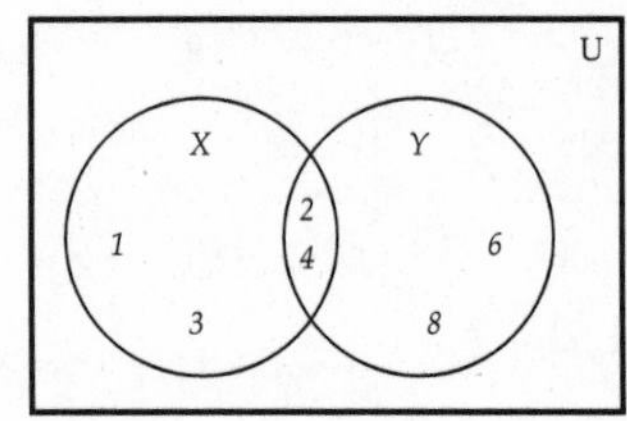

1.170 a.

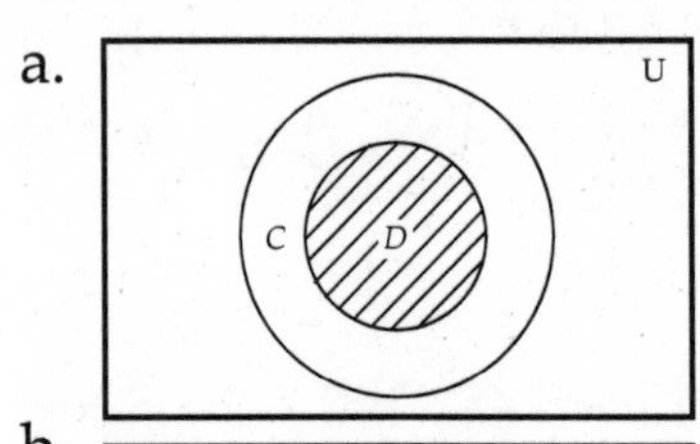

b.

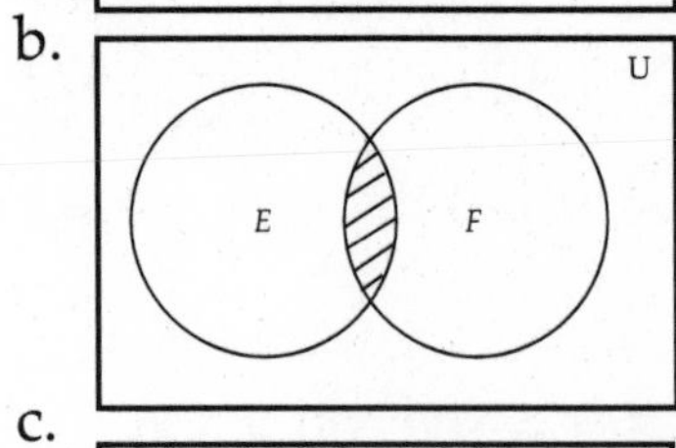

c.

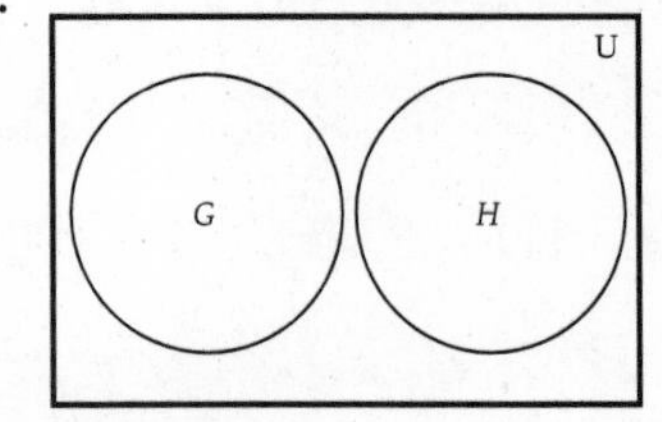

1.171 $15 = 3 \cdot 5$

1.172 $32 = 2 \cdot 2 \cdot 2 \cdot 2 \cdot 2 = 2^5$

1.173 $72 = 2 \cdot 2 \cdot 2 \cdot 3 \cdot 3 = 2^3 \cdot 3^2$

1.174 $150 = 2 \cdot 3 \cdot 5 \cdot 5 = 2 \cdot 3 \cdot 5^2$

1.175 6 is even, so 146 is divisible by 2.

1.176 The last digit is 0, so 1,810 is divisible by 5.

1.177 The sum of the digits = 4 + 1 + 1 + 2 + 1 = 9, so 41,121 is divisible by 9.

1.178 The sum of the digits = 1 + 2 + 1 + 4 + 3 + 1 = 12; 12 is divisible by 3, so 121,431 is divisible by 3.

1.179 $5^2 = 5 \cdot 5 = 25$
$2^5 = 2 \cdot 2 \cdot 2 \cdot 2 \cdot 2 = 32$
$6^3 = 6 \cdot 6 \cdot 6 = 216$

1.180 $7^2 = 7 \cdot 7 = 49$
$8^1 = 8$
$3^4 = 3 \cdot 3 \cdot 3 \cdot 3 = 81$

1.181 $121 = 11 \cdot 11 = 11^2$
$27 = 3 \cdot 3 \cdot 3 = 3^3$
$64 = 2 \cdot 2 \cdot 2 \cdot 2 \cdot 2 \cdot 2 = 2^6$

1.182 $144 = 2 \cdot 2 \cdot 2 \cdot 2 \cdot 3 \cdot 3 = 2^4 \cdot 3^2$
$81 = 3 \cdot 3 \cdot 3 \cdot 3 = 3^4$
$100 = 2 \cdot 2 \cdot 5 \cdot 5 = 2^2 \cdot 5^2$

1.183 $14 = 2 \cdot 7$
$28 = 2 \cdot 2 \cdot 7$
$\text{GCF} = 2 \cdot 7 = 14$

1.184 $27 = 3 \cdot 3 \cdot 3$
$63 = 3 \cdot 3 \cdot 7$
$\text{GCF} = 3 \cdot 3 = 9$

1.185 $13 = 13$
$26 = 2 \cdot 13$
$39 = 3 \cdot 13$
$\text{GCF} = 13$

1.186 $34 = 2 \cdot 17$
$51 = 3 \cdot 17$
$68 = 4 \cdot 17$
$\text{GCF} = 17$

1.187 $\frac{57}{76} = \frac{3 \cdot 19}{4 \cdot 19} = \frac{3}{4}$

1.188 $\frac{85}{102} = \frac{5 \cdot 17}{6 \cdot 17} = \frac{5}{6}$

1.189 $36 = 2 \cdot 2 \cdot 3 \cdot 3$
$54 = 2 \cdot 3 \cdot 3 \cdot 3$
$\text{LCM} = 2 \cdot 2 \cdot 3 \cdot 3 \cdot 3 = 108$

1.190 $12 = 2 \cdot 2 \cdot 3$
$15 = 3 \cdot 5$
$18 = 2 \cdot 3 \cdot 3$
$\text{LCM} = 2 \cdot 2 \cdot 3 \cdot 3 \cdot 5 = 180$

1.191 $4 = 2 \cdot 2$
$5 = 5$
$12 = 2 \cdot 2 \cdot 3$
$20 = 2 \cdot 2 \cdot 5$
$\text{LCM} = 2 \cdot 2 \cdot 3 \cdot 5 = 60$

1.192 $24 = 2 \cdot 2 \cdot 2 \cdot 3$
$30 = 2 \cdot 3 \cdot 5$
$42 = 2 \cdot 3 \cdot 7$
$60 = 2 \cdot 2 \cdot 3 \cdot 5$
$\text{LCM} = 2 \cdot 2 \cdot 2 \cdot 3 \cdot 5 \cdot 7 = 840$

1.193 $5 = 5$
$12 = 2 \cdot 2 \cdot 3$
$\text{LCD} = 2 \cdot 2 \cdot 3 \cdot 5 = 60$

1.194 $24 = 2 \cdot 2 \cdot 2 \cdot 3$
$16 = 2 \cdot 2 \cdot 2 \cdot 2$
$12 = 2 \cdot 2 \cdot 3$
$\text{LCD} = 2 \cdot 2 \cdot 2 \cdot 2 \cdot 3 = 48$

1.195 $\frac{4}{9}$

1.196 Examples: $\frac{2}{6}, \frac{3}{9}, \frac{4}{12}, \frac{5}{15}, \frac{6}{18}, \frac{7}{21}$

1.197 Examples: $\frac{4}{22}, \frac{6}{33}, \frac{8}{44}$

1.198 $\frac{6}{8} = \frac{2 \cdot 3}{2 \cdot 4} = \frac{3}{4}$

$\frac{12}{15} = \frac{3 \cdot 4}{3 \cdot 5} = \frac{4}{5}$

$\frac{14}{21} = \frac{2 \cdot 7}{3 \cdot 7} = \frac{2}{3}$

$\frac{21}{28} = \frac{3 \cdot 7}{4 \cdot 7} = \frac{3}{4}$

$\frac{32}{36} = \frac{4 \cdot 8}{4 \cdot 9} = \frac{8}{9}$

$\frac{15}{45} = \frac{1 \cdot 15}{3 \cdot 15} = \frac{1}{3}$

1.199 a. $\frac{1}{4} = \frac{1 \cdot 2}{4 \cdot 2} = \frac{2}{8}$

b. $\frac{1}{4} = \frac{1 \cdot 3}{4 \cdot 3} = \frac{3}{12}$

c. $\frac{1}{4} = \frac{1 \cdot 5}{4 \cdot 5} = \frac{5}{20}$

d. $\frac{1}{4} = \frac{1 \cdot 7}{4 \cdot 7} = \frac{7}{28}$

1.200 $\frac{8}{3} = 8 \div 3 = 2\frac{2}{3}$

1.201 $\frac{48}{16} = 48 \div 16 = 3$

1.202 a. $\frac{12}{5} = 12 \div 5 = 2\frac{2}{5}$

b. $\frac{18}{11} = 18 \div 11 = 1\frac{7}{11}$

c. $\frac{72}{10} = 72 \div 10 = 7\frac{2}{10} = 7\frac{1}{5}$

1.203 a. $4\frac{1}{3} = \frac{3 \cdot 4 + 1}{3} = \frac{13}{3}$

b. $6\frac{2}{5} = \frac{5 \cdot 6 + 2}{5} = \frac{32}{5}$

c. $8\frac{5}{6} = \frac{6 \cdot 8 + 5}{6} = \frac{53}{6}$

1.204 an infinite amount

1.205 $\frac{5}{2} = 2\frac{1}{2}$

$\frac{14}{5} = 2\frac{4}{5}$

$7\frac{1}{2}, 5, \frac{14}{5}, \frac{5}{2}, 1, \frac{2}{3}, \frac{1}{3}$

1.206 $\frac{12}{7} = 1\frac{5}{7}$

$\frac{4}{5}, 1, \frac{12}{7}, 2, 3\frac{1}{2}, 8$

1.207

0 $\frac{1}{4}$ $\frac{1}{2}$ 1 $1\frac{1}{2}$ 2 $2\frac{1}{2}$ 3 $3\frac{1}{2}$ 4

1.208 twenty-nine and thirteen hundredths

1.209 two hundred fourteen and one hundred forty-six thousandths

1.210 57.89

1.211 136.417

1.212 $(6 \cdot 100) + (7 \cdot 10) + (8 \cdot 1) + (2 \cdot \frac{1}{10})$

$+ (4 \cdot \frac{1}{100}) + (1 \cdot \frac{1}{1{,}000}) = 600 + 70$

$+ 8 + \frac{2}{10} + \frac{4}{100} + \frac{1}{1{,}000} = 678.241$

1.213 5,000,000; 5,000; 5; $\frac{5}{10}$; $\frac{5}{1{,}000}$

1.214 0.03

1.215 $\frac{1}{2} = 1 \div 2 = 0.5$

1.216 $\frac{3}{4} = 3 \div 4 = 0.75$

1.217 $0.05 = \frac{5}{100} = \frac{5 \cdot 1}{5 \cdot 20} = \frac{1}{20}$

1.218 $1.25 = 1\frac{25}{100} = 1\frac{1 \cdot 25}{4 \cdot 25} = 1\frac{1}{4}$

1.219 out of 100; hundredths; divided by 100; per hundred

1.220 To change a percent to a decimal, move the decimal point two places to the left and drop the percent sign. The decimals are 0.16, 0.29, 0.03, 2, and 0.86.

1.221 To change a decimal to a percent, move the decimal point two places to the right and add the percent sign. The percents are 50%, 18%, 6%, 100%, 81%, and 213%.

1.222 $\frac{1}{4} = 0.25 = 25\%$

$\frac{13}{100} = 0.13 = 13\%$

$\frac{1}{2} = 0.5 = 50\%$

$\frac{2}{5} = 0.4 = 40\%$

$\frac{47}{100} = 0.47 = 47\%$

$\frac{3}{4} = 0.75 = 75\%$

1.223 $\frac{156}{1000} = 0.156 = 15.6\%$

1.224 $\frac{773}{1000} = 0.773 = 77.3\%$

1.225 $\frac{57}{100} = 0.57 = 57\%$

1.226 $\frac{24}{36} = \frac{2 \cdot 12}{3 \cdot 12} = \frac{2}{3}$ or 2:3

1.227 $\frac{15}{18} = \frac{3 \cdot 5}{3 \cdot 6} = \frac{5}{6}$ or 5:6

1.228 $\frac{\text{6 nickles}}{\text{5 dimes}} = \frac{6 \cdot 5}{5 \cdot 10} = \frac{30}{50} = \frac{3}{5}$ or 3:5

1.229 $\frac{\text{16 inches}}{\text{4 feet}} = \frac{16 \cdot 1}{4 \cdot 12} = \frac{16}{48} = \frac{1}{3}$ or 1:3

1.230 $\frac{3}{5} = \frac{x}{10}$

$5 \cdot x = 3 \cdot 10$

$5x = 30$

$\frac{5x}{5} = \frac{30}{5}$

$x = 6$

1.231 $\frac{x}{7} = \frac{2}{8}$

$x \cdot 8 = 7 \cdot 2$

$8x = 14$

$\frac{8x}{8} = \frac{14}{8}$

$x = \frac{7}{4}$ or $1\frac{3}{4}$

1.232 $2:3 = 6:9$

$\frac{2}{3} = \frac{6}{9}$

$3 \cdot 6 = 18$

$2 \cdot 9 = 18$

1.233 $3:4 = 9:12$

$\frac{3}{4} = \frac{9}{12}$

$4 \cdot 9 = 36$

$3 \cdot 12 = 36$

1.234 $\frac{CX}{CA} = \frac{CY}{CB}$

$\frac{3}{5} = \frac{6}{CB}$

$3 \cdot CB = 5 \cdot 6$

$3CB = 30$

$\frac{3CB}{3} = \frac{30}{3}$

$CB = 10$

1.235 Let h = height of flagpole

$\frac{h}{40} = \frac{5}{4}$

$h \cdot 4 = 40 \cdot 5$

$4h = 200$

$\frac{4h}{4} = \frac{200}{4}$

$h = 50'$

1.236 Let l = length

$\frac{1}{100} = \frac{l}{900}$

$100 \cdot l = 1 \cdot 900$

$100l = 900$

$\frac{100l}{100} = \frac{900}{100}$

$l = 9$ inches

1.237 a. $1{,}000 = 10 \cdot 10 \cdot 10 = 10^3$

b. $10{,}000 = 10 \cdot 10 \cdot 10 \cdot 10 = 10^4$

1.238 $4{,}000{,}000 = 4 \cdot 1{,}000{,}000 =$
$4 \cdot 10 \cdot 10 \cdot 10 \cdot 10 \cdot 10 \cdot 10 = 4 \cdot 10^6$

1.239 $30 = 3 \cdot 10 = 3 \cdot 10^1$

1.240 $4 \cdot 10^5 = 4 \cdot 10 \cdot 10 \cdot 10 \cdot 10 \cdot 10 =$
$4 \cdot 100{,}000 = 400{,}000$

1.241 $2 \cdot 10^3 = 2 \cdot 10 \cdot 10 \cdot 10 =$
$2 \cdot 1{,}000 = 2{,}000$

1.242 1 m = 10 dm
3 m = $3 \cdot 10 = 30$ dm

1.243 1 cm = 10 mm
5 cm = $5 \cdot 10 = 50$ mm

1.244 1 mm = 0.01 dm
47 mm = $47 \cdot 0.01 = 0.47$ dm

1.245 1 mm = 0.001 m
516 mm = $516 \cdot 0.001 = 0.516$ m

1.246 1 m = 1,000 mm
2m = $2 \cdot 1{,}000 = 2{,}000$ mm

1.247 1 dg = 0.1 g
5 dg = $5 \cdot 0.1 = 0.5$ g

1.248 $0.\overline{4}$

1.249 $0.\overline{6}$

1.250 To change a percent to a decimal, move the decimal point two places to the left and drop the percent sign.
0.4% = 0.004

1.251 0.014% = 0.00014

1.252 To change a decimal to a percent, move the decimal point two places to the right and add the percent sign.
0.006 = 0.6%

1.253 0.000014 = 0.0014%

1.254 1.5 = 150%

1.255 240% = 2.4

1.256 A whole = 100%. Therefore,
161% = 161 – 100 = 61%
more than the whole.

1.257 \$40 – \$35 = \$5
$\frac{5}{40} = \frac{5 \cdot 1}{5 \cdot 8} = \frac{1}{8} =$
0.125 = 12.5%

1.258 $\frac{3}{30} = \frac{3 \cdot 1}{3 \cdot 10} = \frac{1}{10} = 0.1 = 10\%$

II. SECTION TWO

2.1 $\frac{2}{3}$

2.2 $\frac{3}{5}$

2.3 $\frac{8}{12} = \frac{2}{3}$

2.4 $\frac{14}{13} = 1\frac{1}{13}$

2.5 $\frac{23}{23} = 1$

2.6 $\frac{7}{14} = \frac{1}{2}$

2.7 $\frac{42}{52} = \frac{21}{26}$

2.8 $\frac{187}{201}$

2.9 $\frac{9}{8}$ or $1\frac{1}{8}$

2.10 $\frac{10}{16} = \frac{5}{8}$

2.11 $\frac{16}{32} = \frac{1}{2}$

2.12 $\frac{15}{15} = 1$

2.13 $\frac{1}{2} = \frac{2}{4}$
$\frac{1}{4} = \frac{1}{4}$
$\frac{3}{4}$

2.14 LCD = 15
$\frac{2}{5} = \frac{2 \cdot 3}{5 \cdot 3} = \frac{6}{15}$
$\frac{1}{3} = \frac{1 \cdot 5}{3 \cdot 5} = \frac{5}{15}$
$\frac{11}{15}$

2.15 LCD = 10

$$\frac{3}{5} = \frac{6}{10}$$
$$\frac{7}{10} = \frac{7}{10}$$
$$\frac{13}{10} \text{ or } 1\frac{3}{10}$$

2.16 LCD = 72

$$\frac{7}{8} = \frac{7 \cdot 9}{8 \cdot 9} = \frac{63}{72}$$
$$\frac{2}{9} = \frac{2 \cdot 8}{9 \cdot 8} = \frac{16}{72}$$
$$\frac{79}{72} \text{ or } 1\frac{7}{72}$$

2.17 LCD = 12

$$\frac{1}{3} = \frac{1 \cdot 4}{3 \cdot 4} = \frac{4}{12}$$
$$\frac{1}{4} = \frac{1 \cdot 3}{4 \cdot 3} = \frac{3}{12}$$
$$\frac{7}{12}$$

2.18 LCD = 35

$$\frac{3}{7} = \frac{3 \cdot 5}{7 \cdot 5} = \frac{15}{35}$$
$$\frac{2}{5} = \frac{2 \cdot 7}{5 \cdot 7} = \frac{14}{35}$$
$$\frac{29}{35}$$

2.19 LCD = 75

$$\frac{7}{25} = \frac{7 \cdot 3}{25 \cdot 3} = \frac{21}{75}$$
$$\frac{2}{15} = \frac{2 \cdot 5}{15 \cdot 5} = \frac{10}{75}$$
$$\frac{31}{75}$$

2.20 LCD = 108

$$\frac{4}{27} = \frac{4 \cdot 4}{27 \cdot 4} = \frac{16}{108}$$
$$\frac{5}{36} = \frac{5 \cdot 3}{36 \cdot 3} = \frac{15}{108}$$
$$\frac{31}{108}$$

2.21 LCD = 8

$$\frac{1}{8} = \frac{1}{8}$$
$$\frac{1}{4} = \frac{2}{8}$$
$$\frac{1}{2} = \frac{4}{8}$$
$$\frac{7}{8}$$

2.22 LCD = 30

$$\frac{1}{5} = \frac{6}{30}$$
$$\frac{2}{3} = \frac{20}{30}$$
$$\frac{1}{6} = \frac{5}{30}$$
$$\frac{31}{30} \text{ or } 1\frac{1}{30}$$

2.23 LCD = 60

$$\frac{2}{3} = \frac{40}{60}$$
$$\frac{1}{4} = \frac{15}{60}$$
$$\frac{2}{5} = \frac{24}{60}$$
$$\frac{1}{6} = \frac{10}{60}$$
$$\frac{89}{60} \text{ or } 1\frac{29}{60}$$

2.24 LCD = 840

$$\frac{2}{21} = \frac{80}{840}$$
$$\frac{1}{8} = \frac{105}{840}$$
$$\frac{2}{3} = \frac{560}{840}$$
$$\frac{1}{5} = \frac{168}{840}$$
$$\frac{913}{840} \text{ or } 1\frac{73}{840}$$

2.25 LCD = 60

$$\frac{1}{2} = \frac{30}{60}$$
$$\frac{1}{3} = \frac{20}{60}$$
$$\frac{1}{4} = \frac{15}{60}$$
$$\frac{1}{5} = \frac{12}{60}$$
$$\frac{1}{6} = \frac{10}{60}$$
$$\frac{87}{60} = \frac{29}{20} \text{ or } 1\frac{9}{20}$$

2.26 $4\frac{4}{5}$

2.27 $8\frac{3}{5}$

2.28 $14\frac{7}{8}$

2.29 $3\frac{3}{4}$

2.30 $7\frac{2}{3}$

2.31 $23\frac{8}{8} = 24$

2.32 LCD = 12

$$2\frac{1}{3} = 2\frac{4}{12}$$
$$3\frac{1}{4} = 3\frac{3}{12}$$
$$5\frac{7}{12}$$

2.33 LCD = 30

$$4\frac{2}{5} = 4\frac{12}{30}$$
$$5\frac{1}{6} = 5\frac{5}{30}$$
$$9\frac{17}{30}$$

2.34 LCD = 33

$$8\frac{7}{11} = 8\frac{21}{33}$$
$$5\frac{2}{3} = 5\frac{22}{33}$$
$$13\frac{43}{33} = 14\frac{10}{33}$$

2.35 LCD = 6

$$3\frac{1}{6} = 3\frac{1}{6}$$
$$8\frac{2}{3} = 8\frac{4}{6}$$
$$11\frac{5}{6}$$

2.36 LCD = 40

$$10\frac{2}{5} = 10\frac{16}{40}$$
$$8\frac{7}{8} = 8\frac{35}{40}$$
$$18\frac{51}{40} = 19\frac{11}{40}$$

2.37 LCD = 34

$$42\frac{10}{17} = 42\frac{20}{34}$$
$$26\frac{1}{2} = 26\frac{17}{34}$$
$$68\frac{37}{34} = 69\frac{3}{34}$$

2.38 LCD = 84

$$5\frac{1}{4} = 5\frac{21}{84}$$
$$2\frac{2}{3} = 2\frac{56}{84}$$
$$6\frac{4}{7} = 6\frac{48}{84}$$
$$13\frac{125}{84} = 14\frac{41}{84}$$

2.39 LCD = 720

$$48\frac{8}{9} = 48\frac{640}{720}$$
$$1\frac{5}{16} = 1\frac{225}{720}$$
$$12\frac{1}{5} = 12\frac{144}{720}$$
$$61\frac{1009}{720} = 62\frac{289}{720}$$

2.40 LCD = 60

$$181\frac{1}{5} = 181\frac{12}{60}$$
$$116\frac{1}{4} = 116\frac{15}{60}$$
$$814\frac{1}{6} = 814\frac{10}{60}$$
$$1{,}111\frac{37}{60}$$

2.41 LCD = 30

$$18\frac{11}{30} = 18\frac{11}{30}$$
$$8\frac{2}{3} = 8\frac{20}{30}$$
$$15\frac{5}{6} = 15\frac{25}{30}$$
$$41\frac{56}{30} = 42\frac{26}{30} = 42\frac{13}{15}$$

2.42 LCD = 198

$$100\frac{4}{9} = 100\frac{88}{198}$$
$$61\frac{8}{11} = 61\frac{144}{198}$$
$$118\frac{1}{2} = 118\frac{99}{198}$$
$$279\frac{331}{198} = 280\frac{133}{198}$$

2.43 LCD = 54

$$216\frac{5}{18} = 216\frac{15}{54}$$
$$509\frac{4}{9} = 509\frac{24}{54}$$
$$818\frac{14}{27} = 818\frac{28}{54}$$
$$1{,}543\frac{67}{54} = 1{,}544\frac{13}{54}$$

2.44 LCD = 12

$$4\frac{1}{2} = 4\frac{6}{12}$$
$$8\frac{2}{3} = 8\frac{8}{12}$$
$$5\frac{1}{4} = 5\frac{3}{12}$$
$$17\frac{17}{12} = 18\frac{5}{12}$$

2.45 LCD = 24

$$15\frac{1}{8} = 15\frac{3}{24}$$
$$21\frac{2}{3} = 21\frac{16}{24}$$
$$36\frac{19}{24}$$

2.46 LCD = 32

$$8\frac{1}{8} = 8\frac{4}{32}$$
$$9\frac{1}{16} = 9\frac{2}{32}$$
$$10\frac{1}{32} = 10\frac{1}{32}$$
$$27\frac{7}{32} \text{ inches}$$

2.47 LCD = 15

$$4\frac{1}{3} = 4\frac{5}{15}$$
$$5\frac{2}{5} = 5\frac{6}{15}$$
$$9\frac{11}{15} \text{ cups}$$

2.48 $\frac{2}{5}$

2.49 $\frac{4}{6} = \frac{2}{3}$

2.50 $\frac{2}{8} = \frac{1}{4}$

2.51 $\frac{5}{11}$

2.52 LCD = $2 \cdot 3 = 6$

$$\frac{2}{3} = \frac{4}{6}$$
$$\frac{1}{2} = \frac{3}{6}$$
$$\frac{1}{6}$$

2.53 LCD = $13 \cdot 3 = 39$

$$\frac{10}{13} = \frac{30}{39}$$
$$\frac{1}{3} = \frac{13}{39}$$
$$\frac{17}{39}$$

2.54 LCD = $9 \cdot 5 = 45$

$$\frac{8}{9} = \frac{40}{45}$$
$$\frac{1}{5} = \frac{9}{45}$$
$$\frac{31}{45}$$

2.55 LCD = 25

$$\frac{11}{25} = \frac{11}{25}$$
$$\frac{2}{5} = \frac{10}{25}$$
$$\frac{1}{25}$$

2.56 LCD = $3 \cdot 4 = 12$

$$1\frac{1}{3} = 1\frac{4}{12}$$
$$1\frac{1}{4} = 1\frac{3}{12}$$
$$\frac{1}{12}$$

2.57 LCD = $7 \cdot 2 = 14$

$$2\frac{5}{7} = 2\frac{10}{14}$$
$$1\frac{1}{2} = 1\frac{7}{14}$$
$$1\frac{3}{14}$$

2.58 LCD = $9 \cdot 5 = 45$

$$4\frac{7}{9} = 4\frac{35}{45}$$
$$2\frac{3}{5} = 2\frac{27}{45}$$
$$2\frac{8}{45}$$

2.59 LCD = 4

$$5 = 4\frac{4}{4}$$
$$\frac{3}{4} = \frac{3}{4}$$
$$4\frac{1}{4}$$

2.60 LCD = 8

$$7 = 6\frac{8}{8}$$
$$\frac{7}{8} = \frac{7}{8}$$
$$6\frac{1}{8}$$

2.61 LCD = 8

$$10 = 9\frac{8}{8}$$
$$\frac{1}{8} = \frac{1}{8}$$
$$9\frac{7}{8}$$

2.62 LCD = 6

$$4\frac{1}{6} = 4\frac{1}{6} = 3\frac{7}{6}$$
$$3\frac{1}{2} = 3\frac{3}{6} = 3\frac{3}{6}$$
$$\frac{4}{6} = \frac{2}{3}$$

2.63 LCD = $5 \cdot 9 = 45$

$$8\frac{2}{5} = 8\frac{18}{45} = 7\frac{63}{45}$$
$$4\frac{8}{9} = 4\frac{40}{45} = 4\frac{40}{45}$$
$$3\frac{23}{45}$$

2.64 LCD = 12

$$12\frac{1}{12} = 12\frac{1}{12} = 11\frac{13}{12}$$
$$11\frac{2}{3} = 11\frac{8}{12} = 11\frac{8}{12}$$
$$\frac{5}{12}$$

2.65 LCD = 12

$$41\frac{2}{3} = 41\frac{8}{12} = 40\frac{20}{12}$$
$$21\frac{11}{12} = 21\frac{11}{12} = 21\frac{11}{12}$$
$$19\frac{9}{12} = 19\frac{3}{4}$$

2.66 LCD = $25 \cdot 3 = 75$ or $15 \cdot 5 = 75$

$$64\frac{2}{25} = 64\frac{6}{75} = 63\frac{81}{75}$$
$$18\frac{11}{15} = 18\frac{55}{75} = 18\frac{55}{75}$$
$$45\frac{26}{75}$$

2.67 LCD = 21

$$110\frac{8}{21} = 110\frac{8}{21} = 109\frac{29}{21}$$
$$106\frac{5}{7} = 106\frac{15}{21} = 106\frac{15}{21}$$
$$3\frac{14}{21} = 3\frac{2}{3}$$

2.68 LCD = 3

$$12 = 11\frac{3}{3}$$
$$5\frac{2}{3} = 5\frac{2}{3}$$
$$6\frac{1}{3} \text{ ft.}$$

2.69 LCD = 4

$$21\frac{1}{2} = 21\frac{2}{4} = 20\frac{6}{4}$$
$$18\frac{3}{4} = 18\frac{3}{4} = 18\frac{3}{4}$$
$$2\frac{3}{4} \text{ ft.}$$

2.70 LCD = 2 • 3 • 5 = 30

$7\frac{1}{2} = 7\frac{15}{30}$

$+3\frac{2}{3} = 3\frac{20}{30}$

$10\frac{35}{30}$

$-\ 4\frac{3}{5} = 4\frac{18}{30}$

$6\frac{17}{30}$

2.71 91.79

2.71 9.16

2.73 75.066

2.74 75.296

2.75 44.347

2.76 86.1618

2.77 20.1

2.78 15.468

2.79 90.266

2.80 90.9181

2.81 7,135.208

2.82 75.9114

2.83 60.92

2.84 0.909

2.85 0.888

2.86 288.88

2.87 890.82

2.88 90.909

2.89 398.511

2.90 2,229.0704

2.91 4.90184

2.92 178.84

2.93 31.8

2.94 7.094

2.95 8,898.0814

2.96 $984.59

2.97 1,879

2.98 $4.77

2.99 51.4 mm

2.100 99.84 yd.

2.101 $\frac{1}{15}$

2.102 $\frac{1}{12}$

2.103 ${}^{1}\frac{\cancel{2}}{3} \cdot \frac{1}{\cancel{6}_{3}} = \frac{1}{9}$

2.104 $\frac{3}{28}$

2.105 $\frac{5}{{}_{3}\cancel{6}} \cdot \frac{\cancel{2}^{1}}{3} = \frac{5}{9}$

2.106 $\frac{{}^{1}\cancel{4}}{11} \cdot \frac{3}{\cancel{4}_{1}} = \frac{3}{11}$

2.107 $\frac{{}^{1}\cancel{2}}{{}_{1}\cancel{3}} \cdot \frac{1}{\cancel{4}_{2}} \cdot \frac{\cancel{3}^{1}}{5} = \frac{1}{10}$

2.108 $\frac{{}^{1}\cancel{5}}{{}_{3}\cancel{6}} \cdot \frac{1}{\cancel{5}_{1}} \cdot \frac{\cancel{2}^{1}}{7} = \frac{1}{21}$

2.109 $1\frac{1}{2} \cdot \frac{2}{3} = \frac{{}^{1}\cancel{3}}{{}_{1}\cancel{2}} \cdot \frac{\cancel{2}^{1}}{\cancel{3}_{1}} = 1$

2 110 $2\frac{1}{3} \cdot \frac{3}{5} = \frac{7}{{}_{1}\cancel{3}} \cdot \frac{\cancel{3}^{1}}{5} = \frac{7}{5}$ or $1\frac{2}{5}$

2.111 $3\frac{2}{5} \cdot \frac{6}{7} = \frac{17}{5} \cdot \frac{6}{7} = \frac{102}{35}$ or $2\frac{32}{35}$

2.112 $4\frac{1}{4} \cdot \frac{7}{8} = \frac{17}{4} \cdot \frac{7}{8} = \frac{119}{32}$ or $3\frac{23}{32}$

2.113 $5 \cdot \frac{1}{5} = \frac{{}^{1}\cancel{5}}{{}_{1}\cancel{1}} \cdot \frac{\cancel{1}^{1}}{\cancel{5}_{1}} = 1$

2.114 $6 \cdot \frac{2}{3} = \frac{{}^{2}\cancel{6}}{1} \cdot \frac{2}{\cancel{3}_{1}} = \frac{4}{1} = 4$

2.115 $1\frac{1}{4} \cdot 5 = \frac{5}{4} \cdot \frac{5}{1} = \frac{25}{4}$ or $6\frac{1}{4}$

2.116 $2\frac{2}{5} \cdot 7 = \frac{12}{5} \cdot \frac{7}{1} = \frac{84}{5}$ or $16\frac{4}{5}$

2.117 $2\frac{3}{4} \cdot 1\frac{1}{2} = \frac{11}{4} \cdot \frac{3}{2} = \frac{33}{8}$ or $4\frac{1}{8}$

2.118 $3\frac{1}{2} \cdot 2\frac{2}{3} = \frac{7}{{}_1\cancel{2}} \cdot \frac{\cancel{8}^4}{3} = \frac{28}{3}$ or $9\frac{1}{3}$

2.119 $4\frac{7}{8} \cdot 5\frac{2}{5} = \frac{39}{8} \cdot \frac{27}{5} = \frac{1{,}053}{40}$ or $26\frac{13}{40}$

2.120 $6\frac{9}{10} \cdot 7\frac{3}{7} = \frac{69}{10} \cdot \frac{52}{7} = \frac{3{,}588}{70}$ or

$51\frac{18}{70} = 51\frac{9}{35}$

2.121
$$\begin{array}{r} 1.41 \\ \underline{21.1} \\ 141 \\ 141 \\ \underline{282} \\ 29.751 \end{array}$$

2.122
$$\begin{array}{r} 36.1 \\ \underline{0.16} \\ 2166 \\ \underline{361} \\ 5.776 \end{array}$$

2.123
$$\begin{array}{r} 12.38 \\ \underline{68.2} \\ 2476 \\ 9904 \\ \underline{7428} \\ 844.316 \end{array}$$

2.124
$$\begin{array}{r} 3.039 \\ \underline{0.924} \\ 12156 \\ 6078 \\ \underline{27351} \\ 2.808036 \end{array}$$

2.125
$$\begin{array}{r} 567.23 \\ \underline{0.1138} \\ 453784 \\ 170169 \\ 56723 \\ \underline{56723} \\ 64.550774 \end{array}$$

2.126
$$\begin{array}{r} 4.006 \\ \underline{23.8} \\ 32048 \\ 12018 \\ \underline{8012} \\ 95.3428 \end{array}$$

2.127
$$\begin{array}{r} 832 \\ \underline{2.67} \\ 5824 \\ 4992 \\ \underline{1664} \\ 2{,}221.44 \end{array}$$

2.128
$$\begin{array}{r} 4{,}612 \\ \underline{0.306} \\ 27672 \\ \underline{13836} \\ 1{,}411.272 \end{array}$$

2.129 $\frac{1}{2} \div \frac{1}{4} = \frac{1}{{}_1\cancel{2}} \cdot \frac{\cancel{4}^2}{1} = \frac{2}{1} = 2$

2.130 $\frac{2}{3} \div \frac{1}{2} = \frac{2}{3} \cdot \frac{2}{1} = \frac{4}{3}$ or $1\frac{1}{3}$

2.131 $\frac{2}{3} \div \frac{3}{4} = \frac{2}{3} \cdot \frac{4}{3} = \frac{8}{9}$

2.132 $\frac{4}{5} \div \frac{3}{7} = \frac{4}{5} \cdot \frac{7}{3} = \frac{28}{15}$ or $1\frac{13}{15}$

2.133 $\frac{7}{8} \div \frac{2}{3} = \frac{7}{8} \cdot \frac{3}{2} = \frac{21}{16}$ or $1\frac{5}{16}$

2.134 $\frac{8}{9} \div \frac{6}{7} = \frac{{}^4\cancel{8}}{9} \cdot \frac{7}{\cancel{6}_3} = \frac{28}{27}$ or $1\frac{1}{27}$

2.135 $6 \div \frac{2}{3} = \frac{{}^3\cancel{6}}{1} \cdot \frac{3}{\cancel{2}_1} = \frac{9}{1} = 9$

2.136 $5 \div \frac{2}{5} = \frac{5}{1} \cdot \frac{5}{2} = \frac{25}{2}$ or $12\frac{1}{2}$

2.137 $\frac{3}{4} \div 6 = \frac{{}^1\cancel{3}}{4} \cdot \frac{1}{\cancel{6}_2} = \frac{1}{8}$

2.138 $\frac{5}{6} \div 8 = \frac{5}{6} \cdot \frac{1}{8} = \frac{5}{48}$

2.139 $1\frac{1}{2} \div \frac{1}{3} = \frac{3}{2} \cdot \frac{3}{1} = \frac{9}{2}$ or $4\frac{1}{2}$

2.140 $2\frac{2}{3} \div \frac{2}{3} = \frac{{}^4\cancel{8}}{{}_1\cancel{3}} \cdot \frac{\cancel{3}^1}{\cancel{2}_1} = \frac{4}{1} = 4$

2.141 $5 \div 1\frac{2}{3} = \frac{5}{1} \div \frac{5}{3} = \frac{^{1}\cancel{5}}{1} \cdot \frac{3}{\cancel{5}_{1}} = \frac{3}{1} = 3$

2.142 $7 \div \frac{6}{7} = \frac{7}{1} \cdot \frac{7}{6} = \frac{49}{6}$ or $8\frac{1}{6}$

2.143 $2\frac{1}{4} \div 1\frac{4}{5} = \frac{9}{4} \div \frac{9}{5} = \frac{^{1}\cancel{9}}{4} \cdot \frac{5}{\cancel{9}_{1}} = \frac{5}{4}$ or $1\frac{1}{4}$

2.144 $3\frac{2}{5} \div 1\frac{5}{6} = \frac{17}{5} \div \frac{11}{6} = \frac{17}{5} \cdot \frac{6}{11} = \frac{102}{55}$ or $1\frac{47}{55}$

2.145 $5\frac{1}{8} \div 2\frac{7}{9} = \frac{41}{8} \div \frac{25}{9} = \frac{41}{8} \cdot \frac{9}{25} = \frac{369}{200}$ or $1\frac{169}{200}$

2.146 $6\frac{4}{5} \div 4\frac{1}{8} = \frac{34}{5} \div \frac{33}{8} = \frac{34}{5} \cdot \frac{8}{33} = \frac{272}{165}$ or $1\frac{107}{165}$

2.147 $10\frac{4}{7} \div 5\frac{2}{5} = \frac{74}{7} \div \frac{27}{5} = \frac{74}{7} \cdot \frac{5}{27} = \frac{370}{189}$ or $1\frac{181}{189}$

2.148 $14\frac{1}{3} \div 12\frac{2}{5} = \frac{43}{3} \div \frac{62}{5} = \frac{43}{3} \cdot \frac{5}{62} = \frac{215}{186}$ or $1\frac{29}{186}$

2.149
```
            50.
0.007)0.350.0
         35
          0
```

2.150
```
         0.002
0.54 )0.00.108
           108
             0
```

2.151
```
      0.156
45 )7.020
    4 5
    252
    225
     270
     270
       0
```

2.152
```
            5,600.
0.023 )128.800.
        115
         138
         138
           0
```

2.153
```
       1.3 4
3.8 )5.0.9 2
      3 8
      1 29
      1 14
        152
        152
          0
```

2.154
```
          5.8
0.63 )3.65.4
       315
        504
        504
          0
```

For Problems 2.155 through 2.166: to change a decimal to a percent, move the decimal point two places to the right and add the percent sign.

2.155 27%

2.156 18%

2.157 89%

2.158 73%

2.159 3%

2.160 8%

2.161 1%

2.162 120%

2.163 180%

2.164 262%

2.165 319%

2.166 401%

2.167 $\frac{1}{4} = 0.25 = 25\%$

2.168 $\frac{3}{4} = 0.75 = 75\%$

2.169 $\frac{2}{5} = 0.4 = 40\%$

2.170 $\frac{5}{6} = 0.83\frac{1}{3} = 83\frac{1}{3}\%$

2.171 $\frac{4}{5} = 0.8 = 80\%$

2.172 $\frac{1}{8} = 0.125 = 12.5\%$

2.173 $\frac{4}{9} = 0.44\frac{4}{9} = 44\frac{4}{9}\%$

2.174 $\frac{1}{3} = 0.33\frac{1}{3} = 33\frac{1}{3}\%$

2.175 $1\frac{1}{2} = 1.5 = 150\%$

2.176 14% of $60 = 0.14 \cdot 60 = 8.4$

2.177 18% of $46 = 0.18 \cdot 46 = 8.28$

2.178 27% of $110 = 0.27 \cdot 110 = 29.7$

2.179 42% of $72 = 0.42 \cdot 72 = 30.24$

2.180 82% of $24 = 0.82 \cdot 24 = 19.68$

2.181 92% of $246 = 0.92 \cdot 246 = 226.32$

2.182 120% of $81 = 1.2 \cdot 81 = 97.2$

2.183 205% of $11.2 = 2.05 \cdot 11.2 = 22.96$

2.184 256% of $314 = 2.56 \cdot 314 = 803.84$

2.185 314% of $866 = 3.14 \cdot 866 = 2{,}719.24$

2.186 $8\% \cdot B = 64$
$B = 64 \div 8\%$
$B = 64 \div 0.08$
$B = 800$

2.187 $21\% \cdot B = 420$
$B = 420 \div 21\%$
$B = 420 \div 0.21$
$B = 2{,}000$

2.188 $6\% \cdot B = 78$
$B = 78 \div 6\%$
$B = 78 \div 0.06$
$B = 1{,}300$

2.189 $81\% \cdot B = 1{,}620$
$B = 1{,}620 \div 81\%$
$B = 1{,}620 \div 0.81$
$B = 2{,}000$

2.190 $100\% \cdot B = 133$
$B = 133 \div 100\%$
$B = 133 \div 1$
$B = 133$

2.191 $14\% \cdot B = 21$
$B = 21 \div 14\%$
$B = 21 \div 0.14$
$B = 150$

2.192 $2.5\% \cdot B = 10$
$B = 10 \div 2.5\%$
$B = 10 \div 0.025$
$B = 400$

2.193 $150\% \cdot B = 30$
$B = 30 \div 150\%$
$B = 30 \div 1.5$
$B = 20$

2.194 $15 = 10\% \cdot B$
$B \cdot 10\% = 15$
$B = 15 \div 10\%$
$B = 15 \div 0.1$
$B = 150$

2.195 $61 = 13\% \cdot B$
$B \cdot 13\% = 61$
$B = 61 \div 13\%$
$B = 61 \div 0.13$
$B = 469.2$

2.196 $\% \cdot 36 = 9$
$\% = 9 \div 36$
$\% = \frac{1}{4}$
$\frac{1}{4} = 25\%$

2.197
$\% \cdot 72 = 12$
$\% = 12 \div 72$
$\% = \frac{1}{6}$
$\frac{1}{6} = 16\frac{2}{3}\%$

2.198
$\% \cdot 42 = 84$
$\% = 84 \div 42$
$\% = 2$
$2 = 200\%$

2.199
$\% \cdot 250 = 100$
$\% = 100 \div 250$
$\% = \frac{2}{5}$
$\frac{2}{5} = 40\%$

2.200
$\% \cdot 14 = 7$
$\% = 7 \div 14$
$\% = \frac{1}{2}$
$\frac{1}{2} = 50\%$

2.201
$\% \div 7 = 14$
$\% = 14 \div 7$
$\% = 2$
$2 = 200\%$

2.202
$\% \cdot 210 = 210$
$\% = 210 \div 210$
$\% = 1$
$1 = 100\%$

2.203
$15\% \cdot 47 = N$
$0.15 \cdot 47 = N$
$7.05 = N$

2.204
$27\% \cdot B = 54$
$B = 54 \div 27\%$
$B = 54 \div 0.27$
$B = 200$

2.205
$\% \cdot 214 = 36$
$\% = 36 \div 214$
$\% = 0.168$
$0.168 = 16.8\%$

2.206
$18 = \% \cdot 27$
$\% \cdot 27 = 18$
$\% = 18 \div 27$
$\% = \frac{2}{3}$
$= 66\frac{2}{3}\%$

2.207
$110\% \cdot 215 = N$
$1.1 \cdot 215 = N$
$236.5 = N$

2.208
$10 = \% \cdot 100$
$\% \cdot 100 = 10$
$\% = 10 \div 100$
$\% = \frac{1}{10}$
$\frac{1}{10} = 10\%$

2.209
$67 = 30\% \cdot B$
$B \cdot 30\% = 67$
$B = 67 \div 30\%$
$B = 67 \div 0.3$
$B = 223\frac{1}{3}$

2.210
$\% \cdot 18 = 42$
$\% = 42 \div 18$
$\% = 2\frac{1}{3}$
$2\frac{1}{3} = 233\frac{1}{3}\%$

2.211
$19 = \% \cdot 57$
$\% \cdot 57 = 19$
$\% = 19 \div 57$
$\% = \frac{1}{3}$
$\frac{1}{3} = 33\frac{1}{3}\%$

2.212
$3.5\% \cdot 15.1 = N$
$0.035 \cdot 15.1 = N$
$0.5285 = N$

2.213
$6\% \cdot 110{,}000 = N$
$0.06 \cdot 100{,}000 = N$
$\$6{,}600 = N$

2.214 $\% \cdot 56{,}000 = 4{,}480$
$\% = 4{,}480 \div 56{,}000$
$\% = 0.08$
$0.08 = 8\%$

2.215 $5\% \cdot B = 140$
$B = 140 \div 5\%$
$B = 140 \div 0.05$
$B = \$2{,}800$

2.216 $15\% \cdot 200 = N$
$0.15 \cdot 200 = N$
$30 = N$

$20\% \cdot 200 = N$
$0.2 \cdot 200 = N$
$40 = N$

$10\% \cdot 200 = N$
$0.1 \cdot 200 = N$
$20 = N$

The collection contains 30 pennies, 40 nickels, and 20 dimes.

2.217 $12 + 16 + 20 + 24 + 28 = 100$
$\frac{100}{5} = 20.0$

2.218 $6 + 8 + 11 + 14 + 17 = 56$
$\frac{56}{5} = 11\frac{1}{5}$ or 11.2

2.219 $41 + 13 + 6 + 19 + 24 = 103$
$\frac{103}{5} = 20\frac{3}{5}$ or 20.6

2.220 $1 + 9 + 17 + 41 + 60 + 78 = 206$
$\frac{206}{6} = 34\frac{2}{6} = 34\frac{1}{3}$ or 34.3

2.221 $12 + 13 + 13 + 8 + 15 + 17 = 78$
$\frac{78}{6} = 13.0$

2.222 $1.6 + 6 + 8 + 8.1 + 5.7 = 29.4$
$\frac{29.4}{5} = 5.88 = 5.9$

2.223 14, 16, 18, (20), 22, 24, 26; the median is 20.

2.224 5, 6, 7, 8, (9) 10, 11, 12, 13; the median is 9.

2.225 1, 6, 11, (16, 21) 26, 31, 36; the median is $\frac{16 + 21}{2} = \frac{37}{2} = 18\frac{1}{2}$ or 18.5

2.226 21, 30, 39, (46) 57, 57, 86; the median is 46.

2.227 5, 15, 25, (35, 35) 35, 46, 51; the median is 35.

2.228 1, 41, 42, (56) 71, 81, 111; the median is 56.

2.229 4

2.230 7

2.231 1

2.232 91

2.233 2, 4, 4, 6, 8, 9, 11
a. mean $= 2 + 4 + 4 + 6 + 8 + 9 + 11 = 44$
$\frac{44}{7} = 6\frac{2}{7}$ or 6.3
b. median = 6
c. mode = 4

2.234 1, 1, 1, 5, 6, 7, 9, 11
a. mean $= 1 + 1 + 1 + 5 + 6 + 7 + 9 + 11 = 41$
$\frac{41}{8} = 5\frac{1}{8}$ or 5.125
b. median $= \frac{5 + 6}{2} = \frac{11}{2} = 5\frac{1}{2}$ or 5.5
c. mode = 1

2.235 7, 7, 8, 9, 51, 51, 91
a. mean $= 7 + 7 + 8 + 9 + 51 + 51 + 91 = 224$
$\frac{224}{7} = 32$
b. median = 9
c. mode = 7, 51

2.236 12, 16, 16, 17, 18, 18, 18, 19, 20, 20, 22, 22, 28, 30, 30

2.236 cont.

a. mean = 12 + 16 + 16 + 17 +18 + 18 + 18 + 19 + 20 + 20 + 22 + 22 + 28 + 30 + 30 = 306

$\frac{306}{15} = 20\frac{2}{5}$ or 20.4

b. median = 19

c. mode = 18

2.237 spread = 21 – 1 = 20

2.238 spread = 100 – 4 = 96

2.239 spread = 29 – 7 = 22

2.240 spread = 126 – 87 = 39

2.241 spread = 19.6 – 4.7 = 14.9

2.242 spread = 1.1 – 0.006 = 1.094

2.243 3, 6, 6, 6, 6, 7, 7, 7, 7, 7, 7, 8, 8, 8, 9, 9, 9, 11, 11, 14, 17, 17, 19

a. mean = $\frac{209}{23} = 9\frac{2}{23} = 9.1$

b. median = 8

c. mode = 7

d. spread = 19 – 3 = 16

2.244 47, 47, 47, 47, 47, 53, 66, 66, 81, 81, 82, 87, 91, 95, 96, 100, 101, 101 , 101

a. mean = $\frac{1{,}436}{19} = 75\frac{11}{19}$ or 75.6

b. median = 81

c. mode = 47

d. spread = 101– 47 = 54

2.245

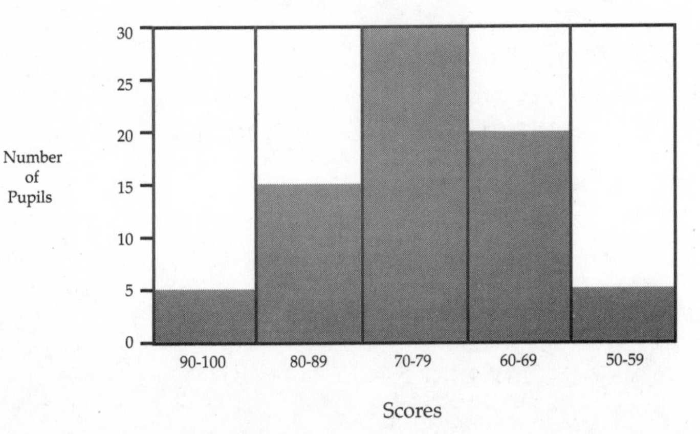

2.246

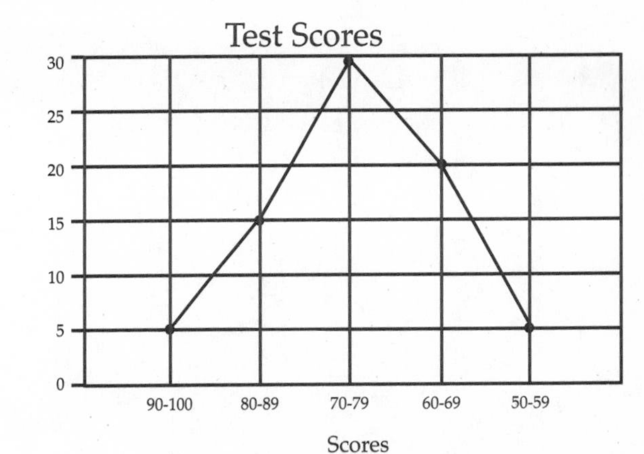

2.247

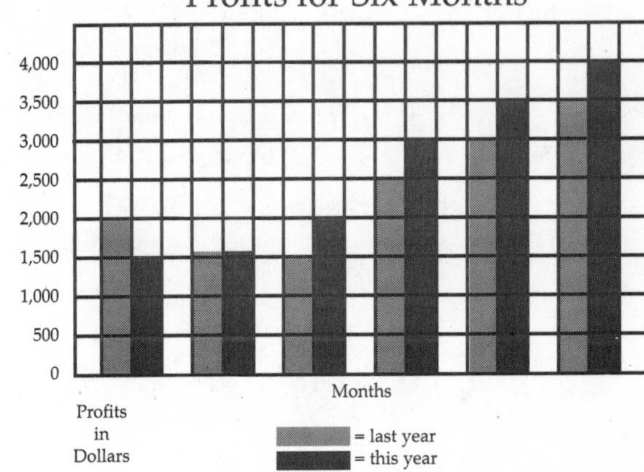

2.248

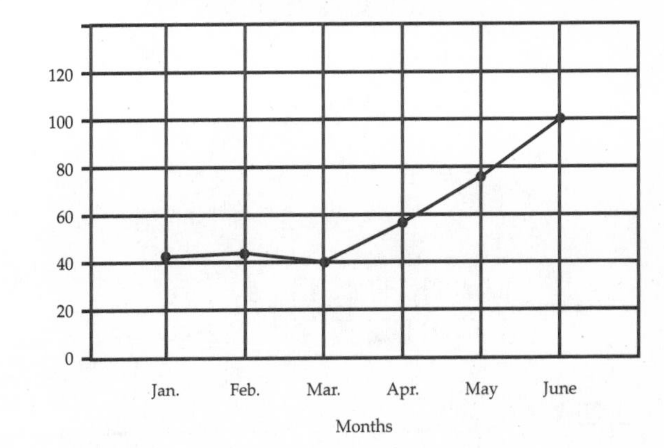

2.249 $f(n) = 3n$

$n = 0$: $f(n) = 3 \cdot 0 = 0$

$n = 2$: $f(n) = 3 \cdot 2 = 6$

$n = 4$: $f(n) = 3 \cdot 4 = 12$

$n = 5$: $f(n) = 3 \cdot 5 = 15$

n	0	1	2	3	4	5
$f(n)$	0	3	6	9	12	15

2.250 $f(n) = 3n - 1$

$n = 4$: $f(n) = 3 \cdot 4 - 1 = 12 - 1 = 11$

$n = 6$: $f(n) = 3 \cdot 6 - 1 = 18 - 1 = 17$

$n = 8$: $f(n) = 3 \cdot 8 - 1 = 24 - 1 = 23$

n	2	4	6	8	10
$f(n)$	5	11	17	23	29

2.251 $f(n) = 5n + 6$

$n = 0$: $f(n) = 5 \cdot 0 + 6 = 0 + 6 = 6$

$n = 2$: $f(n) = 5 \cdot 2 + 6 = 10 + 6 = 16$

$n = 3$: $f(n) = 5 \cdot 3 + 6 = 15 + 6 = 21$

$n = 5$: $f(n) = 5 \cdot 5 + 6 = 25 + 6 = 31$

n	0	1	2	3	4	5
$f(n)$	6	11	16	21	26	31

2.252 $f(n) = n^2 + 1$

$n = 1$: $f(n) = 1^2 + 1 = 1 + 1 = 2$

$n = 3$: $f(n) = 3^2 + 1 = 9 + 1 = 10$

$n = 4$: $f(n) = 4^2 + 1 = 16 + 1 = 17$

$n = 5$: $f(n) = 5^2 + 1 = 25 + 1 = 26$

n	0	1	2	3	4	5
$f(n)$	1	2	5	10	17	26

2.253

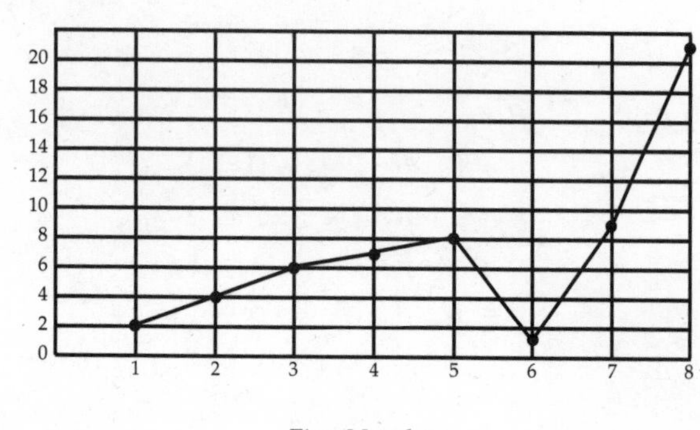

2.254 a. $P(\text{green}) = \frac{2}{16} = \frac{1}{8}$

b. $P\ (\text{red}) = \frac{3}{16}$

c. $P(\text{white}) = \frac{4}{16} = \frac{1}{4}$

d. $P\ (\text{black}) = \frac{7}{16}$

2.255 $P\ (\text{even}) = \frac{6}{12} = \frac{1}{2}$

2.256 The composite whole numbers between 2 and 15 are 4, 6, 8, 9, 10, 12, 14, and 15.
$P\ (\text{composite}) = \frac{8}{14} = \frac{4}{7}$

2.257 Sample space = {2, 3, 4, 5, 6, 7, 8, 9, 10, 11, 12, 13, 14, 15}

$P\ (\text{number} < 5) = \frac{3}{14}$

$P\ (\text{number} > 10) = \frac{5}{14}$

$P\ (\text{number} < 5 \text{ and } > 10) = \frac{3}{14} + \frac{5}{14}$

$\frac{8}{14} = \frac{4}{7}$

2.258 $P\ (\text{red}) = \frac{4}{25}$

$P\ (\text{black}) = \frac{6}{25}$

$P\ (\text{red or black}) = \frac{4}{25} + \frac{6}{25} = \frac{10}{25} = \frac{2}{5}$

2.259 Sample space A = {1, 3}.
Sample space B = {9, 11, 13}.

$P(A) = \frac{2}{7}$

$P(B) = \frac{3}{7}$

$P(A \text{ and } B) = P(A) \cdot P(B) = \frac{2}{7} \cdot \frac{3}{7} = \frac{6}{49}$

2.260 $P(A) = \frac{5}{25} = \frac{1}{5}$

$P(B) = \frac{6}{25}$

$P(A \text{ and } B) = P(A) \cdot P(B) = \frac{1}{5} \cdot \frac{6}{25} = \frac{6}{125}$

III. SECTION THREE

3.1 -1, -6, 9, 4, -16, 0

3.2 97, -1, 8, 1, -97, -6, -5, 7

3.3 -1, -2, -3, -4, -5, -6 . . .

3.4 0, 1, 2, 3, 4, 5 . . .

3.5 1, 3, 5, 7, 9 . . .

3.6 -5, -4, -3, -2, -1, 0, 1, 2, 3

3.7 1

3.8 -112

3.9 7, 3, 0, -1, -2, -3, -5, -9

3.10 -7, -3, 0, 1, 2, 3, 5, 9

3.11 $\frac{1}{5}$

3.12 -112

3.13
a. 8
b. 7
c. 14
d. 314
e. 321
f. 0

3.14
a. =
b. <
c. >
d. =

3.15 $|8| + |-9| + |-6| = 8 + 9 + 6 = 23$

3.16 $|-7| + |-10| + 9 = 7 + 10 + 9 = 26$

3.17 $|-31| - |-6| + |-2| =$
$31 - 6 + 2 = 27$

3.18 $|4 + 6| + |5 + 7| =$
$|10| - |12| =$
$10 - 12 = -2$

3.19 $|-5 + 1| + |5 - 1| - |-6| =$
$|-4| + |4| - |-6| =$
$4 + 4 - 6 = 2$

3.20 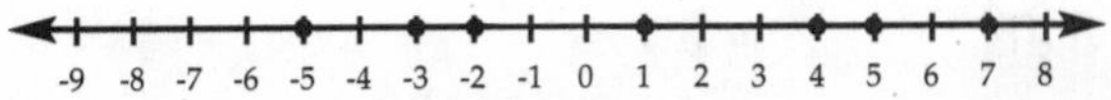 -9 -8 -7 -6 -5 -4 -3 -2 -1 0 1 2 3 4 5 6 7 8

3.21 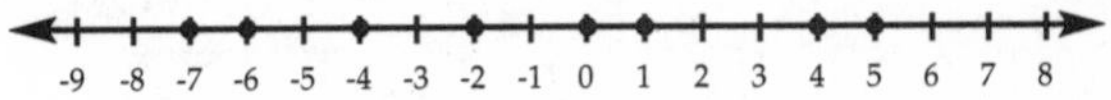 -9 -8 -7 -6 -5 -4 -3 -2 -1 0 1 2 3 4 5 6 7 8

3.22  -9 -8 -7 -6 -5 -4 -3 -2 -1 0 1 2 3 4 5 6 7 8

3.23 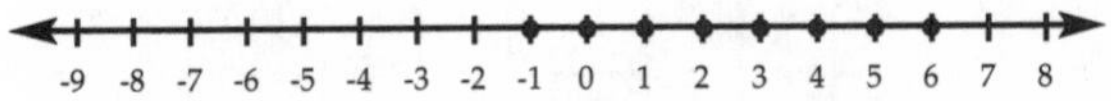 -9 -8 -7 -6 -5 -4 -3 -2 -1 0 1 2 3 4 5 6 7 8

3.24 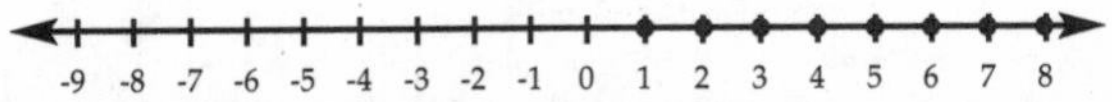 -9 -8 -7 -6 -5 -4 -3 -2 -1 0 1 2 3 4 5 6 7 8

3.25 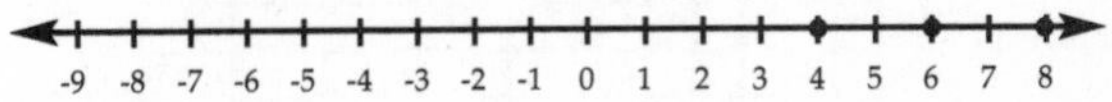 -9 -8 -7 -6 -5 -4 -3 -2 -1 0 1 2 3 4 5 6 7 8

3.26 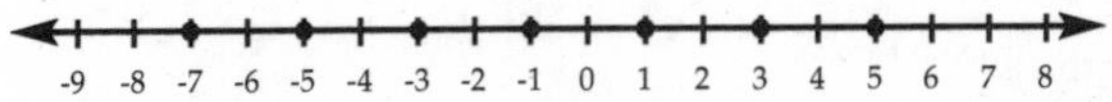 -9 -8 -7 -6 -5 -4 -3 -2 -1 0 1 2 3 4 5 6 7 8

3.27 -9 -8 -7 -6 -5 -4 -3 -2 -1 0 1 2 3 4 5 6 7 8

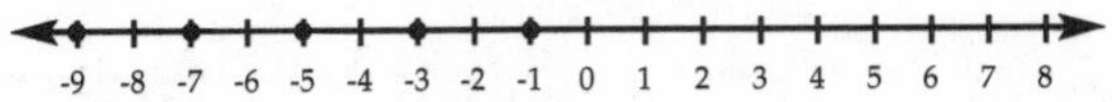

3.28 through 3.37

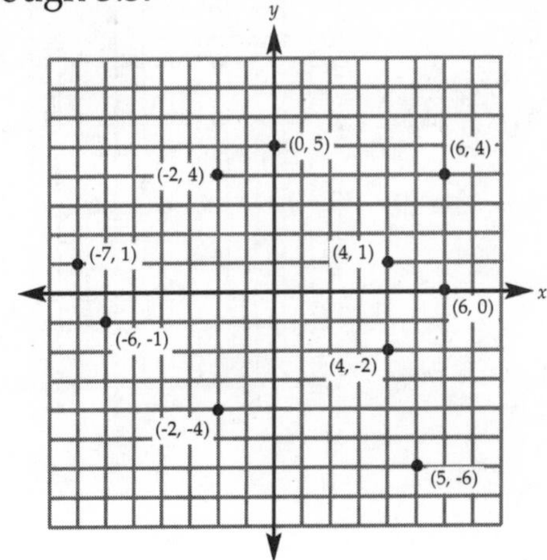

3.38 (1, 5)

3.39 (5, 0)

3.40 (2, -1)

3.41 (0, -5)

3.42 (-1, -4)

3.43 (-3, -2)

3.44 (-7 , 0)

3.45 (-5, 1)

3.46 (-2, 4)

3.47 (0, 3)

3.48 $A = b \cdot h$
$A = 10 \cdot 10 = 100$

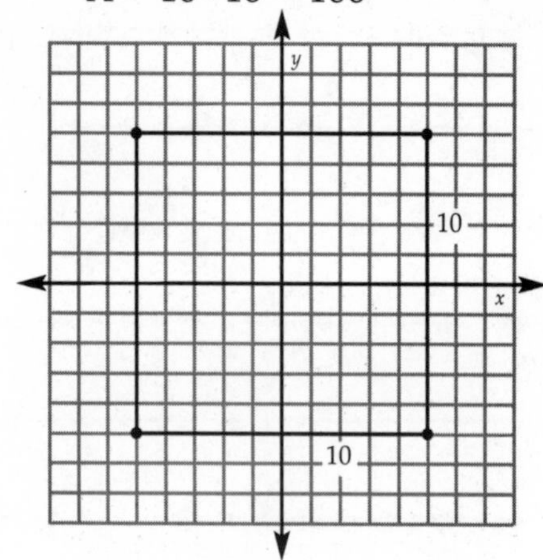

3.49

$A = \frac{b \cdot h}{2}$

$A = \frac{7 \cdot 8}{2}$

$A = \frac{56}{2} = 28$

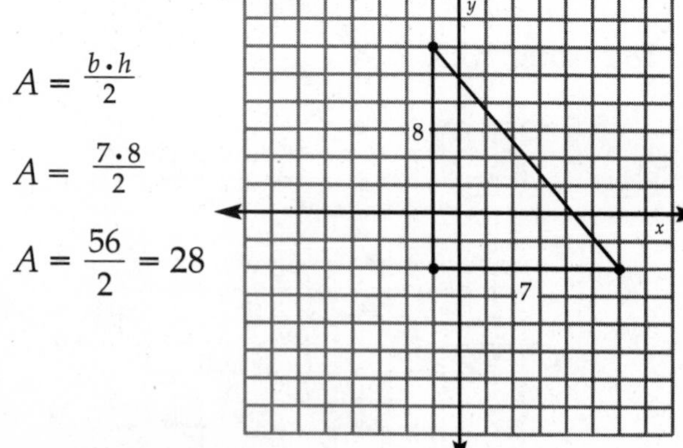

3.50 $A = s^2$
$A = 5^2 = 25$

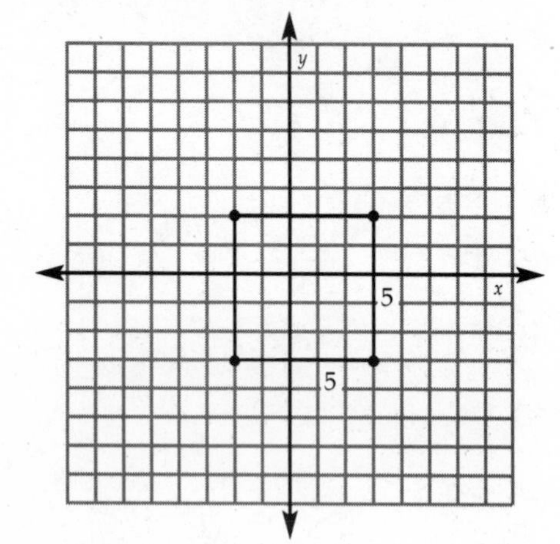

3.51 12

3.52 6 + 7 + (-2) =
6 + 7 – 2 = 11

3.53 14 + (-7) + 8 =
14 – 7 + 8 = 15

3.54 11 + 1 + (-8) =
11 + 1 – 8 = 4

3.55 2 + (-7) + (-8) + 9 + 10 =
2 – 7 – 8 + 9 + 10 = 6

3.56 (-1) + (-6) + (-7) + (-9) =
-1 – 6 – 7 – 9 = -23

3.57 2 + 3 + (-5) + (-6) + 7 =
2 + 3 – 5 – 6 + 7 = 1

3.58 12 + (-7) + 13 + (-11) + 6 =
12 – 7 + 13 – 11 + 6 = 13

3.59 (-19) + (-8) + 14 + 12 =
-19 – 8 + 14 + 12 = -1

3.60 40 + (-6) + 56 + (-54) =
40 – 6 + 56 – 54 = 36

3.61 100 + (-100) + 0 + (-5) + 5 =
100 – 100 + 0 + -5 + 5 = 0

3.62 (-216) + (-106) + (-400) =
-216 – 106 – -400 = -722

3.63 4

3.64 5

3.65 16

3.66 8

3.67 -47

3.68 -5

3.69 14 – (-2) = 14 + 2 = 16

3.70 27 – (-8) = 27 + 8 = 35

3.71 -81

3.72 -10 – (-26) = -10 + 26 = 16

3.73 3 – (-1) – 3 = 3 + 1 – 3 = 1

3.74 -6 – (-5) – 1 = -6 + 5 – 1 = -2

3.75 0 – (-6) = 0 + 6 = 6

3.76 $14 - (0) - (-8) = 14 - 0 + 8 = 22$

3.77 $6 - (-3) - 8 - (-7) - 10 =$
$6 + 3 - 8 + 7 - 10 = -2$

3.78 $-14 - 6 - (-19) - (-21) =$
$-14 - 6 + 19 + 21 = 20$

3.79 36

3.80 88

3.81 48

3.82 117

3.83 0

3.84 -126

3.85 -105

3.86 -86

3.87 -232

3.88 -498

3.89 0

3.90 432

3.91 -32

3.92 -576

3.93 $2^3 = 2 \cdot 2 \cdot 2 = 8$

3.94 $4^2 = 4 \cdot 4 = 16$

3.95 $5^4 = 5 \cdot 5 \cdot 5 \cdot 5 = 625$

3.96 $2^6 = 2 \cdot 2 \cdot 2 \cdot 2 \cdot 2 \cdot 2 = 64$

3.97 $(-6)^2 = (-6) \cdot (-6) = 36$

3.98 $(-7)^3 = (-7) \cdot (-7) \cdot (-7) = -343$

3.99 $(-2)^7 = (-2) \cdot (-2) \cdot (-2) \cdot (-2) \cdot (-2) \cdot (-2) \cdot (-2)$
$= -128$

3.100 $(-3)^4 = (-3) \cdot (-3) \cdot (-3) \cdot (-3) = 81$

3.101 $4^1 = 4$

3.102 $5^2 = 5 \cdot 5 = 25$

3.103 $6^2 = 6 \cdot 6 = 36$

3.104 $2^3 = 2 \cdot 2 \cdot 2 = 8$

3.105 $15^0 = 1$

3.106 $6^0 = 1$

3.107 $3^3 = 3 \cdot 3 \cdot 3 = 27$

3.108 5

3.109 -7

3.110 -4

3.111 0

3.112 8

3.113 -5

3.114 -1

3.115 -20

3.116 9

3.117 575

3.118 19

3.119 -30

3.120 $4a = 4 \cdot 1 = 4$

3.121 $5b = 5 \cdot 2 = 10$

3.122 $6c = 6(-3) = -18$

3.123 $7d = 7(-4) = -28$

3.124 $-2x = -2(0) = 0$

3.125 $8y = 8(-1) = -8$

3.126 $a - b = 1 - 2 = -1$

3.127 $x - y = 0 - (-1) = 0 + 1 = 1$

3.128 $c - d = -3 - (-4) = -3 + 4 = 1$

3.129 $2a - 3 = 2 \cdot 1 - 3 = 2 - 3 = -1$

3.130 $4d - x = 4(-4) - 0 = -16$

3.131 $6a + 2d = 6(1) + 2(-4) = 6 - 8 = -2$

3.132 $a + b - c = 1 + 2 - (-3) =$
$1 + 2 + 3 = 6$

3.133 $x + y - c =$
$0 + (-1) - (-3) =$
$-1 + 3 = 2$

3.134 $abc = 1 \cdot 2 \cdot (-3) = -6$

3.135 $cdxy = (-3) \cdot (-4) \cdot (0) \cdot (-1) = 0$

3.136 $\frac{a+b}{2} = \frac{1+2}{2} = \frac{3}{2}$ or $1\frac{1}{2}$

3.137 $\frac{c-d}{3} = \frac{-3-(-4)}{3} = \frac{-3+4}{3} = \frac{1}{3}$

3.138 $\frac{a}{b} = \frac{1}{2}$

3.139 $\frac{x}{y} = \frac{0}{-1} = 0$

3.140 $\frac{a+b}{c+d} = \frac{1+2}{-3+(-4)} = \frac{3}{-3-4} = \frac{3}{-7} = -\frac{3}{7}$

3.141 $x + y = 4$
Examples:
$x = 2$: $2 + y = 4$
$y = 4 - 2$
$y = 2$
$(2, 2)$
$x = 3$: $3 + y = 4$
$y = 4 - 3$
$y = 1$
$(3, 1)$
$x = 1$: $1 + y = 4$
$y = 4 - 1$
$y = 3$
$(1, 3)$
$x = 5$: $5 + y = 4$
$y = 4 - 5$
$y = -1$
$(5, -1)$

3.142 $x - y = 2$
Examples:
$x = 4$: $4 - y = 2$
$-y = 2 - 4$
$y = -2$
$y = 2$
$(4, 2)$
$x = 6$: $6 - y = 2$
$-y = 2 - 6$
$-y = -4$
$y = 4$
$(6, 4)$
$x = 1$: $1 - y = 2$
$-y = 2 - 1$
$-y = 1$
$y = -1$
$(1, -1)$
$x = -2$: $-2 - y = 2$
$-y = 2 + 2$
$y = 4$
$y = -4$
$(-2, -4)$

3.143 $x = y$
Examples:
$x = 1$: $1 = y$
$(1, 1)$
$x = -3$: $-3 = y$
$(-3, -3)$
$x = 5$: $5 = y$
$(5, 5)$
$x = -6$: $-6 = y$
$(-6, -6)$

3.144 $3x + 2y = 8$
Examples:
$x = 0$: $3 \cdot 0 + 2y = 8$
$2y = 8$
$\frac{2y}{2} = \frac{8}{2}$
$y = 4$
$(0, 4)$
$x = 2$: $3 \cdot 2 + 2y = 8$
$6 + 2y = 8$
$2y = 8 - 6$
$2y = 2$
$\frac{2y}{2} = \frac{2}{2}$
$y = 1$
$(2, 1)$
$x = 4$: $3 \cdot 4 + 2y = 8$
$12 + 2y = 8$
$2y = 8 - 12$
$2y = -4$

3.144 cont.

$\frac{2y}{2} = \frac{-4}{2}$

$y = -2$

(4, -2)

$x = 6$: $3 \cdot 6 + 2y = 8$

$18 + 2y = 8$

$2y = 8 - 18$

$2y = -10$

$\frac{2y}{2} = \frac{-10}{2}$

$y = -5$

(6, -5)

3.145 $3x + y = 6$

Examples:

$x = 2$: $3 \cdot 2 + y = 6$

$6 + y = 6$

$y = 6 - 6$

$y = 0$

(2, 0)

$x = 0$: $3 \cdot 0 + y = 6$

$y = 6$

(0, 6)

$x = 4$: $3 \cdot 4 + y = 6$

$12 + y = 6$

$y = 6 - 12$

$y = -6$

(4, - 6)

$x = 1$: $3 \cdot 1 + y = 6$

$3 + y = 6$

$y = 6 - 3$

$y = 3$

(1, 3)

3.146 $x = y^2$

Examples:

$y = 1$: $x = 1^2$

$x = 1$

(1, 1)

$y = 2$: $x = 2^2$

$x = 4$

(4, 2)

$y = 3$: $x = 3^2$

$x = 9$

(9, 3)

$y = 4$ $x = 4^2$

$x = 16$

(16, 4)

3.147 $x = 4y$

Examples:

$y = 1$: $x = 4 \cdot 1$

$x = 4$

(4, 1)

$y = 2$: $x = 4 \cdot 2$

$x = 8$

(8, 2)

$y = \frac{1}{2}$: $x = 4 \cdot \frac{1}{2}$

$x = 2$

$(2, \frac{1}{2})$

$y = 3$: $x = 4 \cdot 3$

$x = 12$

(12, 3)

3.148 $2x + 3y = 12$

Examples:

$x = 3$: $2 \cdot 3 + 3y = 12$

$6 + 3y = 12$

$3y = 12 - 6$

$3y = 6$

$\frac{3y}{3} = \frac{6}{3}$

$y = 2$

(3, 2)

$x = 6$: $2 \cdot 6 + 3y = 12$

$12 + 3y = 12$

$3y = 12 - 12$

$3y = 0$

$y = 0$

(6, 0)

$x = 12$: $2 \cdot 12 + 3y = 12$

$24 + 3y = 12$

$3y = 12 - 24$

$3y = -12$

$\frac{3y}{3} = \frac{-12}{3}$

$y = -4$

(12, -4)

$x = -3$: $2(-3) + 3y = 12$

$-6 + 3y = 12$

$3y = 12 + 6$

3.148 cont.

$$3y = 18$$
$$\frac{3y}{3} = \frac{18}{3}$$
$$y = 6$$
$$(-3, 6)$$

3.149 2

3.150 3

3.151 4

3.152 5

3.153 6

3.154 7

3.155 8

3.156 9

3.157 10

3.158 11

3.159 12

3.160 13

3.161

```
       1  4
     √1  96
      1
 24   0  96
         96
          0
```

$\sqrt{196} = 14$

3.162

```
           1  4 .  1   4   2
         √2  00 . 00  00  00
          1
   24     1  00
             96
   281       4   00
             2   81
   2,824     1   19   00
             1   12   96
   28,282             6   04   00
                      5   65   64
                          38   36
```

$\sqrt{200} = 14.14$

3.163

```
              1   0 .  4   8   8
            √1   10 . 00  00  00
              1
   20         0  10
                  0
   204           10   00
                  8   16
   2,088          1   84  00
                  1   67  04
   20,968             16  96  00
                      16  77  44
                          18  56
```

$\sqrt{110} = 10.49$

3.164 $C = 2 \cdot \pi \cdot r$
$C = 2 \cdot 3.14 \cdot 12$
$C = 75.36$ inches

3.165 $A = \pi \cdot r^2$
$A = 3.14 \cdot 8^2$
$A = 200.96$ sq. in.

3.166 $P = a + b + c$
$P = 15 + 18 + 21$
$P = 54$ inches

3.167 $A = \frac{1}{2} \cdot b \cdot h$
$A = \frac{1}{2} \cdot 18.6 \cdot 12$
$A = 111.6 \text{ mm}^2$

3.168 $a^2 + b^2 = c^2$
$6^2 + 8^2 = c^2$
$36 + 64 = c^2$
$100 = c^2$
$10 = c$

3.169 $P = 2l + 2w$
$P = 2 \cdot 14.2 + 2 \cdot 8.6$
$P = 28.4 + 17.2$
$P = 45.6$ cm

3.170 $A = b \cdot h$
$A = 11.16 \cdot 6.75$
$A = 75.33 \text{ m}^2$

3.171
$A = \frac{1}{2} \cdot h \cdot (a + b)$
$A = \frac{1}{2} \cdot 4.2 \cdot (6 + 12)$
$A = 2.1 \cdot (18)$
$A = 37.8 \text{ in.}^2$

3.172
$A = s^2$
$A = 8.1^2$
$A = 65.61$ sq. yds.
$P = 4s$
$P = 4 \cdot 8.1 = 32.4$ yds.

3.173
$P = 4 + 5 + 2 + 5 + 4 + 8$
$P = 28$

3.174
$V = l \cdot w \cdot h$
$V = 8 \cdot 4 \cdot 7$
$V = 224$ cu. in. or in.^{3}

3.175
$P = 2l + 2w$
$P = 2 \cdot 5.8 + 2 \cdot 3.9$
$P = 11.6 + 7.8$
$P = 19.4$ yd.
$LA = P \cdot h$
$LA = 19.4 \cdot 4.3$
$LA = 83.42$ sq. yds.

3.176
$TA = P \cdot h + 2 \cdot B$
$TA = (2 \cdot l + 2 \cdot w) \cdot h + 2 \cdot l \cdot w$
$TA = (2 \cdot 12 + 2 \cdot 9) \cdot 10 + 2 \cdot 12 \cdot 9$
$TA = (42) \cdot 10 + 2 \cdot 12 \cdot 9$
$TA = 420 + 216 = 636$ sq. cm
$V = l \cdot w \cdot h$
$V = 12 \cdot 9 \cdot 10$
$V = 1{,}080$ cu. cm

3.177
$V = l \cdot w \cdot h$
$V = 10 \cdot 12 \cdot 15$
$V = 1{,}800$ cu. ft.
$V = \frac{1800}{27} = 66\frac{2}{3}$ cu. yds. or yds^3.

3.178
$P = 4 \cdot s$
$P = 4 \cdot 11$
$P = 44$ cm
$LA = \frac{1}{2} \cdot P \cdot l$
$LA = \frac{1}{2} \cdot 44 \cdot 8.4$
$LA = 184.8$ sq. cm.

3.179
$B = s^2$
$B = 10^2$
$B = 100$
$V = \frac{1}{3} \cdot B \cdot h$
$V = \frac{1}{3} \cdot 100 \cdot 12$
$V = 400$ cu. m.

3.180
$P = 4 \cdot s$
$P = 4 \cdot 20 = 80$ in.
$LA = \frac{1}{2} \cdot P \cdot l$
$LA = \frac{1}{2} \cdot 80 \cdot 22$
$LA = 880$ sq. in.
$B = s^2 = 20^2 = 400$ sq. in.
$TA = LA + B$
$TA = 880 + 400$
$TA = 1{,}280$ sq. in.
$V = \frac{1}{3} \cdot B \cdot h$
$V = \frac{1}{3} \cdot 400 \cdot 18$
$V = 2{,}400$ cu. in.

3.181
$V = \frac{1}{3} \cdot B \cdot h$
$V = \frac{1}{3} \cdot 39 \cdot 6$
$V = 78$ cu. ft.

3.182
$C = 2 \cdot \pi \cdot r$
$C = 2 \cdot 3.14 \cdot 7$
$C = 43.96$ m

3.183
$L.A. = 2 \cdot \pi \cdot r \cdot h$
$L.A. = 2 \cdot 3.14 \cdot 6 \cdot 15$
$L.A. = 565.2$ sq. ft.
$T.A. = 2 \cdot \pi \cdot r \cdot (h + r)$
$T.A. = 2 \cdot 3.14 \cdot 6 \cdot (15 + 6)$
$T.A. = 37.68 \cdot (21)$
$T.A. = 791.28$ sq. ft.

3.184 $V = \pi \cdot r^2 \cdot h$
$V = 3.14 \cdot 3^2 \cdot 6$
$V = 169.56$ cu. yds.

3.185 $L.A. = 2 \cdot \pi \cdot r \cdot h$
$L.A. = 2 \cdot 3.14 \cdot 6 \cdot 10$
$L.A. = 376.8$ sq. ft.
$T.A. = 2 \cdot \pi \cdot r \cdot (h + r)$
$T.A. = 2 \cdot 3.14 \cdot 6 \cdot (10 + 6)$
$T.A. = 37.68 \cdot (16)$
$T.A. = 602.88$ sq. ft.
$V = \pi \cdot r^2 \cdot h$
$V = 3.14 \cdot 6^2 \cdot 10$
$V = 1{,}130.4$ cu. ft.

3.186 $L.A. = \pi \cdot r \cdot l$
$L.A. = 3.14 \cdot 5 \cdot 13$
$L.A. = 204.1$ sq. m.

3.187 $L.A. = \pi \cdot r \cdot l$
$L.A. = 3.14 \cdot 8 \cdot 14$
$L.A. = 351.68$ sq. in.
$T.A. = \pi \cdot r \cdot (l + r)$
$T.A. = 3.14 \cdot 8 \cdot (14 + 8)$
$T.A. = 25.12 \cdot (22)$
$T.A. = 552.64$ sq. in.

3.188 $V = \frac{1}{3} \cdot \pi \cdot r^2 \cdot h$
$V = \frac{1}{3} \cdot 3.14 \cdot 4^2 \cdot 12$
$V = \frac{1}{3} \cdot 3.14 \cdot 16 \cdot 12$
$V = 200.96$ cu. cm.

3.189 $L.A. = \pi \cdot r \cdot l$
$L.A. = 3.14 \cdot 9 \cdot 15$
$L.A. = 423.9$ sq. in.
$T.A. = \pi \cdot r \cdot (l + r)$
$T.A. = 3.14 \cdot 9 \cdot (15 + 9)$
$T.A. = 28.26 \cdot (24)$
$T.A. = 678.24$ sq. in.
$V = \frac{1}{3} \cdot \pi \cdot r^2 \cdot h$
$V = \frac{1}{3} \cdot 3.14 \cdot 9^2 \cdot 12$
$V = \frac{1}{3} \cdot 3.14 \cdot 81 \cdot 12$
$V = 1{,}017.36$ cu. in.

3.190 $S = 4 \cdot \pi \cdot r^2$
$S = 4 \cdot 3.14 \cdot 11^2$
$S = 4 \cdot 3.14 \cdot 121$
$S = 1{,}519.76$ sq. ft.

3.191 $V = \frac{4}{3} \cdot \pi \cdot r^3$
$V = \frac{4}{3} \cdot 3.14 \cdot 4^3$
$V = \frac{4}{3} \cdot 3.14 \cdot 64$
$V = 267.95$ cu. cm.

3.192 $S = 4 \cdot \pi \cdot r^2$
$S = 4 \cdot 3.14 \cdot 12^2$
$S = 4 \cdot 3.14 \cdot 144$
$S = 1.808.64$ sq. yds.
$V = \frac{4}{3} \cdot \pi \cdot r^3$
$V = \frac{4}{3} \cdot 3.14 \cdot 12^3$
$V = \frac{4}{3} \cdot 3.14 \cdot 1{,}728$
$V = 7{,}234.56$ cu. yds.

3.193 $V = \frac{1}{2} \cdot \frac{4}{3} \cdot \pi \cdot r^3$
$V = \frac{1}{2} \cdot \frac{4}{3} \cdot 3.14 \cdot 10^3$
$V = \frac{1}{2} \cdot \frac{4}{3} \cdot 3.14 \cdot 1{,}000$
$V = 2{,}093.3$ cu. m.

3.194 Since $11 + 2 = 13$ and
$2 + 11 = 13$,
$11 + 2 = 2 + 11$

3.195 Since $18 + 91 = 109$ and
$91 + 18 = 109$,
$18 + 91 = 91 + 18$

3.196 $a + b = b + a$

3.197 Since $1 + 5 = 6$ and
$5 + 1 = 6$,
$1 + 5 = 5 + 1$

3.198 Since $104 + 116 = 220$ and
$116 + 104 = 220$,
$104 + 116 = 116 + 104$

3.199 Since $409 + 747 = 1{,}156$ and $747 + 409 = 1{,}156$, $409 + 747 = 747 + 409$.

3.200 Since $5 \cdot 8 = 40$ and $8 \cdot 5 = 40$, so $5 \cdot 8 = 8 \cdot 5$.

3.201 Since $14 \cdot 71 = 994$ and $71 \cdot 14 = 994$, $14 \cdot 71 = 71 \cdot 14$

3.202 Since $96 \cdot 11 = 1{,}056$ and $11 \cdot 96 = 1{,}056$, $96 \cdot 11 = 11 \cdot 96$.

3.203 Since $(5 + 6) + 9 = 11 + 9 = 20$ and $5 + (6 + 9) = 5 + 15 = 20$, $(5 + 6) + 9 = 5 + (6 + 9)$.

3.204 Since $(15 + 18) + 2 = 33 + 2 = 35$ and $15 + (18 + 2) = 15 + 20 = 35$, $(15 + 18) + 2 = 15 + (18 + 2)$.

3.205 Since $(41 + 66) + 89 = 107 + 89 = 196$ and $41 + (66 + 89) = 41 + 155 = 196$, $(41 + 66) + 89 = 41 + (66 + 89)$.

3.206 Since $(x + y) + z = x + y + z$ and $x + (y + z) = x + y + z$, $(x + y) + z = x + (y + z)$.

3.207 Since $(4 + d) + 10 = 4 + d + 10 = 14 + d$ and $4 + (d + 10) = 4 + d + 10 = 14 + d$, $(4 + d) + 10 = 4 + (d + 10)$.

3.208 Since $(514 + 216) + 89 = 730 + 89 = 819$ and $514 + (216 + 89) = 514 + 305 = 819$, $(514 + 216) + 89 = 514 + (216 + 89)$.

3.209 Since $(17 \cdot 45) \cdot 8 = 765 \cdot 8 = 6{,}120$ and $17 \cdot (45 \cdot 8) = 17 \cdot 360 = 6{,}120$, $(17 \cdot 48) \cdot 8 = 17 \cdot (45 \cdot 8)$.

3.210 Since $(89 \cdot 34) \cdot 12 = 3{,}026 \cdot 12 = 36{,}312$ and $89 \cdot (34 \cdot 12) = 89 \cdot 408 = 36{,}312$, $(89 \cdot 34) \cdot 12 = 89 \cdot (34 \cdot 12)$.

3.211 Since $(125 \cdot 97) \cdot 57 = 12{,}125 \cdot 57 = 691{,}125$ and $125 \cdot (97 \cdot 57) = 125 \cdot 5{,}529 = 691{,}125$, $(125 \cdot 97) \cdot 57 = 125 \cdot (97 \cdot 57)$

3.212 Since $(a \cdot b) \cdot c = abc$ and $a \cdot (b \cdot c) = abc$, $(a \cdot b) \cdot c = a \cdot (b \cdot c)$.

3.213 -3, -2, -1, 0, 1, 2,

3.214 9

3.215 -14

3.216 -19 and 5

3.217 14

3.218 30

3.219 A double negative is a positive.

3.220 $-6t$

3.221 0

3.222 $-12pq$

3.223 $-c + 4d$

3.224 LCD = 12

$$\frac{1}{2}x + \frac{1}{3}y - \frac{1}{4}x + \frac{1}{6}y - \frac{2}{3}x =$$

$$\frac{6}{12}x + \frac{4}{12}y - \frac{3}{12}x + \frac{2}{12}y - \frac{8}{12}x =$$

$$-\frac{5}{12}x + \frac{6}{12}y = -\frac{5}{12}x + \frac{1}{2}y$$

3.225 $0.5r + 3.3s$

3.226 4, 15, 5, 47, 1, $\frac{1}{2}$

3.227 $4a - 28$

3.228 $5x - 5$

3.229 $12b + 18$

3.230 $28x + 14y$

3.231 $12a - 16b$

3.232 $24y - 8$

3.233 $40c - 48d + 32e$

3.234 $60p + 90q - 45r$

3.235 $5a^3 + 5a^2 - 35a$

3.236 $8r^5 + 8r^4 - 48r^3 + 8r^2$

3.237 $(x-2)\cdot(x-3) = x(x-3) - 2$
$(x-3) = x^2 - 3x - 2x + 6 =$
$x^2 - 5x + 6$

3.238 $(2x+3)\cdot(3x-1) = 2x(3x-1) +$
$3(3x-1) = 6x^2 - 2x + 9x - 3 =$
$6x^2 + 7x - 3$

3.239 $(3y-1)\cdot(2y+3) = 3y(2y+3) -$
$1(2y+3) = 6y^2 + 9y - 2y - 3 =$
$6y^2 + 7y - 3$

3.240 $(t+6)\cdot(t-2) = t(t-2) +$
$6(t-2) = t^2 - 2t + 6t - 12 =$
$t^2 + 4t - 12$

3.241 $(x+y)\cdot(x-y) = x(x-y) +$
$y(x-y) = x^2 - xy + xy - y^2 =$
$x^2 - y^2$

3.242 $(c-d)\cdot(a-b) = c(a-b) - d(a-b) =$
$ac - bc - ad + bd$

3.243 $(g+3)\cdot(g+3) = g(g+3) +$
$3(g+3) = g^2 + 3g + 3g + 9 =$
$g^2 + 6g + 9$

3.244 $5a + 10 = 5\cdot a + 5\cdot 2 =$
$5\cdot(a+2)$

3.245 $6x + 4 = 2\cdot 3x + 2\cdot 2 =$
$2\cdot(3x+2)$

3.246 $ab + a = a\cdot b + a\cdot 1 =$
$a\cdot(b+1)$

3.247 $a^2 + a = a\cdot a + a\cdot 1$
$a\cdot(a+1)$

3.248 $5a^2 + 10b^2 - 15 = 5\cdot a^2 +$
$5\cdot 2b^2 - 5\cdot 3 = 5\cdot(a^2 + 2b^2 - 3)$

3.249 $t^3 + 2t^2 - 3t = t\cdot t^2 + t\cdot 2t -$
$t\cdot 3 = t\cdot(t^2 + 2t - 3)$

3.250 $x - 4 = 11$
$4 = 4$
$x = 15$
Check:
$15 - 4 = 11$
$11 = 11$

3.251 $x + 7 = 10$
$-7 = -7$
$x = 3$
Check:
$3 + 7 = 10$
$10 = 10$

3.252 $3x = 15$
$\frac{3x}{3} = \frac{15}{3}$
$x = 5$
Check:
$3\cdot 5 = 15$
$15 = 15$

3.253 $\frac{x}{8} = 5$
$8\frac{x}{8} = 8\cdot 5$
$x = 40$
Check:
$\frac{40}{8} = 5$
$5 = 5$

3.254 $2x - 1 = 7$
$1 = 1$
$2x = 8$
$\frac{2x}{2} = \frac{8}{2}$
$x = 4$
Check:
$2\cdot 4 - 1 = 7$
$8 - 1 = 7$
$7 = 7$

3.255 $3y + 2 = 11$
$-2 = -2$
$3y = 9$
$\frac{3y}{3} = \frac{9}{3}$
$y = 3$
Check:
$3\cdot 3 + 2 = 11$
$9 + 2 = 11$
$11 = 11$

3.256
$$4y = y + 6$$
$$-y = -y$$
$$3y = 6$$
$$\frac{3y}{3} = \frac{6}{3}$$
$$y = 2$$
Check:
$$4 \cdot 2 = 2 + 6$$
$$8 = 2 + 6$$
$$8 = 8$$

3.257
$$5x - 9 = 1$$
$$9 = 9$$
$$5x = 10$$
$$\frac{5x}{5} = \frac{10}{5}$$
$$x = 2$$
Check:
$$5 \cdot 2 - 9 = 1$$
$$10 - 9 = 1$$
$$1 = 1$$

3.258
$$2a + a = 4 + 5$$
$$3a = 9$$
$$\frac{3a}{3} = \frac{9}{3}$$
$$a = 3$$
Check:
$$2 \cdot 3 + 3 = 4 + 5$$
$$6 + 3 = 9$$
$$9 = 9$$

3.259
$$6x + 3 = 6 + 3x$$
$$-3 = -3$$
$$6x = 3 + 3x$$
$$-3x = -3x$$
$$3x = 3$$
$$\frac{3x}{3} = \frac{3}{3}$$
$$x = 1$$
Check:
$$6 \cdot 1 + 3 = 6 + 3 \cdot 1$$
$$6 + 3 = 6 + 3$$
$$9 = 9$$

3.260
$$5a + 1 = 2a - 2$$
$$-1 = -1$$
$$5a = 2a - 3$$
$$-2a = -2a$$
$$3a = -3$$
$$\frac{3a}{3} = \frac{-3}{3}$$
$$a = -1$$
Check:
$$5(-1) + 1 = 2(-1) - 2$$
$$-5 + 1 = -2 - 2$$
$$-4 = -4$$

3.261
$$a + 6a = 8a - 4$$
$$7a = 8a - 4$$
$$-8a = 8a$$
$$-a = -4$$
$$a = 4$$
Check:
$$4 + 6 \cdot 4 = 8 \cdot 4 - 4$$
$$4 + 24 = 32 - 4$$
$$28 = 28$$

3.262
$$10p - 14 - p = 4p - 9$$
$$9p - 14 = 4p - 9$$
$$14 = 14$$
$$9p = 4p + 5$$
$$-4p = -4p$$
$$5p = 5$$
$$\frac{5p}{5} = \frac{5}{5}$$
$$p = 1$$
Check:
$$10 \cdot 1 - 14 - 1 = 4 \cdot 1 - 9$$
$$10 - 14 - 1 = 4 - 9$$
$$-5 = -5$$

3.263
$$6x + 1 - 2x = 3x - 3$$
$$4x + 1 = 3x - 3$$
$$-1 = -1$$
$$4x = 3x - 4$$
$$-3x = -3x$$
$$x = -4$$
Check:
$$6(-4) + 1 - 2(-4) = 3(-4) - 3$$
$$-24 + 1 + 8 = -12 - 3$$
$$-15 = -15$$

3.264
$$\begin{aligned} b - 8b + 7 &= 2b - 11 \\ -7b + 7 &= 2b - 11 \\ -7 &= \quad -7 \\ \hline -7b &= 2b - 18 \\ -2b &= -2b \\ \hline -9b &= -18 \\ \frac{-9b}{9} &= \frac{-18}{9} \\ b &= 2 \end{aligned}$$
Check:
$$\begin{aligned} 2 - 8 \cdot 2 + 7 &= 2 \cdot 2 - 11 \\ 2 - 16 + 7 &= 4 - 11 \\ -7 &= -7 \end{aligned}$$

3.265
$$\begin{aligned} 10x - 14 - x &= 4x - 9 \\ 9x - 14 &= 4x - 9 \\ 14 &= \quad 14 \\ \hline 9x &= 4x + 5 \\ -4x &= -4x \\ \hline 5x &= 5 \\ \frac{5x}{5} &= \frac{5}{5} \\ x &= 1 \end{aligned}$$
Check:
$$\begin{aligned} 10 \cdot 1 - 14 - 1 &= 4 \cdot 1 - 9 \\ 10 - 14 - 1 &= 4 - 9 \\ -5 &= -5 \end{aligned}$$

3.266
$$\begin{aligned} 4(a + 2) &= 2a - 4 \\ 4a + 8 &= 2a - 4 \\ -8 &= \quad -8 \\ \hline 4a &= 2a - 12 \\ -2a &= -2a \\ \hline 2a &= -12 \\ \frac{2a}{2} &= \frac{-12}{2} \\ a &= -6 \end{aligned}$$
Check:
$$\begin{aligned} 4(-6 + 2) &= 2(-6) - 4 \\ 4(-4) &= -12 - 4 \\ -16 &= -16 \end{aligned}$$

3.267
$$\begin{aligned} 5(2x - 3) &= 3x - 1 \\ 10x - 15 &= 3x - 1 \\ 15 &= \quad 15 \\ \hline 10x &= 3x + 14 \\ -3x &= -3x \\ \hline 7x &= 14 \\ \frac{7x}{7} &= \frac{14}{7} \\ x &= 2 \end{aligned}$$
Check:
$$\begin{aligned} 5(2 \cdot 2 - 3) &= 3 \cdot 2 - 1 \\ 5 \cdot 1 &= 6 - 1 \\ 5 &= 5 \end{aligned}$$

3.268
$$\begin{aligned} 7(3a + 1) - 6 &= 2a - 18 \\ 21a + 7 - 6 &= 2a - 18 \\ -1 &= \quad -1 \\ \hline 21a &= 2a - 19 \\ -2a &= -2a \\ \hline 19a &= -19 \\ \frac{19a}{19} &= \frac{-19}{19} \\ a &= -1 \end{aligned}$$
Check:
$$\begin{aligned} 7[3(-1) + 1] - 6 &= 2(-1) - 18 \\ -14 - 6 &= -2 - 18 \\ -20 &= -20 \end{aligned}$$

3.269 $n - 5$

3.270 $8n$

3.271 $4 + 2n$ or $2n + 4$

3.272 $7 + x$

3.273 $10(2n)$

3.274 $7 + 12x$

3.275 20% of n or 0.20 of n

3.276 $2y + 2$

3.277 $14 - n$

3.278 $\frac{2n}{5n}$

3.279 one more than a number *or* a number plus one

3.280 a number minus six

3.281 twice a number

3.282 one less than four times a number *or* four times a number minus one

3.283 five times a number divided by six

3.284 six more than five times a number *or* five times a number plus six

3.285 fifteen minus twice a number

3.286 seven more than six divided by a number *or* six divided by a number plus seven

3.287 three times a number plus five *or* five more than three times a number

3.288 the sum of five plus a number minus six

3.289 $n - 2 = 8$

3.290 $\frac{7}{n} = 10$

3.291 $2x - 1 = 14$

3.292 $5 + 2x = 4$

3.293 $20 - 2n = \frac{1}{2}n$

3.294 $7 + \frac{3}{4}x = x - 6$

3.295 $\frac{1}{2} = 3n + \frac{1}{4}$

3.296 $\frac{7}{m} = m - 8$

3.297 $n = 2n + 1$

3.298 $3x = x - 3$

3.299 Two less than a number is seven *or* a number minus two is seven.

3.300 Four times a number is eight.

3.301 A number minus six is twice the number.

3.302 Five divided by a number is two.

3.303 Two plus five times a number is six.

3.304 One less than one-half of a number is twice the number.

3.305 Five less than twice a number is seventeen.

3.306 The sum of three times a number and five is two less than the number.

3.307 Three-fifths of a number is two less than the number *or* three-fifths of a number is the number minus two.

3.308 Four times a number plus seven is three times the number minus five.

3.309

$$\begin{array}{r} n + 5 = 7 \\ -5 = -5 \\ \hline n = 2 \end{array}$$

3.310

$$\begin{array}{r} 3n = n - 4 \\ -n = -n \\ \hline 2n = -4 \\ \frac{2n}{2} = \frac{-4}{2} \\ n = -2 \end{array}$$

3.311

$$\frac{n}{8} = 5$$

$$8 \cdot \frac{n}{8} = 8 \cdot 5$$

$$n = 40$$

3.312

$$\begin{array}{r} 7 = 2x - 1 \\ 1 = \quad 1 \\ \hline 8 = 2x \\ \frac{8}{2} = \frac{2x}{2} \\ 4 = x \\ x = 4 \end{array}$$

3.313

$$\begin{array}{r} a + 9 = 10 \\ -9 = -9 \\ \hline a = 1 \end{array}$$

3.314

$$\begin{array}{r} 3y + 2 = 11 \\ -2 = -2 \\ \hline 3y = 9 \\ \frac{3y}{3} = \frac{9}{3} \\ y = 3 \end{array}$$

3.315 Let n = the number

$3n = 69$

$\frac{3n}{3} = \frac{69}{3}$

$n = 23$

3.316 Let n = the number

$n + 29 = 96$

$-29 = -29$

$n = 67$

3.317 Let s = George's salary

$\frac{1}{5}s = 217$

$5 \cdot \frac{1}{5}s = 5 \cdot 217$

$s = \$1{,}085$

3.318 Let n = one integer

$n = 1$ = next consecutive integer

$n + n + 1 = 21$

$2n + 1 = 21$

$-1 = -1$

$2n = 20$

$\frac{2n}{2} = \frac{20}{2}$

$n = 10$

$n + 1 = 11$

3.319 Let n = one integer

$n + 1$ = 2nd consecutive integer

$n + 2$ = 3rd consecutive integer

$n + n + 1 + n + 2 = 18$

$3n + 3 = 18$

$-3 = -3$

$3n = 15$

$\frac{3n}{3} = \frac{15}{3}$

$n = 5$

$n + 1 = 6$

$n + 2 = 7$

3.320 Let j = Jane's age

$\frac{6j}{6} = \frac{72}{6}$

$j = 12$ years

3.321 Let a = Jack's age now

$7a$ = Jim's age now

$a + 6$ = Jack's age in 6 years

$7a + 6$ = Jim's age in 6 years

$7a + 6 = 5(a + 6)$

$7a + 6 = 5a + 30$

$-6 \quad -6$

$7a = 5a + 24$

$-5a = -5a$

$2a = 24$

$\frac{2a}{2} = \frac{24}{2}$

$a = 12$ years

$7a = 84$ years

3.322 Let d = the distance

$\frac{300}{7} = \frac{d}{15}$

$7 \cdot d = 300 \cdot 15$

$7d = 4{,}500$

$\frac{7d}{7} = \frac{4{,}500}{7}$

$d = 642.9$ or 643 miles

3.323 Let x = the cost

$\frac{12}{360} = \frac{50}{x}$

$12 \cdot x = 360 \cdot 50$

$12x = 18{,}000$

$\frac{12x}{12} = \frac{18{,}000}{12}$

$x = \$1{,}500$

3.324 Let a = largest angle

$a - 67 = 35$

$67 = 67$

$a = 102°$

SELF TEST 1

1.01 c

1.02 a

1.03 e

1.04 g

1.05 b

1.06 f

1.07 d

1.08 four hundred ten

1.09 two thousand, four hundred twenty-six

1.010 57

1.011 4,006

1.012 60,000 + 1,000 + 500 + 40 + 7

1.013 62,000

1.014 61,547,000,000,000

1.015 sixty-one trillion, five hundred forty-seven billion

1.016 a. 650
b. 600

1.017 a. 1,450
b. 1,400

1.018 a. 14,680
b. 14,700

1.019
Monday
Tuesday
Wednesday
Thursday
Friday

each represents 1 can

1.020 vertical

1.021 Either order:
a. Matthew
b. Acts

1.022 Either order:
a. Galatians
b. Ephesians

1.023 10

1.024 16

1.025 Mark, Romans, I Corinthians

1.026
0 5 10 15 20 25 30 35 40 45

SELF TEST 2

2.01 f

2.02 e

2.03 b

2.04 i

2.05 k

2.06 g

2.07 h

2.08 m

2.09 l

2.010 a

2.011 c

2.012 j

2.013
```
 51
 47
 98
```

2.014
```
  7
 13
 26
 46
```

2.015 1,405

2.016 951

2.017 19,301

2.018 nineteen thousand, three hundred one

2.019 22

2.020 111

2.021 88

2.022 1,888

2.023
```
 314,281
 126,882
 187,399
```

2.024 187,000

2.025 498

2.026
```
    93
    39
   837
  279
 3,627
```

2.027
```
     591
     321
     591
   1182
  1773
 189,711
```

2.028
```
   2,007
     286
   12042
  16056
  4014
 574,002
```

2.029
```
      3,428
      5,867
      23996
     20568
    27424
   17140
 20,112,076
```

2.030 591; 11,820; 177,300

2.031
```
     7
 8 )56
    56
     0
```

2.032
```
    14
 7 )98
    7
    28
    28
     0
```

2.033
```
      14
 23 )322
     23
      92
      92
       0
```

2.034 $186\overline{)9{,}118} = 49 = 49\frac{4}{186} = 49\frac{2}{93}$
```
         49
 186 )9,118
       744
       1678
       1674
          4
```

2.035
```
      4,156
238 )989,128
     952
      371
      238
     1332
     1190
      1428
      1428
         0
```

2.036
```
    3,282,641,718
2 )6,565,283,436
   6
   05
    4
    16
    16
     05
      4
      12
      12
       08
        8
        03
         2
         14
         14
          03
           2
          16
          16
           0
```

2.037 30,048

2.038 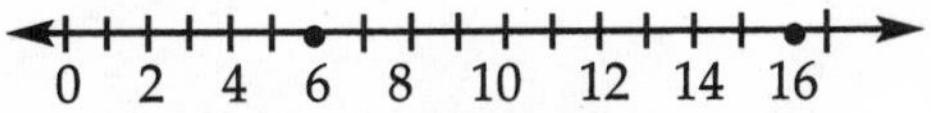

2.039 Any two in any order:
a. and b. pictographs,
bar graphs,
number lines

2.040 a. minuend
b. minus sign
c. subtrahend
d. equal sign
e. difference

2.041 Any order:
a. addition
b. subtraction
c. multiplication
d. division

SELF TEST 3

3.01 b

3.02 e

3.03 a

3.04 g

3.05 f

3.06 d

3.07 c

3.08 5,800

3.09 6,000

3.010 5,780

3.011 5,100,000

3.012 8,788

3.013 6,628,941

3.014 2,585

3.015 2,864

3.016
```
     572
     378
    4576
   4004
  1716
 216,216
```

3.017
```
     7,284
       525
     36420
    14568
  36420
 3,824,100
```

3.018
```
       75
27 )2,025
    189
     135
     135
       0
```

3.019

```
        475
112 )53,200
     448
      840
      784
       560
       560
         0
```

3.020 perimeter = 5 + 7 + 12
perimeter = 24 yards

3.021

```
  440
    3
1,320 feet
```

3.022

```
 56
 81
 49
 75
 92
353 points
```

3.023 a.

```
  1,205
+ 1,461
  2,666 students
```

b.

```
  1,461
- 1,205
    256 more boys
```

3.024

```
  24           72 nickels
x 15       5 )360
 120           35
 24             10
 360¢           10
                 0
```

3.025

```
   78 pieces of silver each
3 )234
   21
    24
    24
     0
```

3.026 fifteen million, four hundred sixty-seven thousand, five hundred ten

3.027 2,000,027,063

3.028

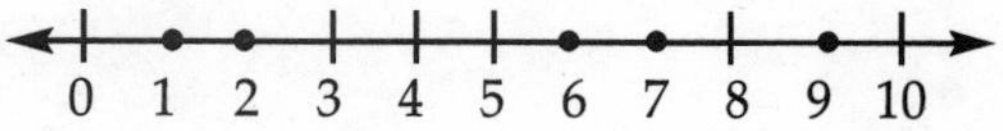

SELF TEST 1

1.01
a. $4 \times 3 = 12$
b. $24 \div 2 \times 1 = 12$
c. $6 \times 2 \times 0 = 0$
d. $18 - 6 = 12$
c is the correct answer.

1.02
a. $(3 + 5) \times 2 = 8 \times 2 = 16$
b. $(3 + 5) \div 2 = 8 \div 2 = 4$
c. $(3 \times 5) + 2 = 15 + 2 = 17$
d. $(3 \div 5) \times 2 = \frac{3}{5} \times 2 = \frac{6}{5}$
c is the correct answer.

1.03
a. 9||| is the Egyptian numeral for 103.
b. 12_2 does not exist; base 2 has only 0's and 1's.
c. XII = 12 in Roman numerals.
d. ~~||||~~ ~~||||~~ ~~||||~~ = 15
c is the correct choice.

1.04 d

1.05 b

1.06 c

1.07 b

1.08 d

1.09 d

1.010 b

1.011 Eight million, six hundred forty-two thousand, seven hundred ninety-eight

1.012 Seventy-six million, three hundred thousand, six hundred forty

1.013 11111_2

1.014 10000011_2

1.015 $11101_2 = (1 \times 2^4) + (1 \times 2^3) + (1 \times 2^2) + (1 \times 1)$
$= 16 + 8 + 4 + 1$
$= 29$

1.016 $110111_2 = (1 \times 2^5) + (1 \times 2^4) + (1 \times 2^2) + (1 \times 2^1) + (1 \times 1)$
$= 32 + 16 + 4 + 2 + 1 = 55$

1.017 $11001_2 = (1 \times 2^4) + (1 \times 2^3) + (1 \times 1) = 16 + 8 + 1 = 25$
$1011_2 = (1 \times 2^3) + (1 \times 2^1) + (1 \times 1) = 8 + 2 + 1 = 11$
$25 + 11 = 36$

1.018

+	0	1	2	3
0	0	1	2	3
1	1	2	3	0
2	2	3	0	1
3	3	0	1	2

$2 + 2 = 0$

1.019 Example:
Base 2 allows indefinite counting, while Mod 2 is a finite arithmetic.

1.020 Example:
The numeral is the symbol actually written. A number is an idea.

1.021
a. 100111110_2
b. $(1 \times 2^8) + (1 \times 2^5) + (1 \times 2^4) + (1 \times 2^3) + (1 \times 2^2) + (1 \times 2^1)$

SELF TEST 2

2.01 i

2.02 a

2.03 e

2.04 j

2.05 c

2.06 h

2.07 f

2.08 b

2.09 g

2.010 l

2.011 Either order:
- a. addition
- b. multiplication

2.012 Either order:
- a. subtraction
  Example: $5 - 2 \neq 2 - 5$
- b. division
  Example: $5 \div 2 \neq 2 \div 5$

2.013 Either order:
- a. by listing
- b. by rule

2.014 Either order:
- a. intersection ∩
- b. union ∪

2.015 {O, P}

2.016 {K, L, M, N, O, P, Q, R}

2.017 ∅

2.018 {K, L, M, N, O, P, H, I, J}

2.019 yes

2.020 b

2.021 a

2.022 c

2.023 true

2.024 true

2.025 true

256 128 64 32 16 eights fours twos ones

2.026 110010000_2

2.027 10001_2

2.028 $1111111_2 = (1 \times 2^6) + (1 \times 2^5) + (1 \times 2^4) + (1 \times 2^3) + (1 \times 2^2) + (1 \times 2^1) + (1 \times 1) = 64 + 32 + 16 + 8 + 4 + 2 + 1 = 127$

2.029 $10101_2 = (1 \times 2^4) + (1 \times 2^2) + (1 \times 1) = 16 + 4 + 1 = 21$

2.030 sixty-six thousand, six hundred sixty-six

2.031 six million, six hundred thousand, sixty

2.032 sixty-six million, six

2.033 six billion, six hundred thousand, six hundred

SELF TEST 3

3.01 j

3.02 c

3.03 f

3.04 i

3.05 e

3.06 g

3.07 d

3.08 a

3.09 h

3.010 $684 = 2 \times 2 \times 3 \times 3 \times 19$
$= 2^2 \times 3^2 \times 19$

3.011 $546 = 2 \times 3 \times 7 \times 13$

3.012 $378 = 2 \times 3 \times 3 \times 3 \times 7$
$= 2 \times 3^3 \times 7$

3.013 $12 = 2^2 \times 3$
$18 = 2 \times 3^2$
$GCF = 2 \times 3 = 6$

3.014 $114 = 2 \times 3 \times 19$
$190 = 2 \times 5 \times 19$
$GCF = 2 \times 19 = 38$

3.015 $15 = 3 \times 5$
$20 = 2^2 \times 5$
$GCF = 5$

3.016 $12 = 2^2 \times 3$
$18 = 2 \times 3^2$
$LCM = 2^2 \times 3 \times 3 = 4 \times 9 = 36$

3.017 $14 = 2 \times 7$
$52 = 2^2 \times 13$
$LCM = 2 \times 7 \times 2 \times 13 = 14 \times 26$
$= 364$

3.018 $15 = 3 \times 5$
$20 = 2^2 \times 5$
$LCM = 3 \times 5 \times 2^2 = 60$

3.019 $8 = 2^3$
$9 = 3^2$
$4 = 2^2$
$LCD = 2^3 \times 3^2 = 8 \times 9 = 72$

3.020 $3 = 1 \times 3$
$4 = 2^2$
$6 = 2 \times 3$
$LCD = 3 \times 2^2 = 3 \times 4 = 12$

3.021 $10 = 2 \times 5$
$20 = 2^2 \times 5$
$25 = 5^2$
$LCD = 2 \times 5 \times 2 \times 5 = 10 \times 10$
$= 100$

3.022 {1, 2, 3, 4, 5, 6, 7, 9}

3.023 {1, 2, 3, 4, 5, 6, 8, 10}

3.024 { } or ∅

3.025 {1, 3, 5}

3.026 a

3.027 c

3.028 b

3.029 d

3.030 b

3.031 d

3.032 c

3.033 c

SELF TEST 1

1.01 e

1.02 f

1.03 g

1.04 h

1.05 d

1.06 c

1.07 a

1.08 b

1.09 $\frac{7}{9}$

1.010 $\frac{3}{8}$

1.011 $\frac{5}{10} = \frac{5 \div 5}{10 \div 5} = \frac{1}{2}$

$\frac{4}{8} = \frac{4 \div 4}{8 \div 4} = \frac{1}{2}$

1.012 a. $\frac{20}{35} = \frac{20 \div 5}{35 \div 5} = \frac{4}{7}$

b. $\frac{28}{42} = \frac{28 \div 14}{42 \div 14} = \frac{2}{3}$

1.013 $28 \div 7 = 4$

$\frac{5}{7} = \frac{5 \times 4}{7 \times 4} = \frac{20}{28}$

1.014 $35 \div 5 = 7$

$\frac{5}{9} = \frac{5 \times 7}{9 \times 7} = \frac{35}{63}$

1.015 $\frac{12}{7} = 12 \div 7 = 1\frac{5}{7}$

1.016 divisor

1.017 $\frac{7}{1}$

1.018 $\frac{200}{25} = 200 \div 25 = 8$

1.019 $\frac{4}{1} = \frac{4 \times 3}{1 \times 3} = \frac{12}{3}$

1.020 $\frac{31}{6} = 31 \div 6 = 5\frac{1}{6}$

1.021 a. $1\frac{1}{2}$

b. $4\frac{2}{3}$

1.022 $4\frac{1}{3} = \frac{4 \times 3 + 1}{3} = \frac{12 + 1}{3} = \frac{13}{3}$

1.023 $\frac{8}{3} = 2\frac{2}{3}$

$\frac{6}{2} = 3$

$\frac{8}{5} = 1\frac{3}{5}$

$\frac{10}{5} = 2$

$4\frac{1}{5}, \frac{6}{2}, \frac{8}{3}, \frac{10}{5}, \frac{8}{5}, 1\frac{1}{4}, \frac{1}{2}, 0$

1.024 Example:

$\frac{72}{10} = \frac{72 \div 2}{10 \div 2} = \frac{36}{5}$

$\frac{72}{10} = \frac{72 \times 2}{10 \times 2} = \frac{144}{20}$

$\frac{72}{10} = \frac{36 \times 3}{5 \times 3} = \frac{108}{15}$

$\frac{72}{10} = \frac{36 \times 5}{5 \times 5} = \frac{180}{25}$

$\frac{72}{10} = \frac{72 \times 3}{10 \times 3} = \frac{216}{30}$

$\frac{72}{10} = 72 \div 10 = 7\frac{2}{10} = 7\frac{1}{5}$

1.025 a

1.026 c

1.027 a

1.028 d

$\frac{7}{3} = \frac{7 \times 2}{3 \times 2} = \frac{14}{6}$

1.029 c

$\frac{42}{70} = \frac{42 \div 14}{70 \div 14} = \frac{3}{5}$

SELF TEST 2

2.01 e

2.02 g

2.03 f

2.04 a

2.05 d

2.06 c

2.07 b

2.08 twelve and twenty-six hundredths

2.09 56.106

2.010 a. 37.47
b. thirty-seven and forty-seven hundredths

2.011 $\frac{211}{100} = 2\frac{11}{100} = 2.11$

2.012 a. 5 is 5 ones
b. 1 is 1 tenth
c. 3 is 3 hundredths

2.013 $40 + 2 + \frac{5}{10} + \frac{6}{100} + \frac{7}{1{,}000} =$
$42 + 0.5 + 0.06 + 0.007 =$
42.567

2.014 $36 \div 3 = 12$
$\frac{1}{3} = \frac{1 \times 12}{3 \times 12} = \frac{12}{36}$

2.015 $\frac{24}{64} = \frac{24 \div 8}{64 \div 8} = \frac{3}{8}$

2.016 $\frac{58}{19} = 58 \div 19 = 3\frac{1}{19}$

2.017 $\frac{1}{4}, \frac{1}{2}, \frac{2}{3}, \frac{5}{7}, \frac{5}{6}, 1\frac{1}{4}, 8$

2.018 5.6 is larger

2.019 0.25

2.020 $0.06 = \frac{6}{100} = \frac{6 \div 2}{100 \div 2} = \frac{3}{50}$

2.021 Examples:
out of 100, hundredths, divided by 100, per 100, in each 100, parts of 100

2.022 Drop the percent sign and move the decimal point two places to the left.
3% = 0.03

2.023 7 out of 100 or $\frac{7}{100}$ or 0.07

2.024 Move the decimal point two places to the right and add the percent sign.
0.31 = 31%

2.025 Drop the percent sign and write 100 as the denominator.
$20\% = \frac{20}{100} = \frac{20 \div 20}{100 \div 20} = \frac{1}{5}$

2.026 Examples:

$\frac{3}{5} = \frac{3 \times 2}{5 \times 2} = \frac{6}{10} = 0.6$

$\frac{3}{5} = 0.6 = 60\%$

$\frac{3}{5} = \frac{3 \times 2}{5 \times 2} = \frac{6}{10}$

$\frac{3}{5} = \frac{3 \times 3}{5 \times 3} = \frac{9}{15}$

$\frac{3}{5} = \frac{3 \times 4}{5 \times 4} = \frac{12}{20}$

$\frac{3}{5} = \frac{3 \times 5}{5 \times 5} = \frac{15}{25}$

$\frac{3}{5} = \frac{3 \times 6}{5 \times 6} = \frac{18}{30}$

$\frac{3}{5} = \frac{3 \times 7}{5 \times 7} = \frac{21}{35}$

$\frac{3}{5} = \frac{3 \times 8}{5 \times 8} = \frac{24}{40}$

$\frac{3}{5} = \frac{3 \times 9}{5 \times 9} = \frac{27}{45}$

$\frac{3}{5} = \frac{3 \times 10}{5 \times 10} = \frac{30}{50}$

$\frac{3}{5} = \frac{3 \times 16}{5 \times 16} = \frac{48}{80}$

2.027 b

2.028 b

$\frac{3}{5} = \frac{3 \times 2}{5 \times 2} = \frac{6}{10}$ (a)

$\frac{3}{5} = \frac{3 \times 3}{5 \times 3} = \frac{9}{15}$ (c)

$\frac{3}{5} = \frac{3 \times 4}{5 \times 4} = \frac{12}{20}$ (d)

2.029 c

2.030 c

$\frac{7}{10} = \frac{7 \times 10}{10 \times 10} = \frac{70}{100} = 70\%$

2.031 c

SELF TEST 3

3.01 c

3.02 g

3.03 d

3.04 b

3.05 k

3.06 e

3.07 f

3.08 j

3.09 i

3.010 a

3.011 5:16 or $\frac{5}{16}$

3.012 The means are 7 and 10; the extremes are 5 and 14.
7 x 10 = 70 and 5 x 14 = 70;
7 x 10 = 5 x 14 is true.

3.013 $\frac{22}{5} = 22 \div 5 = 4\frac{2}{5}$

3.014 1:100 = ?:650
100 x ? = 1 x 650
100 x ? = 650
? = 650 ÷ 100
? = 6.5 inches

3.015 10^6

3.016 0.211

3.017 15.19

3.018 1 m = 100 cm, so
9 m = 9 x 100 = 900 cm

3.019 $\frac{1}{8} = 1 \div 8 = 0.125$ or 0.1250

3.020 35 ÷ 7 = 5

$\frac{1}{7} = \frac{1 \times 5}{7 \times 5} = \frac{5}{35}$

3.021 $\frac{7}{9} = 9\overline{)7.000}$ = 0.777 . . . = $0.\overline{7}$

```
   0.777...
9 )7.000
   63
    70
    63
     70
     63
      7
```

3.022 Drop the % sign and move the decimal point two places to the left.
0.12% = 0.0012

3.023 Move the decimal point to the right two places and add the % sign.
2.16 = 216%

3.024 $\frac{10}{11}$

3.025
```
  5.18
+ 0.019
  5.199
```

3.026

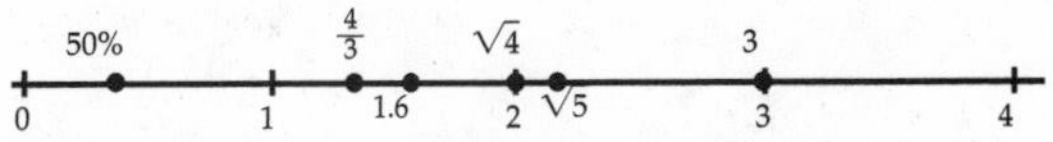

3.027 b

3.028 a

3.029 c

3.030 b
2.81 = 281%

3.031 b

3.032 b

SELF TEST 1

1.01 a. 4
b. 3

1.02 mixed

1.03 unlike

1.04 improper

1.05 cannot

1.06 true

1.07 equivalent fractions

1.08 $\frac{6}{8} = \frac{3}{4}$

1.09 $17\frac{2}{3}$

1.010 $7\frac{4}{4} = 8$

1.011 $\frac{7}{8} = \frac{21}{24}$

$\frac{1}{6} = \frac{4}{24}$

$\frac{25}{24} = 1\frac{1}{24}$

1.012 $\frac{3}{4} = \frac{15}{20}$

$\frac{4}{5} = \frac{16}{20}$

$\frac{31}{20} = 1\frac{11}{20}$

1.013 $\frac{3}{8} = \frac{12}{32}$

$\frac{6}{32} = \frac{6}{32}$

$\frac{18}{32} = \frac{18}{32} \div \frac{2}{2} = \frac{9}{16}$

1.014 $25\frac{3}{4} = 25\frac{18}{24}$

$16\frac{5}{6} = 16\frac{20}{24}$

$41\frac{38}{24} = 41\frac{38 \div 2}{24 \div 2} = 41\frac{19}{12} = 42\frac{7}{12}$

1.015 $74\frac{7}{11} = 74\frac{21}{33}$

$16\frac{2}{3} = 16\frac{22}{33}$

$90\frac{43}{33} = 91\frac{10}{33}$

1.016 $365\frac{1}{2} = 365\frac{6}{12}$

$253\frac{7}{12} = 253\frac{7}{12}$

$618\frac{13}{12} = 619\frac{1}{12}$

1.017 $16 = 2 \cdot 2 \cdot 2 \cdot 2$
$6 = 2 \cdot 3$
$3 = 3$
$LCD = 2 \cdot 2 \cdot 2 \cdot 2 \cdot 3 = 48$

$\frac{7}{16} = \frac{21}{48}$

$\frac{5}{6} = \frac{40}{48}$

$\frac{2}{3} = \frac{32}{48}$

$\frac{93}{48} = 1\frac{45}{48} = 1\frac{45 \div 3}{48 \div 3} = 1\frac{15}{16}$

1.018 $6 = 2 \cdot 3$
$5 = 5$
$4 = 2 \cdot 2$
$LCD = 2 \cdot 2 \cdot 3 \cdot 5 = 60$

$44\frac{5}{6} = 44\frac{50}{60}$

$32\frac{3}{5} = 32\frac{36}{60}$

$67\frac{3}{4} = 67\frac{45}{60}$

$143\frac{131}{60} = 145\frac{11}{60}$

1.019 $8 = 2 \cdot 2 \cdot 2$
$6 = 2 \cdot 3$
$9 = 3 \cdot 3$
$LCD = 2 \cdot 2 \cdot 2 \cdot 3 \cdot 3 = 72$

$354\frac{7}{8} = 354\frac{63}{72}$

$223\frac{5}{6} = 223\frac{60}{72}$

$151\frac{7}{9} = 151\frac{56}{72}$

$728\frac{179}{72} = 730\frac{35}{72}$

1.020
$$4\frac{5}{8} = 4\frac{10}{16}$$
$$3\frac{1}{8} = 3\frac{2}{16}$$
$$6\frac{7}{16} = 6\frac{7}{16}$$
$$13\frac{19}{16} = 14\frac{3}{16}$$

SELF TEST 2

2.01 $\frac{5}{6}$

2.02 proper

2.03 $25\frac{6}{8} = 25\frac{3}{4}$

2.04 $4\frac{2}{7}$

2.05
$$17 = 16\frac{8}{8}$$
$$-\,4\frac{5}{8} = 4\frac{5}{8}$$
$$12\frac{3}{8}$$

2.06
$$3\frac{1}{7} = 3\frac{4}{28}$$
$$+\,3\frac{3}{4} = 3\frac{21}{28}$$
$$6\frac{25}{28}$$

2.07
$$16\frac{5}{8} = 16\frac{15}{24}$$
$$-\,6\frac{1}{3} = 6\frac{8}{24}$$
$$10\frac{7}{24}$$

2.08
$$7\frac{2}{5} = 7\frac{6}{15} = 6\frac{21}{15}$$
$$-\,3\frac{7}{15} = 3\frac{7}{15}$$
$$3\frac{14}{15}$$

2.09
$$6 = 2 \cdot 3$$
$$4 = 2 \cdot 2$$
$$5 = 5$$
$$\text{LCD} = 2 \cdot 2 \cdot 3 \cdot 5 = 60$$
$$3\frac{5}{6} = 3\frac{50}{60}$$
$$2\frac{3}{4} = 2\frac{45}{60}$$
$$+\,4\frac{4}{5} = 4\frac{48}{60}$$
$$9\frac{143}{60} = 11\frac{23}{60}$$

2.010
$$7 = 7$$
$$4 = 2 \cdot 2$$
$$14 = 2 \cdot 7$$
$$\text{LCD} = 2 \cdot 2 \cdot 7 = 28$$
$$5\frac{6}{7} = 5\frac{24}{28} \qquad 22\frac{45}{28} = 22\frac{45}{28}$$
$$+\,17\frac{3}{4} = 17\frac{21}{28} \qquad -\,21\frac{9}{14} = 21\frac{18}{28}$$
$$22\frac{45}{28} \qquad 1\frac{27}{28}$$

2.011
$$4\frac{15}{16} = 4\frac{15}{16}$$
$$\frac{1}{8} = \frac{2}{16}$$
$$+\,\frac{1}{8} = \frac{2}{16}$$
$$4\frac{19}{16} = 5\frac{3}{16} \text{ inches}$$

SELF TEST 3

3.01 21.21

3.02 $\frac{10}{8} = \frac{5}{4} = 1\frac{1}{4}$

3.03 $4.58 (1 point off for missing $ sign)

3.04 344.396

3.05 657.75

3.06
$$17\frac{3}{4} = 17\frac{15}{20} = 16\frac{35}{20}$$
$$-\,9\frac{9}{10} = 9\frac{18}{20} = 9\frac{18}{20}$$
$$7\frac{17}{20}$$

3.07 322.925

3.08 718.67

3.09 2.58111

3.010 553.438

3.011
$$3\frac{1}{4} = 3\frac{4}{16}$$
$$+\ 19\frac{3}{16} = 19\frac{3}{16}$$
$$22\frac{7}{16}$$

3.012
$8 = 2 \cdot 2 \cdot 2$
$5 = 5$
$10 = 2 \cdot 5$
$\text{LCD} = 2 \cdot 2 \cdot 2 \cdot 5 = 40$
$$75\frac{7}{8} = 75\frac{35}{40}$$
$$666\frac{3}{5} = 666\frac{24}{40}$$
$$+\ 27\frac{9}{10} = 27\frac{36}{40}$$
$$768\frac{95}{40} = 770\frac{15}{40} = 770\frac{3}{8}$$

3.013 denominators

3.014 equivalent

3.015
$20.00
– 16.75
$3.25

3.016 subtraction

3.017 3.77 – 2.53 = 1.24 meters

3.018
$$1\frac{1}{2} = 1\frac{3}{6}$$
$$+\ 2\frac{1}{3} = 2\frac{2}{6}$$
$$3\frac{5}{6}$$

3.019
$$24\frac{3}{8} = 24\frac{6}{16} = 23\frac{22}{16}$$
$$-\ 17\frac{7}{16} = 17\frac{7}{16}$$
$$6\frac{15}{16}$$

3.020
Joe:
$ 7.25
+ 14.32
$21.57

Sam:
$12.35
+ 9.85
$22.20
– 21.57
$0.63 or 63¢

a. Sam
b. 63¢
c. Joe

SELF TEST 4

4.01
$$\frac{3}{4} = \frac{12}{16}$$
$$\frac{7}{16} = \frac{7}{16}$$
$$\frac{19}{16} = 1\frac{3}{16}$$

4.02
$$4\frac{7}{8} = 4\frac{35}{40}$$
$$+\ 9\frac{7}{10} = 9\frac{28}{40}$$
$$13\frac{63}{40} = 14\frac{23}{40}$$

4.03
$$16\frac{4}{7} = 16\frac{12}{21}$$
$$+\ 25\frac{2}{3} = 25\frac{14}{21}$$
$$41\frac{26}{21} = 42\frac{5}{21}$$

4.04
$$\frac{3}{4} = \frac{21}{28}$$
$$\frac{7}{14} = \frac{14}{28}$$
$$\frac{7}{28} = \frac{1}{4}$$

4.05
$$1\frac{4}{7} = 1\frac{44}{77} = \frac{121}{77}$$
$$-\ \frac{7}{11} = \frac{49}{77} = \frac{49}{77}$$
$$\frac{72}{77}$$

4.06

$$\begin{array}{r} 452\frac{1}{6} = 452\frac{4}{24} = 451\frac{28}{24} \\ -426\frac{3}{8} = 426\frac{9}{24} = 426\frac{9}{24} \\ \hline 25\frac{19}{24} \end{array}$$

4.07 368.92

4.08 10,499.04

4.09 69.074

4.010 1,728.616

4.011

$$\begin{array}{r} 14.32 \\ +\ 10.68 \\ \hline 25.00 \\ -\ 24.45 \\ \hline 0.55 \end{array}$$

4.012 d

4.013 c

4.014 b

4.015 e

4.016 a

4.017
a. 452.357
b. 452.4
c. 452
d. 500

4.018
a. 7,749.991
b. 7,750.0*
c. 7,750
d. 7,700

4.019
a. 85,456.342
b. 85,456.3
c. 85,456
d. 85,500

*Count off 1 point if the zero after the decimal point is not written.

SELF TEST 1

1.01 Multiply the numerators to obtain the numerator of the product. Multiply the denominators to obtain the denominator of the product.

1.02 First count the number of digits after the decimal point of the two numbers being multiplied. Then place the decimal point in the product where the correct number of digits will follow the decimal point.

1.03 do not

1.04 improper

1.05 true

1.06 $\frac{3}{{}_2\cancel{8}} \times \frac{\cancel{4}^1}{5} = \frac{3}{10}$

1.07 $\frac{16}{35}$ Note: the 4's must not be canceled, since they are both in the numerator.

1.08 $\frac{{}^1\cancel{7}}{{}_4\cancel{16}} \times \frac{\cancel{4}^1}{\cancel{21}_3} = \frac{1}{12}$

1.09 $\frac{{}^7\cancel{21}}{5} \times \frac{23}{\cancel{3}_1} = \frac{161}{5} = 32\frac{1}{5}$

1.010 $\frac{53}{8} \times \frac{13}{5} = \frac{689}{40} =$

$$\begin{array}{r} 17 \\ 40\overline{)689} \\ \underline{40} \\ 289 \\ \underline{280} \\ 9 \end{array} = 17\frac{9}{40}$$

1.011 $\frac{{}^1\cancel{7}}{{}_3\cancel{6}} \times \frac{\cancel{88}^{44}}{\cancel{21}_3} = \frac{44}{9} = 4\frac{8}{9}$

1.012 $\frac{4}{1} \times \frac{16}{5} = \frac{64}{5} = 12\frac{4}{5}$

1.013 $\frac{{}^{13}\cancel{52}}{{}_1\cancel{3}} \times \frac{\cancel{213}^{71}}{\cancel{16}_4} = \frac{923}{4} =$

$$\begin{array}{r} 230 \\ 4\overline{)923} \\ \underline{8} \\ 12 \\ \underline{12} \\ 03 \end{array} = 230\frac{3}{4}$$

1.014 $\frac{{}^{12}\cancel{84}}{{}_1\cancel{5}} \times \frac{\cancel{160}^{32}}{\cancel{7}_1} = \frac{384}{1} = 384$

1.015 0.009

1.016

$$\begin{array}{r} 4.32 \\ \underline{\times\ 0.44} \\ 1728 \\ \underline{1728} \\ 1.9008 \end{array}$$

1.017

$$\begin{array}{r} 352 \\ \underline{\times\ 1.34} \\ 1408 \\ 1056 \\ \underline{352} \\ 471.68 \end{array}$$

1.018

$$\begin{array}{r} 371.4 \\ \underline{\times\ 4.04} \\ 14856 \\ \underline{14856} \\ 1{,}500.456 \end{array}$$

1.019

$$\begin{array}{r} 1.65 \\ \underline{\times\ 23} \\ 495 \\ \underline{330} \\ \$37.95 \end{array}$$

1.020 $2 \times 3 \times 1.5 = 9.0 = 9$ yards

1.021 $\frac{3}{4} \times \frac{7}{1} = \frac{21}{4} = 5\frac{1}{4}$

$5\frac{1}{4} + 5 = 10\frac{1}{4}$

1.022 $\frac{2}{3} \times \frac{7}{1} = \frac{14}{3} = 4\frac{2}{3}$

$4\frac{2}{3} + 5 = 9\frac{2}{3}$

1.023 $47.5 \times 7 = 332.5$

$332.5 + 5 = 337.5$

1.024 $0.326 \times 7 = 2.282$
$2.282 + 5 = 7.282$

1.025 $0 \times 7 = 0$
$0 + 5 = 5$

SELF TEST 2

2.01 Multiply the numerators of the fractions to get the numerator of the product. Multiply the denominators of the fractions to get the denominator of the product.

2.02 a. quotient
b. divisor
c. dividend

2.03 Either order:
a. divisor
b. quotient

2.04 multiply

2.05 inverted *or* changed

2.06 $\frac{4}{3}$ or $1\frac{1}{3}$

2.07 $\frac{8}{7}$ or $1\frac{1}{7}$

2.08 $1\frac{1}{3} = \frac{4}{3}$; reciprocal is $\frac{3}{4}$

2.09 $\frac{5}{2}$ or $2\frac{1}{2}$

2.010 $\frac{3}{16}$

2.011 $\frac{\not{4}^{1}}{5} \times \frac{7}{\not{8}_{2}} = \frac{7}{10}$

2.012 $\frac{21}{5} \div \frac{7}{10} =$

$\frac{{}^{3}\not{21}}{{}_{1}\not{5}} \times \frac{\not{10}^{2}}{\not{7}_{1}} = \frac{6}{1} = 6$

2.013 $\frac{{}^{9}\not{27}}{4} \times \frac{31}{\not{6}_{2}} = \frac{279}{8} =$

$279 \div 8 = 34\frac{7}{8}$

```
   34
8)279
  24
   39
   32
    7
```

2.014
```
   4.23
  x 7.3
   1269
  2961
 30.879
```

2.015 $\frac{23}{3} \div \frac{13}{6} =$

$\frac{23}{{}_{1}\not{3}} \times \frac{\not{6}^{2}}{13} = \frac{46}{13} = 3\frac{7}{13}$

2.016 13.0312 → 13.031
```
     13.0312
32)417.0000
   32
    97
    96
     100
      96
       40
       32
        80
        64
```

2.017 23.3333 → 23.333
```
     23.3333
75)1750.0000
   150
    250
    225
     250
     225
      250
      225
       250
       225
        250
        225
```

2.018 0.2230 → 0.223
```
        0.2230
3204)714.5000
     6408
      7370
      6408
       9620
       9612
          80
```

2.019

$$745\overline{)56.5437} = 0.0758 \rightarrow 0.076$$

5215
4393
3725
6687
5960

2.020

$$307\overline{)155071.84} = 505.12 \rightarrow 505.120$$

1535
1571
1535
368
307
614
614
0

2.021

$$157\overline{)2650.00} = 16.87 \rightarrow 16.9 \text{ mpg}$$

157
1080
942
1380
1256
1240
1099

SELF TEST 3

3.01 % x *B* = *N* or, the percent times the base equals the number.

3.02 improper

3.03 $\frac{7}{_2\cancel{8}} \times \frac{\cancel{4}^1}{3} = \frac{7}{6} = 1\frac{1}{6}$

3.04 $\frac{^1\cancel{4}}{5} \times \frac{7}{\cancel{8}_2} = \frac{7}{10}$

3.05 $\frac{37}{8} \div \frac{27}{4} =$

$\frac{37}{_2\cancel{8}} \times \frac{\cancel{4}^1}{27} = \frac{37}{54}$

3.06 $\frac{3}{_2\cancel{8}} \times \frac{\cancel{4}^1}{5} = \frac{3}{10}$

3.07 $\frac{8}{3} \div 4 = \frac{^2\cancel{8}}{3} \times \frac{1}{\cancel{4}_1} = \frac{2}{3}$

3.08 $\frac{50}{7} \div \frac{21}{5} =$

$\frac{50}{7} \times \frac{5}{21} = \frac{250}{147} = 1\frac{103}{147}$

3.09

$$43\overline{)19440.3} = 452.1$$

172
224
215
90
86
43
43
0

3.010

$$471\overline{)621.72} = 1.32$$

471
1507
1413
942
942
0

3.011

$$147\overline{)632835} = 4,305$$

588
448
441
735
735
0

3.012

34.01
x 62.3
10203
6802
20406
2,118.823

3.013 0.06

3.014

654
x 0.0041
654
2616
2.6814

3.015

$214\overline{)1730.000}$ = 8.084 → 8.08

1712
1800
1712
880
856

3.016

$315\overline{)6773.500}$ = 21.503 → 21.50

630
473
315
1585
1575
1000
945

3.017 $\frac{7}{25} = \frac{28}{100} = 0.28$

3.018 $\frac{7}{16} = 16\overline{)7.0000}$ = 0.4375

6 4
60
48
120
112
80
80
0

3.019 $0.34 \times \frac{100}{100} = \frac{34}{100} = \frac{34 \div 2}{100 \div 2} = \frac{17}{50}$

3.020 $0.625 \times \frac{1,000}{1,000} = \frac{625}{1,000} =$

$\frac{625 \div 125}{1,000 \div 125} = \frac{5}{8}$

3.021 $0.45 \times \frac{100}{100} = \frac{45}{100} = 45\%$

3.022 $1.4 \times \frac{100}{100} = \frac{140}{100} = 140\%$

3.023 $0.004 \times \frac{100}{100} = \frac{0.4}{100} = 0.4\%$

3.024 $\frac{3}{5} = 0.6$

$0.6 \times \frac{100}{100} = \frac{60}{100} = 60\%$

3.025 0.34

3.026 1.0

3.027 0.02

3.028 0.19

3.029 $0.27 \times 42 = N$

0.27
x 42
54
108
$11.34 = N$

3.030 $R \times 24 = 6.6$

$R = 6.6 \div 24$

$240\overline{)66.000}$ = 0.275

480
1800
1680
1200
1200
0

$R = 27.5\%$

3.031 $0.09 \times B = 180$

$B = 180 \div 0.09$

$B = 2,000$

3.032 $125 - 115 = 10$

$\frac{10}{125} = 0.08 = 8\%$ decrease

3.033 $1.54 \times \text{B} = 549.78$

$\text{B} = 549.78 \div 1.54$

$154\overline{)54978}$ = 357

462
877
770
1078
1078
0

3.034 $0.0009 \times 3,482 = N$

$3.1338 = N$

3.035 $0.32 \times B = 6.4$

$\text{B} = 6.4 \div 0.32$

$\text{B} = 20$

3.036 $34 - 25 = 9$

$\frac{9}{25} = 0.36 = 36\%$ increase

3.037 $R \times 128 = 16$

$R = 16 \div 128$

$$\begin{array}{r} 0.125 \\ 128\overline{)16.000} \\ \underline{128} \\ 320 \\ \underline{256} \\ 640 \\ \underline{640} \\ 0 \end{array}$$

$R = 12.5\%$

3.038 $0.004 \times 78 = N$

$0.312 = N$

3.039 $R \times 72 = 18$

$R = 18 \div 72$

$$\begin{array}{r} 0.25 \\ 72\overline{)18.00} \\ \underline{144} \\ 360 \\ \underline{360} \\ 0 \end{array}$$

R = 25%

3.040 $150 - 120 = 30$

$\frac{30}{120} = \frac{1}{4} = 0.25 = 25\%$ increase

SELF TEST 1

1.01 $\frac{118}{6} = 19\frac{2}{3} \rightarrow 20$

1.02 $\frac{19.9}{5} = 3\frac{4.9}{5} \rightarrow 4$

1.03 $\frac{470}{6} = 78\frac{1}{3} \rightarrow 78$

1.04 2, 3, 4, 5, 6, 7, 8, 9

median $\frac{4+5}{2} = \frac{9}{2} = 4.5$

1.05 1.02, 1.4, 1.8, 2.5, 3.05, 3.1

median $= \frac{1.8 + 2.5}{2} = \frac{4.3}{2} = 2.15$

1.06 70, 70, 75, 75, 80, 90, 95

median = 75

1.07 2

1.08 90

1.09 3.1 – 1.02 = 2.08

1.010 mean $= \frac{40}{9} = 4\frac{4}{9} \rightarrow 4$

Score	Mean	Diff.
2	4	2
2	4	2
2	4	2
3	4	1
5	4	1
5	4	1
6	4	2
7	4	3
8	4	4
40	36	18

average spread $= \frac{18}{9} = 2$

1.011 mean $= \frac{189}{19} = 9\frac{9}{19} \rightarrow 10$

E	F	E x F	D	D x F
6	1	6	4	4
7	3	21	3	9
8	2	16	2	4
9	0	0	1	0
10	5	50	0	0
11	3	33	1	3
12	4	48	2	8
13	0	0	3	0
14	0	0	4	0
15	1	15	5	5
	19	189		33

1.012 mean $= \frac{3,860}{50} = 77\frac{1}{5} \rightarrow 77$

I	E	F	E x F	D	D x F
50-59	55	6	55 x 6 = 330	77 – 55 = 22	132
60-69	65	11	65 x 11 = 715	77 – 65 = 12	132
70-79	75	9	75 x 9 = 675	77 – 75 = 2	18
80-89	85	14	85 x 14 = 1,190	85 – 77 = 8	112
90-99	95	10	95 x 10 = 950	95 – 77 = 18	180
		50	3,860		574

median interval: the twenty-fifth and twenty-sixth numbers are in the 70-79 interval; therefore, the median interval is 70-79.

mode interval = 80-89

spread = 98 – 50 = 48

average spread $= \frac{574}{50} = 11.5$

SELF TEST 2

2.01 $\frac{3,030}{40} = 75\frac{3}{4} \rightarrow 76$

2.02 55, 55, 58, 59, 60, 61, 65, 65, 68, 69, 69, 69, 71, 71, 72, 72, 73, 75, 75, 75, 76, 76, 76, 78, 78, 79, 82, 82, 82, 82, 85, 86, 87, 87, 88, 90, 92, 93, 93, 100

$$\text{median} = \frac{75 + 76}{2} = \frac{151}{2} = 75.5$$

2.03 82

2.04 $100 - 55 = 45$

2.05

I	E	F	E x F	D	D x F
50-59	55	4	55 x 4 = 220	76 – 55 = 21	84
60-69	65	8	65 x 8 = 520	76 – 65 = 11	88
70-79	75	14	75 x 14 = 1,050	76 – 75 = 1	14
80-89	85	9	85 x 9 = 765	85 – 76 = 9	81
90-100	95	5	95 x 5 = 475	95 – 76 = 19	95
		40	3,030		362

2.06 $\frac{362}{40} = 9\frac{1}{20} \rightarrow 9$

2.07

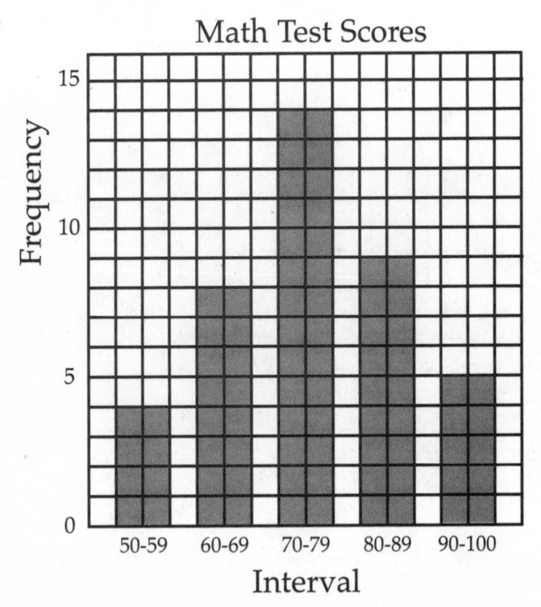

2.08

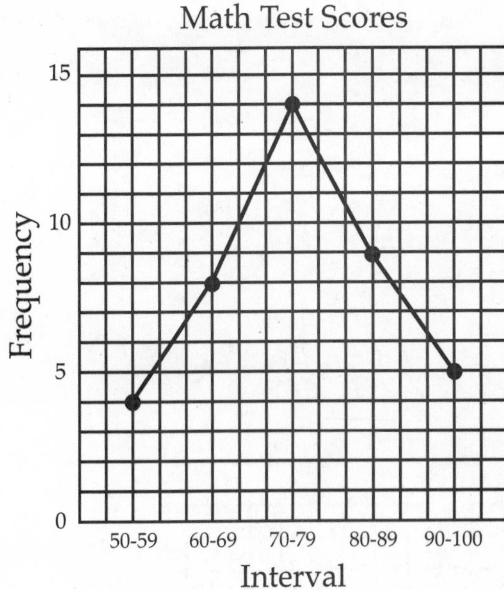

2.09 $\frac{712}{7} = 102$

2.010 $\frac{535}{7} = 76$

2.011

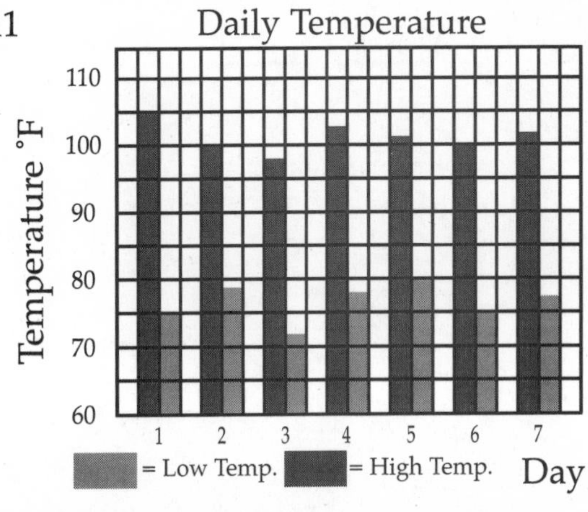

2.012 Monthly Sales, Voss Co.

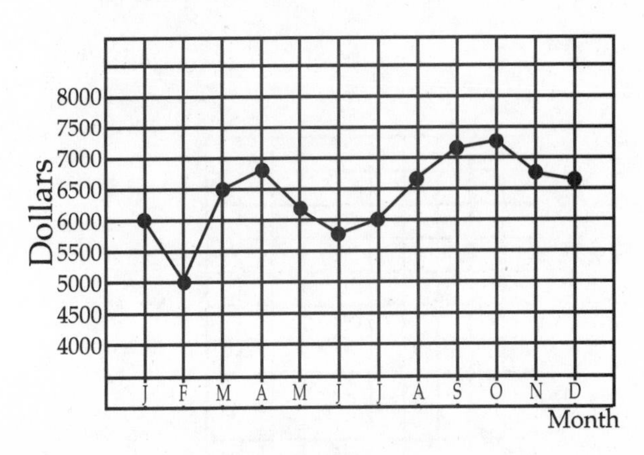

2.013 a. $f(n) = 2(2) - 2$
$= 4 - 2$
$= 2$

b. $f(n) = 2(3) - 2$
$= 6 - 2$
$= 4$

c. $f(n) = 2\,(4) - 2$
$= 8 - 2$
$= 6$

d. $f(n) = 2(5) - 2$
$= 10 - 2$
$= 8$

2.014

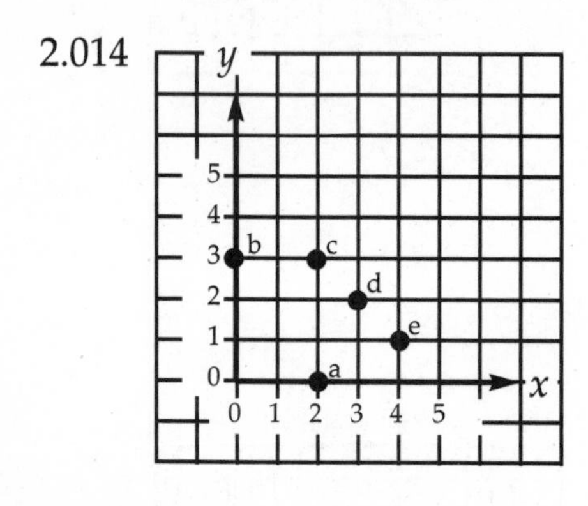

2.015 a. Examples;

f(n) = 5 – 2(0)
= 5 – 0
= 5
(0, 5)

f(n) = 5 – 2(1)
= 3
(1, 3)

f(n) = 5 – 2(2)
= 5 – 4
= 1
(2, 1)

b.

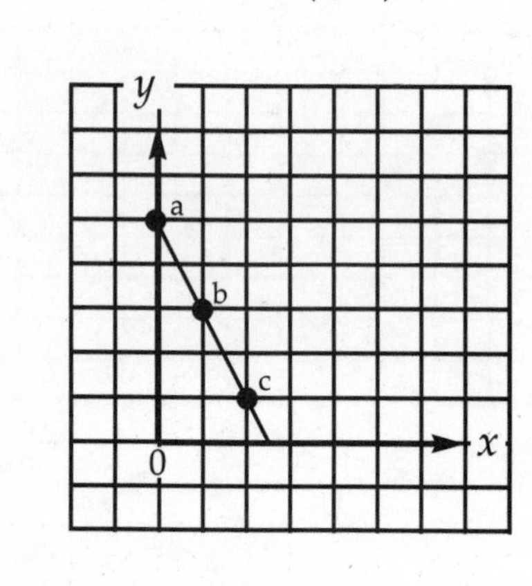

SELF TEST 3

3.01 $\frac{115}{7} = 16\frac{3}{7} \rightarrow 16$

3.02 17, 22, 23, 31, 54, 60, 92
median = 31

3.03 150 – 43 = 107

3.04 mean = $\frac{202}{13} = 15\frac{7}{13} \rightarrow 16$

Score	Mean	Diff.
1	16	15
9	16	7
10	16	6
10	16	6
12	16	4
15	16	1
16	16	0
17	16	1
17	16	1
18	16	2
22	16	6
25	16	9
30	16	14
202	208	72

average spread = $\frac{72}{13} = 5.5$

3.05 4

3.06

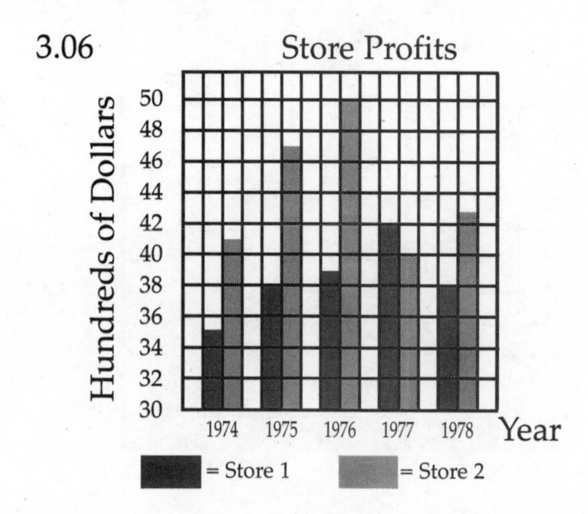

3.07

Interval	E	F	E x F	D	F x D
50-59	55	4	220	25	100
60-69	65	6	390	15	90
70-79	75	12	900	5	60
80-89	85	16	1360	5	80
90-99	95	12	1140	15	180
		50	4010		510

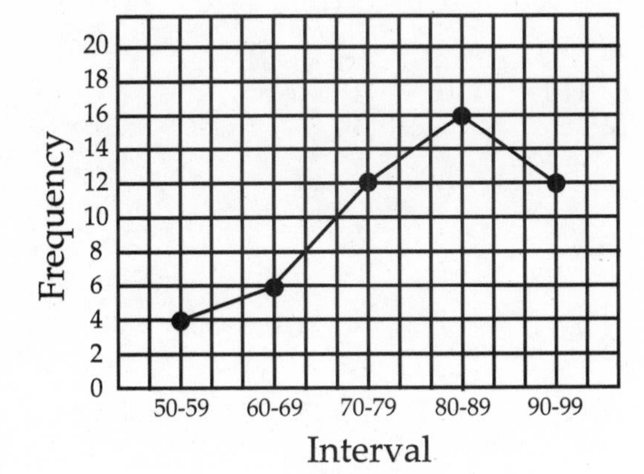

mean = $\frac{4010}{50} = 80.2 \rightarrow 80$

median interval: the twenty-fifth and twenty-sixth numbers are in the 80-89 interval; therefore, the median interval is 80-89.

average spread = $\frac{510}{50} = 10.2$

3.08 a. $f(n) = 3(1) - 2 = 3 - 2 = 1$
(1, 1)
$f(n) = 3(2) - 2 = 6 - 2 = 4$
(2, 4)
$f(n) = 3(3) - 2 = 9 - 2 = 7$
(3, 7)
$f(n) = 3(4) - 2 = 12 - 2 = 10$
(4, 10)

b.

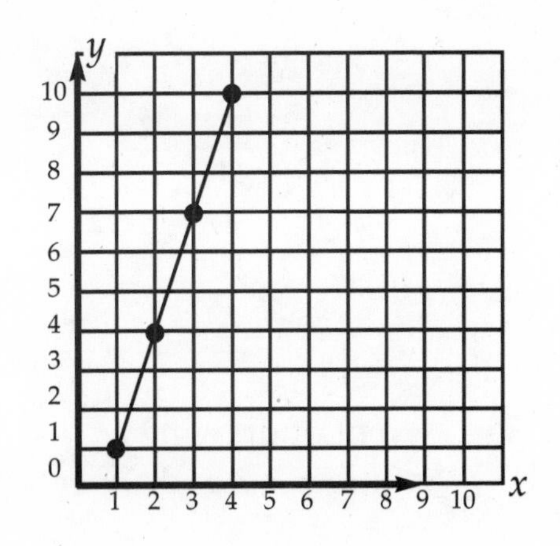

3.09 $f(n) = 10 - 2n$

3.010 $\frac{1}{11}$

3.011 $\frac{1}{11} + \frac{1}{11} = \frac{2}{11}$

3.012 sample space (prime) = {2, 3, 5, 7}
P(6 or prime number) =
P(6) + P(prime number) =
$\frac{1}{11} + \frac{4}{11} = \frac{5}{11}$

3.013 sample space = {0, 2, 4, 6, 8, 10}
P(even) = $\frac{6}{11}$

3.014 sample space = {2, 3, 5, 7}
P (even) = $\frac{4}{11}$

3.015 sample space = {0, 1, 4, 6, 8, 9, 10}
P(not prime) = $\frac{7}{11}$

3.016 Let A = event of drawing a number > 8 and
B = event of drawing a number < 2.

sample space A = {9, 10}

sample space B = {0, 1}

3.016 (cont.)

$$P(A \text{ and } B) = \frac{2}{11} \cdot \frac{2}{10} = \frac{2}{11} \cdot \frac{1}{5} = \frac{2}{55}$$

3.017 Let A = event of drawing a number < 3 and

B = event of drawing a number > 8.

sample space = {0, 1, 2}

sample space = {9, 10}

$$P(A \text{ and } B) = \frac{3}{11} \cdot \frac{2}{10} = \frac{3}{11} \cdot \frac{1}{5} = \frac{3}{55}$$

3.018 P (sum of two numbers = 5) =
P(0 and 5) or P(5 and 0) or
P(1 and 4) or P(4 and 1) or
P(2 and 3) or P(3 and 2) =

$$\frac{1}{11} \cdot \frac{1}{10} + \frac{1}{11} \cdot \frac{1}{10} + \frac{1}{11} \cdot \frac{1}{10} + \frac{1}{11} \cdot \frac{1}{10} + \frac{1}{11} \cdot \frac{1}{10} + \frac{1}{11} \cdot \frac{1}{10} =$$

$$\frac{1}{110} + \frac{1}{110} + \frac{1}{110} + \frac{1}{110} + \frac{1}{110} + \frac{1}{110} = \frac{6}{110} = \frac{3}{55}$$

3.019 P(sum of two numbers = 5 or 6) =
P(sum of two numbers = 5) or
P(sum of two numbers = 6) =
P(0 and 5) or P(5 and 0) or
P(1 and 4) or P(4 and 1) or
P(2 and 3) or P(3 and 2) or
P(0 and 6) or P(6 and 0) or
P(1 and 5) or P(5 and 1) or
P(2 and 4) or P(4 and 2) =

(answer to 3.018)

$$\frac{6}{110} + \frac{1}{11} \cdot \frac{1}{10} + \frac{1}{11} \cdot \frac{1}{10} + \frac{1}{11} \cdot \frac{1}{10} + \frac{1}{11} \cdot \frac{1}{10} + \frac{1}{11} \cdot \frac{1}{10} + \frac{1}{11} \cdot \frac{1}{10}$$

$$= \frac{6}{110} + \frac{1}{110} + \frac{1}{110} + \frac{1}{110} + \frac{1}{110} + \frac{1}{110} + \frac{1}{110} + \frac{1}{110} = \frac{12}{110} = \frac{6}{55}$$

3.020 $0.350 = \frac{350}{1,000} = \frac{7}{20}$

SELF TEST 1

1.01 12

1.02 -400

1.03 false

1.04 -4, -3, -2, -1

1.05 0, 2, 4

1.06 The non negative integers include zero; the positive integers do not include zero.

1.07 $-51 < -15 < 0 < 15 < 51$

1.08 $A = bh$
$A = 5 \cdot 3 = 15$

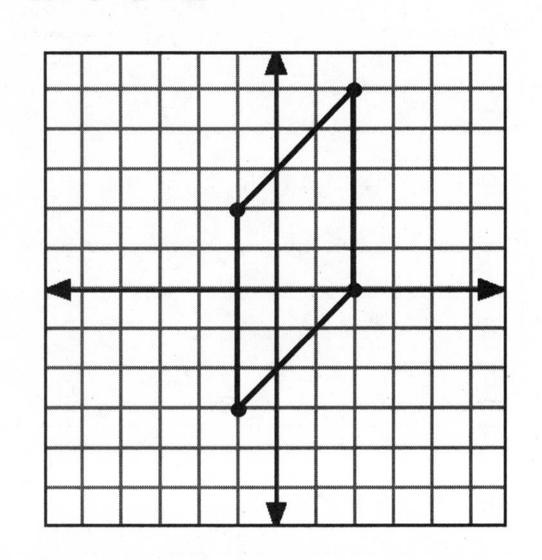

1.09 >

1.010 >

1.011 >

1.012 <

1.013 13

1.014 8

1.015 0

1.016 $9 < 11$

1.017 $0 < 7$

1.018 $14 = 14$

1.019 $2 + 3$ ____ $2 + 3$
$5 = 5$

1.020

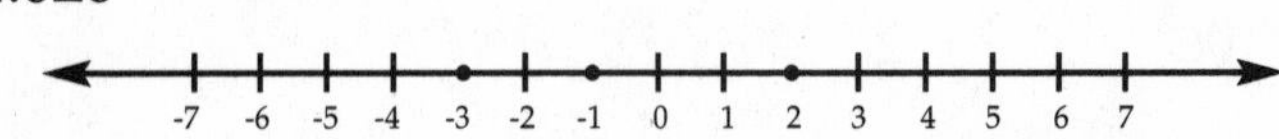

1.021

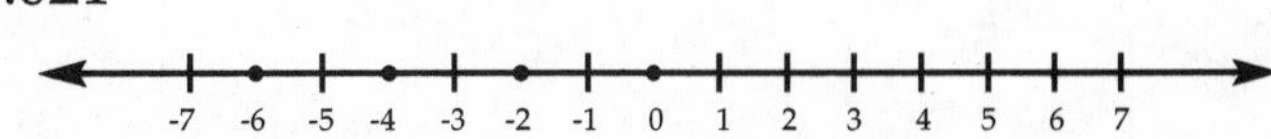

1.022

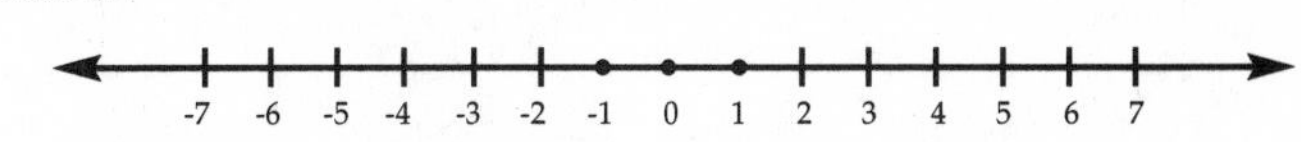

1.023 through 1.026

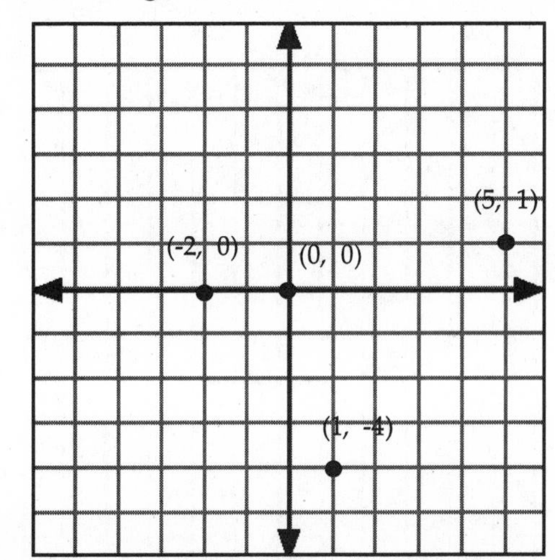

1.027 (-2, 3)

1.028 (4, -1)

1.029 (-4, -3)

1.030 (0, 1)

SELF TEST 2

2.01 132

2.02 0, 1, 2, 3

2.03 <

2.04 >

2.05 >

2.06 $2 = 2$

2.07 $-3 < 4$

2.08 5 ___ 1 + 4
$5 = 5$

2.09

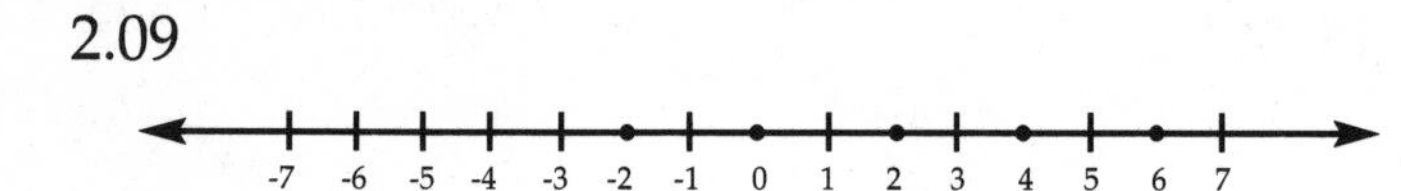

2.010 $A = lw$
$A = 4 \cdot 3 = 12$

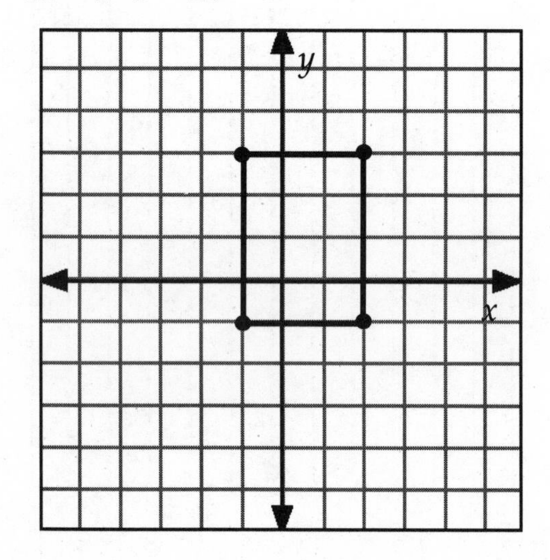

2.011 (-2, 4)

2.012 (2, 0)

2.013 (-1, -2)

2.014 7, 6, 5, 4

2.015 2, 1, 0, -1

2.016

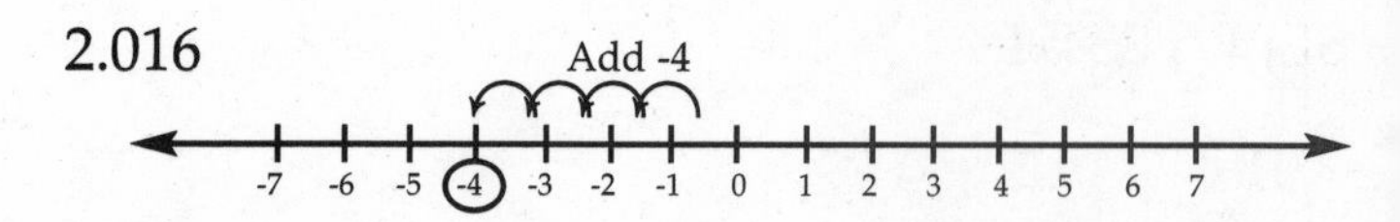

2.017

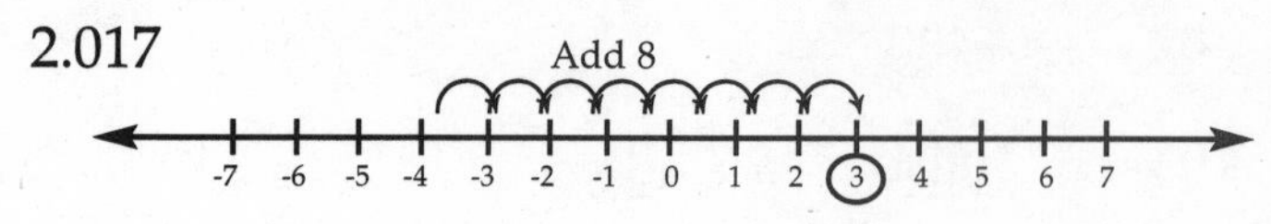

2.018

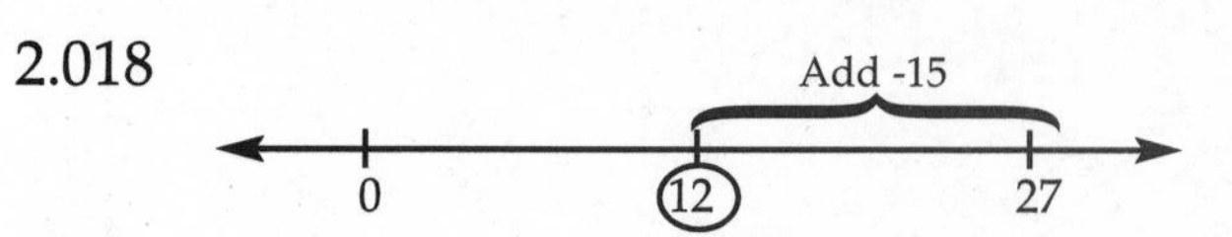

2.019 -10

2.020 -3

2.021
$-25 + 17 + 8 + (-6) + 15$
$= -8 + 8 + (-6) + 15$
$= 0 + (-6) + 15$
$= -6 + 15$
$= 9$

2.022 3, 4, 5, 6

2.023 1, 0, -1, -2

2.024

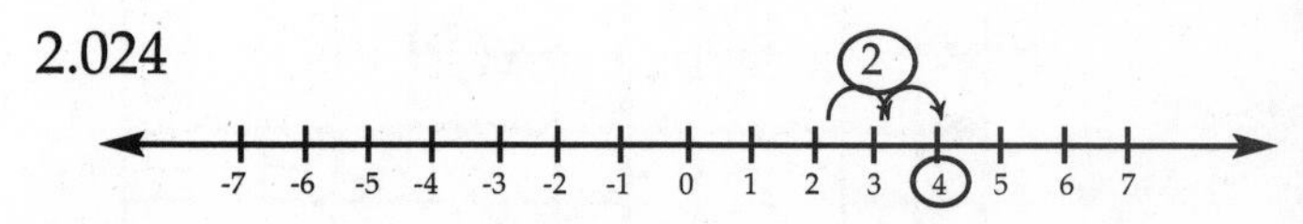

2.025

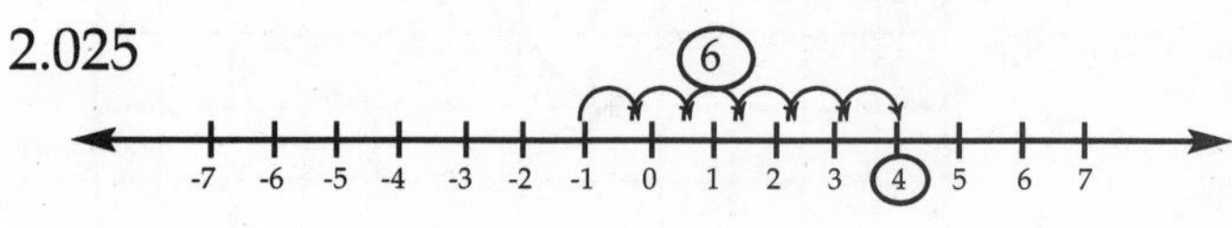

2.026

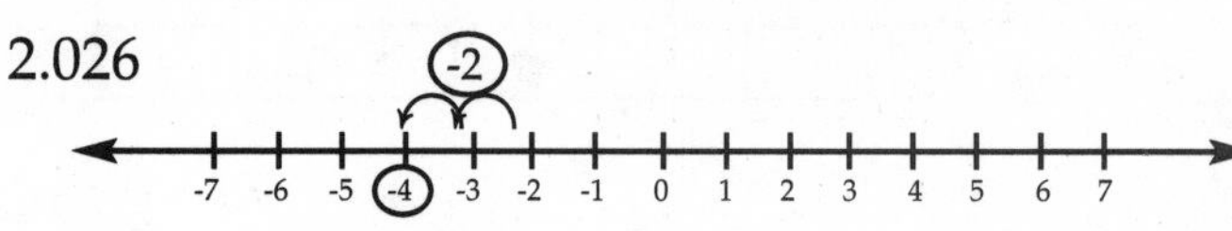

2.027 -15

2.028 100

2.029
$5 - (-4) - 3 - (-2) - 1$
$= 5 + 4 + (-3) + 2 + (-1)$
$= 9 + (-3) + 2 + (-1)$
$= 6 + 2 + (-1)$
$= 8 + (-1)$
$= 7$

2.030 2, 1, 0, -1

2.031 15, 5, -5, -15

2.032 14

2.033 0

2.034
$$\begin{aligned} & \underbrace{3 \cdot (-2)} \cdot (-7) \cdot (-10) \cdot 1 \\ = & \underbrace{-6 \cdot (-7)} \cdot (-10) \cdot 1 \\ = & \underbrace{42 \cdot (-10)} \cdot 1 \\ = & \underbrace{-420 \cdot 1} \\ & -420 \end{aligned}$$

2.035 $6^3 = 6 \cdot 6 \cdot 6 = 216$

2.036 $(-13)^2 = (-13) \cdot (-13) = 169$

2.037 1

2.038 $\frac{1}{8}$

2.039 -5

2.040 0

2.041 21

2.042 undefined

2.043 -1

2.044 17

2.045 addition and multiplication

SELF TEST 3

3.01 <

3.02 >

3.03 48 = 48

3.04

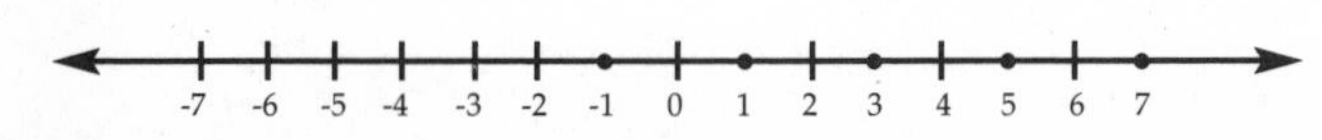

3.05

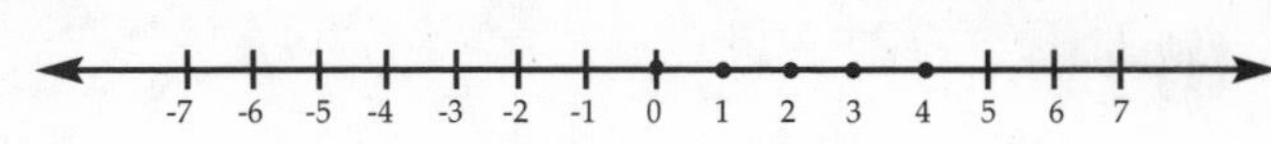

3.06 (4, -3)

3.07 (-3, 0)

3.08 (0, 0)

3.09 -12

3.010 0

3.011 58

3.012 -5

3.013 -5

3.014 -70

3.015 56

3.016 1

3.017 -729

3.018 -10

3.019 2

3.020 undefined

3.021 $5 \cdot (-4) = -20$

3.022
$$\begin{aligned} & 3 + \underbrace{2 \cdot (-1)} = \\ & \underbrace{3 + (-2)} = \\ & 1 \end{aligned}$$

3.023
$$\begin{aligned} & -4 + 3 - (-1) = \\ & \underbrace{-4 + 3} + 1 = \\ & \underbrace{-1 + 1} = \\ & 0 \end{aligned}$$

3.024 $\frac{-4 + 3}{-1} = \frac{-1}{-1} = -1 \div (-1) = 1$

3.025 $-4 \cdot 3 \cdot (-1)$
$-12 \cdot (-1)$
12

3.026 $3^3 = 27$

3.027 $14 \cdot (-1)^2 =$
$14 \cdot 1 =$
14

3.028 $-4^3 = -64$

3.029 $10^{-1} = \frac{1}{10^1} = \frac{1}{10}$

3.030 $(-4)^2 - 3 \cdot (-1) =$
$16 - \underbrace{3 \cdot (-1)} =$
$16 - (-3) =$
$\underbrace{16 + 3} =$
19

3.031 through 3.033 Examples:

3.031 $x = 4$: $4 + y = 4$
$y = 4 - 4$
$y = 0$
$(4, 0)$
$x = 1$: $1 + y = 4$
$y = 4 - 1$
$y = 3$
$(1, 3)$

3.032 $x = 5$: $5 - y = 4$
$-y = 4 - 5$
$-y = -1$
$y = 1$
$(5, 1)$
$x = 7$: $7 - y = 4$
$-y = 4 - 7$
$-y = -3$
$-y = 3$
$(7, 3)$

3.033 $x = -4$: $-4 + y = -4$
$y = -4 + 4$
$y = 0$
$(-4, 0)$
$x = -2$: $-2 + y = -4$
$y = -4 + 2$
$y = -2$
$(-2, -2)$

3.034

a	0	3
$b = a + 1$	$0 + 1$	$3 + 1$
b	1	4

b	-2
$a = b - 1$	$-2 - 1$
a	-3

(0, 1), (3, 4), (-3, -2)

3.035

p	4	-2
$5p$	5 4	5 (-2)
q	20	-10

q	0
$p = \frac{q}{5}$	$\frac{0}{5}$ $0 \div 5$
p	0

(4, 20), (-2, -10), (0, 0)

3.036

x	-2	0	3
x^3	$(-2)^3$	0^3	3^3
y	-8	0	27

(-2, -8), (0, 0), (3, 27)

3.037

m	-1	1	2
$4m^2 - 5$	$4 \cdot (-1)^2 - 5$ $4 \cdot 1 - 5$ $4 - 5$	$4 \cdot 1^2 - 5$ $4 \cdot 1 - 5$ $4 - 5$	$4 \cdot 2^2 - 5$ $4 \cdot 4 - 5$ $16 - 5$
n	-1	-1	11

(-1, -1), (1, -1), (2, 11)

3.038

a	1	2	3	4	5
$b = 3 - a$	$3 - 1$	$3 - 2$	$3 - 3$	$3 - 4$	$3 - 5$
b	2	1	0	-1	-2

3.038 cont.

(1, 2), (2, 1), (3, 0), (4, -1), (5, -2)

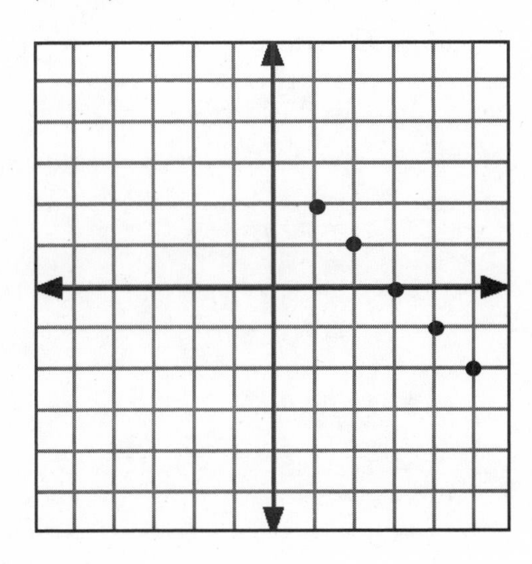

3.039

m	-1	0
$3m - 1$	$3 \cdot (-1) - 1$ $-3 \;\; - 1$	$3 \cdot 0 - 1$ $0 \;\; - 1$
n	-4	-1

1	2
$3 \cdot 1 - 1$ $3 \;\; - 1$	$3 \cdot 2 - 1$ $6 \;\; - 1$
2	5

(-1, -4), (0, -1), (1, 2), (2, 5)

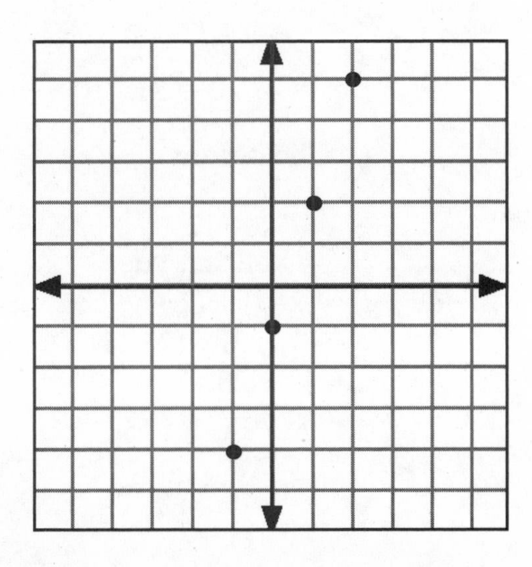

3.040

x	-2	-1
$x^2 - 2$	$(-2)^2 - 2$ $4 \;\; - 2$	$(-1)^2 - 2$ $1 \;\; - 2$
y	2	-1

0	1	2
$0^2 - 2$ $0 \;\; - 2$	$1^2 \;\; - 2$ $1 \;\; - 2$	$2^2 \;\; - 2$ $4 \;\; - 2$
-2	-1	2

(-2, 2), (-1, -1), (0, -2), (1, -1), (2, 2)

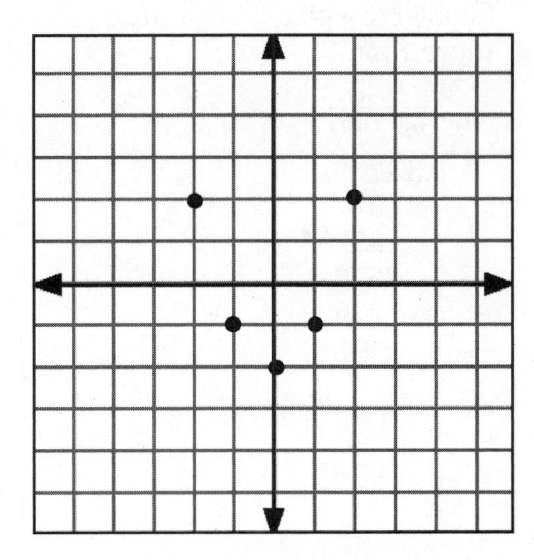

SELF TEST 1

1.01 Arithmetic method:

```
          9.     2     1 → 9.2
        √85.    00    00
          81
18 2       4    00
           3    64
184 1           36    00
                18    41
                17    59
```

Divide-and-average method:
$9^2 = 81$ is too small.
$10^2 = 100$ is too large.
Guess 9.3 since 85 is a little more than 81.

```
       9.13
  9.3)85.000
      83 7
       1 30
         93
        370
        279
         91
```

$\frac{9.3 + 9.13}{2} = \frac{18.43}{2} = 9.215$

Round to 9.2.

1.02 Arithmetic method:

```
        6.     4     8 → 6.5
      √42.    00    00
       36
12 4    6     00
        4     96
128 8   1     04    00
        1     03    04
                    96
```

Divide-and-average method:
$6^2 = 36$ is too small.
$7^2 = 49$ is too large.
Guess 6.5.

```
       6.46
  6.5)42.0.00
      39 0
       3 00
       2 60
         400
         390
          10
```

1.02 cont. $\frac{6.5 + 6.46}{2} = \frac{12.96}{2} = 6.48$

Round to 6.5.

1.03 a. $C = \pi \cdot d$
$= 3.14 \cdot 8$
$= 25.12$ ft.

b. $r = \frac{1}{2} \cdot d = \frac{1}{2} \cdot 8 = 4$ ft.

$A = \pi \cdot r^2$
$= 3.14 \cdot 4^2$
$= 3.14 \cdot 16$
$= 50.24$ ft.^{2}

1.04 $A = \frac{1}{2} \cdot b \cdot h$

$= \frac{1}{2} \cdot 5 \cdot 8$

$= \frac{1}{2} \cdot 40$

$= 20$ cm^2

1.05 $a = 7$
$b = 9$
$c =$ unknown
$7^2 + 9^2 = c^2$
$49 + 81 = c^2$
$130 = c^2$
$\sqrt{c^2} = \sqrt{130}$
$c \doteq 11.40 \doteq 11.4$

1.06 $a = 4$
$b = 11$
$c =$ unknown
$4^2 + 11^2 = c^2$
$16 + 121 = c^2$
$137 = c^2$
$\sqrt{c^2} = \sqrt{137}$
$c \doteq 11.70 \doteq 11.7$ ft.

1.07 a. parallelogram
b. $A = b \cdot h$
c. $= 20 \cdot 12$
$= 240$ sq. yds.
d. $P = (2 \cdot l) + (2 \bullet w)$
$= (2 \cdot 20) + (2 \cdot 14)$
$= 40 + 28$
$= 68$ yds.

1.08
- a. trapezoid: one set of parallel sides
- b. $A = \frac{1}{2} \cdot h \cdot (b_1 = b_2)$
- c. $A = \frac{1}{2} \cdot 1.7 \cdot (8 + 6)$
  $= \frac{1}{2} \cdot 1.7 \cdot 14$
  $= 11.9 \text{ m}^2$
- d. $P = 2 + 6 + 2 + 8 = 18$ m

1.09
- a. rhombus: four sides equal
- b. $A = b \cdot h$
- c. $A = 78 \cdot 70$
  $= 5{,}460 \text{ cm}^2$
- d. $P = 78 + 78 + 78 + 78$ or 4.78
  $= 312$ cm

1.010
- a. rectangle
- b. $A = l \cdot w$ or $A = b \cdot h$
- c. $A = 250 \cdot 150$
  $= 37{,}500$ sq. ft.
- d. $P = (2 \cdot l) + (2 \cdot w)$
  $= (2 \cdot 250) + (2 \cdot 150)$
  $= 500 + 300$
  $= 800$ ft.

1.011
- a. square
- b. $A = s^2$
- c. $A = 12^2$
  $= 144 \text{ ft.}^2$ or sq. ft.
- d. $P = 4 \cdot s$
  $= 4 \cdot 12$
  $= 48$ ft.

1.012 $A = l \cdot w$
$= 24 \cdot 18$
$= 432 \text{ ft.}^2$

$432 \text{ ft.}^2 \cdot \frac{1 \text{ yd.}^2}{9 \text{ ft.}^2} =$

$432 \text{ ft.}^2 \cdot \frac{1 \text{ yd.}^2}{9 \text{ ft.}^2} =$

$\frac{432 \text{ yd.}}{9} = 48 \text{ yd.}^2$

$48 \cdot \$18.50 = \$888$

SELF TEST 2

2.01 5

2.02 Arithmetic method:

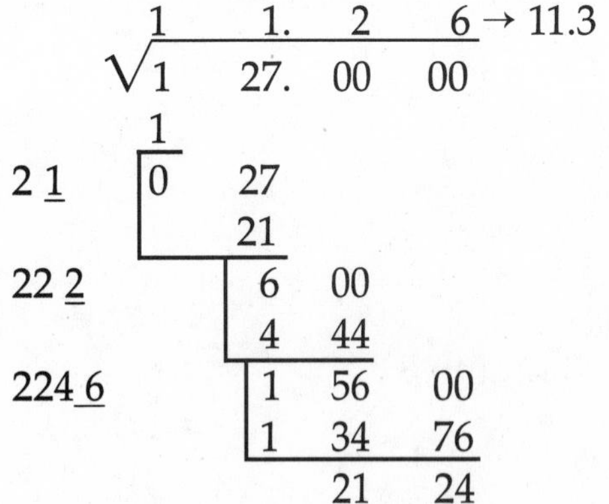

Divide-and-average method:
$11^2 = 121$ is too small.
$12^2 = 144$ is too large.
Guess 11.3 since 127 is a little more than 121.

$127 \div 11.3 = 11.23$: $1270.00 - 1130 = 140$; $140 - 113 = 27$; $270 - 226 = 44$; $440 - 339 = 101$

$\frac{11.3 + 11.23}{2} = \frac{22.53}{2} = 11.265$

Round to 11.3.

2.03 Arithmetic method:

$\sqrt{69.\ 00\ 00} = 8.\ 3\ 0 \rightarrow 8.3$
- 64
- $16\underline{3}$: 5 00 − 4 89 = 11
- $166\underline{0}$: 11 00 − 0 = 0

Divide-and-average method:
$8^2 = 64$ is too small.
$9^2 = 81$ is too large.
Guess 8.2 since 69 is a little more than 64.

2.03 cont.

$$8.2\overline{)69.000} = 8.41$$

656
340
328
120
82
38

$\frac{8.2 + 8.41}{2} = \frac{16.61}{2} = 8.305$

Round to 8.3.

2.04 a. $A = \pi \cdot r^2$
$= 3.14 \cdot 40$
$= 3.14 \cdot 1{,}600$
$= 5{,}024$ ft.2

b. $C = 2 \cdot \pi \cdot r$
$= 2 \cdot 3.14 \cdot 40$
$= 6.28 \cdot 40$
$= 251.2$ ft.

2.05 a. $P = 4 + 5 + 6$
$= 15"$

b. $A = \frac{1}{2} \cdot b \cdot h$
$= \frac{1}{2} \cdot 6 \cdot 3.4$
$= 3 \cdot 3.4$
$= 10.2 \text{ in.}^2$

2.06 $a = 2$
$b = 4$
c = unknown
$2^2 + 4^2 = c^2$
$4 + 16 = c^2$
$20 = c^2$
$\sqrt{c^2} = \sqrt{20}$
$c \doteq 4.47 \doteq 4.5\text{m}$

2.07 a. parallelogram

b. $A = b \cdot h$
$= 12 \cdot 7$
$= 84 \text{ cm}^2$

c. $P = 8 + 12 + 8 + 12$
$= 40$ cm

2.08 a. trapezoid

b. $A = \frac{1}{2} \cdot h \cdot (b_1 + b_2)$
$= \frac{1}{2} \cdot 5 \cdot (14 + 8)$
$= \frac{1}{2} \cdot 5 \cdot 22$
$= 55 \text{ ft.}^2$

2.09 a. rhombus

b. $A = b \cdot h$
$= 16 \cdot 14$
$= 224 \text{ in.}^2$

2.010 a. $A = l \cdot w$
$= 20 \cdot 18$
$= 360 \text{ ft }^2$

b. $V = l \cdot w \cdot h$
$= 20 \cdot 18 \cdot 8$
$= 360 \cdot 8$
$= 2{,}880$ ft.

c. $360 \text{ ft.}^2 \cdot \frac{1 \text{ yd.}^2}{9 \text{ ft.}^2}$

$\frac{360 \text{ yd.}^2}{9} = 40 \text{ yd.}^2$

2.011 a. $V = \frac{1}{2} \cdot b_{\triangle} \cdot h_{\triangle} \cdot h_p$
$= \frac{1}{2} \cdot 18 \cdot 12 \cdot 40$
$= \frac{1}{2} \cdot 12 \cdot 40$
$= 108 \cdot 40$
$= 4{,}320 \text{ ft.}^3$

b. area of 2 rectangular surfaces $= 2 \cdot 40 \cdot 15$
$= 80 \cdot 15$
$= 1{,}200 \text{ ft.}^2$

area of 2 ends $= 2 \cdot \frac{1}{2} \cdot b \cdot h$
$= 2 \cdot \frac{1}{2} \cdot 18 \cdot 12$
$= 1 \cdot 18 \cdot 12$
$= 216 \text{ ft.}^2$

area of bottom $= 40 \cdot 18$
$= 720 \text{ ft.}^2$

total area $= 1{,}200 + 216 + 720$
$= 2{,}136 \text{ ft.}^2$

2.012 a. $A = s^2$
$= 14^2$
$= 196 \text{ ft.}^2$

2.012 cont. b. $L.A. = 4 \cdot \frac{1}{2} \cdot b \cdot l$

$= 4 \cdot \frac{1}{2} \cdot 14 \cdot 10$
$= 2 \cdot 14 \cdot 10$
$= 28 \cdot 10$
$= 280 \text{ ft.}^2$

c. total area = $L.A.$ + area of base
$= 280 + 196$
$= 476 \text{ ft.}^2$

2.013 $C = \pi \cdot d$
$6 = 3.14 \cdot d$
$6 \div 3.14 = d$
$1.91 \doteq d$
$d \doteq 1.9$ ft.

2.014 $r = \frac{1}{2} \cdot d = \frac{1}{2} \cdot 1.9 = 0.95$ ft.

$A = \pi \cdot r^2$
$= 3.14 \cdot 0.95^2$
$= 3.14 \cdot 0.9025$
$= 2.83385 \text{ ft.}^2 \doteq 2.8 \text{ ft.}^2$

SELF TEST 3

3.01 Arithmetic method:

```
        1    9.  2
     √3    68.  64
 29    1
       2   68
       2   61
 382        7  64
            7  64
                0
```

Divide-and-average method:
$19^2 = 361$ is too small.
$20^2 = 400$ is too large.
Guess 19.2 since 368 is a little more than 361.

```
        19.16
19.2)368.0,00
     192
     1760
     1728
       320
       192
       1280
       1152
        128
```

3.01 cont. $\frac{19.2 + 19.16}{2} = \frac{38.36}{2} = 19.18$

Round to 19.2.

3.02 $r = \frac{1}{2} \cdot d = \frac{1}{2} \cdot 1.8 = 0.9$ cm

$A = \pi \cdot r^2$
$= 3.14 \cdot 0.9^2$
$= 3.14 \cdot 0.81$
$= 2.5434 \text{ cm}^2$

3.03 $C = \pi \cdot d$
$= 3.14 \cdot 1.8$
$= 5.652$ cm

3.04 $A = \frac{1}{2} \cdot b \cdot h$
$= \frac{1}{2} \cdot 21 \cdot 13$
$= \frac{1}{2} \cdot 273$
$= 136.5 \text{ cm}^2$

3.05 $a = 5$
$b = 8$
c = unknown
$5^2 + 8^2 = c^2$
$25 + 64 = c^2$
$89 = c^2$
$\sqrt{c^2} = \sqrt{89}$
$c \doteq 9.13 \doteq 9.4$ yd.

3.06 a. parallelogram
b. $A = b \cdot h$
$= 13 \cdot 5$
$= 65 \text{ in.}^2$

3.07 a. trapezoid
b. $A = \frac{1}{2} \cdot h \cdot (b_1 + b_2)$
$= \frac{1}{2} \cdot 4.6 \cdot (9.2 + 5.8)$
$= \frac{1}{2} \cdot 4.6 \cdot 15$
$= 2.3 \cdot 15$
$= 34.5 \text{ cm}^2$

3.08 $A = l \cdot w$ or $A = b \cdot h$
$= 60 \cdot 42$
$= 2{,}520 \text{ ft.}^2$

3.09 $2{,}520 \text{ ft.}^2 \cdot \frac{1 \text{ yd.}^2}{9 \text{ ft.}^2}$

$\frac{2{,}520 \text{ yd.}^2}{9} = 280 \text{ yd.}^2$

3.010 $280 \div 100 = 2.8$ bags
Round to whole number: 3 bags.
$3 \cdot \$6.50 = \$19.50$

3.011 $V = l \cdot w \cdot h$
$= 18 \cdot 9.5 \cdot 11$
$= 171 \cdot 11$
$= 1{,}881 \text{ in.}^3$

3.012 a. prism
b. $V = \frac{1}{2} \cdot b \cdot_{\Delta} h \cdot_{\Delta} h_p$
$= \frac{1}{2} \cdot 10 \cdot 8 \cdot 36$
$= 5 \cdot 8 \cdot 36$
$= 40 \cdot 36$
$= 1{,}440 \text{ in.}^3$

3.013 a. pyramid
b. $V = \frac{1}{3} \cdot B \cdot h$
$= \frac{1}{3} \cdot s^2 \cdot h$
$= \frac{1}{3} \cdot 11^2 \cdot 4$
$= \frac{1}{3} \cdot 121 \cdot 4$
$= \frac{1}{3} \cdot 484$
$= 161.\overline{3} \text{ ft.}^3$

3.014 $r = \frac{1}{2} \cdot d = \frac{1}{2} \cdot 3 = \frac{3}{2} = 1.5''$
$V = \pi \cdot r^2 \cdot h$
$= 3.14 \cdot 1.5^2 \cdot 6.5$
$= 3.14 \cdot 2.25 \cdot 6.5$
$= 7.065 \cdot 6.5$
$= 45.9225 \text{ in.}^3$

3.015 $TA = 2 \cdot \pi \cdot r \cdot (r + h)$
$= 2 \cdot 3.14 \cdot 1.5 \cdot (1.5 + 6.5)$
$= 2 \cdot 3.14 \cdot 1.5 \cdot 8$
$= 6.28 \cdot 1.5 \cdot 8$
$= 9.42 \cdot 8$
$= 75.36 \text{ in.}^2$

3.016 $V = \frac{1}{3} \cdot \pi \cdot r^2 \cdot h$
$= \frac{1}{3} \cdot 3.14 \cdot 4^2 \cdot 7$
$= \frac{1}{3} \cdot 3.14 \cdot 16 \cdot 7$
$= \frac{1}{3} \cdot 50.24 \cdot 7$

3.016 cont. $= \frac{1}{3} \cdot 351.68$
$= 117.22\overline{6}$
$\doteq 117.2 \text{ in.}^3$

3.017 $r = \frac{1}{2} \cdot d = \frac{1}{2} \cdot 18 = 9$ ft.
$V = \frac{4}{3} \cdot \pi \cdot r^3$
$= \frac{4}{3} \cdot 3.14 \cdot 9^3$
$= \frac{4}{3} \cdot 3.14 \cdot 729$
$= \frac{4}{3} \cdot 2{,}289.06$
$= 3{,}052.08 \text{ ft.}^3$

3.018 area of top:
$4.5 \cdot 3 = 13.5 \text{ ft}^2$
area of bottom:
$4.5 \cdot 3 = 13.5 \text{ ft.}^2$
area of left side:
$3 \cdot 2 = 6.0 \text{ ft.}^2$
area of right side:
$3 \cdot 2 = 6.0 \text{ ft.}^2$
area of front:
$4.5 \cdot 2 = 9.0 \text{ ft.}^2$
area of back:
$4.5 \cdot 2 = 9.0 \text{ ft.}^2$
total area 57.0 ft.^2

3.019 Rectangle:
$A = l \cdot w$
$= 5 \text{ in.} \cdot 7 \text{ in.}$
$= 35 \text{ in.}^2$
Circle:
$C = 2 \cdot \pi \cdot r$
$21.98 = 2 \cdot 3.14 \cdot r$
$\frac{21.98}{6.28} = \frac{6.28 \cdot r}{6.28}$
$3.5 = r$

Now with r, find area.
$A = \pi \cdot r^2$
$= 3.14 \cdot 3.5^2$
$= 3.14 \cdot 12.25$
$= 38.465 \text{ in.}^2$
A_{circle} : 38.465
$A_{rect.}$: $-\ 35$
3.465 in.
Circle is larger than rectangle by 3.465 square inches.

3.020 area of large square:

$A = s^2$

$= 30^2$

$= 900 \text{ ft.}^2$

area of small square:

$A = s^2$

$= 10^2$

$= 100 \text{ ft.}^2$

$900 \text{ ft.}^2 - 100 \text{ ft.}^2 = 800 \text{ ft.}^2$

SELF TEST 1

1.01 j

1.02 d

1.03 h

1.04 i

1.05 k

1.06 a

1.07 b

1.08 e

1.09 f

1.010 c

1.011 c

1.012 a

1.013 c

1.014 a

1.015 b

1.016 c

1.017 a

1.018 d

1.019 b

1.020 a

1.021 $5 \cdot 6 = 30$ and $6 \cdot 5 = 30$, so $5 \cdot 6 = 6 \cdot 5$ is true.

1.022 14

1.023 $(x + b) + c$

1.024 $i = prt$

$i = 500 \cdot 0.09 \cdot 2 = 90$

1.025 6

1.026 13

1.027 $-4a$

1.028 $-9xy$

1.029 $3x + 8y - a - 4a + 6x - 9y =$

$3x + 6x + 8y - 9y - a - 4a =$

$9x - y - 5a$

1.030 76

1.031 $7x(a + c) = 7x \cdot a + 7x \cdot c =$

$7ax + 7cx$

1.032 $3x(a + 2b + 3c) =$

$3x \cdot a + 3x \cdot 2b + 3x \cdot 3c =$

$3ax + 6bx + 9cx$

1.033 $7x - 14x = 7x \cdot y - 7x \cdot 2 =$

$7x(y - 2)$

1.034 $(x - 2)(x - 3) = x^2 - 3x - 2x + 6$

The missing terms are $3x$ and $2x$.

1.035 No, because $17x - y \neq y - 17x$.

SELF TEST 2

2.01 k

2.02 g

2.03 i

2.04 c

2.05 b

2.06 a

2.07 e

2.08 d

2.09 f

2.010 h

2.011 b

2.012 c

2.013 c

2.014 a

2.015 b

2.016 a

2.017 d

2.018 b

2.019 d

2.020 c

2.021 d

2.022

$c - 4 = 7$
$4 = 4$
$c = 11$
Check:
$11 - 4 = 7$
$7 = 7$

2.023

$5x = 45$
$\frac{5x}{5}$ $\frac{45}{5}$
$x = 9$
Check:
$5 \cdot 9 = 45$
$45 = 45$

2.024

$2y + 1 = 5$
$-1 = -1$
$2y = 4$
$\frac{2y}{2}$ $\frac{4}{2}$
$y = 2$

Check:
$2 \cdot 2 + 1 = 5$
$5 = 5$

2.025

$\frac{a}{6} = 12$
$6 \cdot \frac{a}{6} = 6 \cdot 12$
$a = 72$
Check:
$\frac{72}{16} = 12$
$12 = 12$

2.026

$\frac{1}{2}h + 2 = 2$
$-2 = -2$
$\frac{1}{2}h = 0$
$2 \cdot \frac{1}{2}h = 2 \cdot 0$
$h = 0$
Check:
$\frac{1}{2} \cdot 0 + 2 = 2$
$0 + 2 = 2$
$2 = 2$

2.027

$3g - 1 = 2g + 3$
$1 = 1$
$3g = 2g + 4$
$-2g = -2g$
$g = 4$
Check:
$3 \cdot 4 - 1 = 2 \cdot 4 + 3$
$12 - 1 = 8 + 3$
$11 = 11$

2.028

$5x - 2x + 6 = x - 2$
$3x + 6 = x - 2$
$-6 = -6$
$3x = x - 8$
$-x = -x$
$2x = -8$
$\frac{2x}{2} = \frac{-8}{2}$
$x = -4$

2.028 cont.

Check:

$5(-4) - 2(-4) + 6 = -4 - 2$

$-20 + 8 + 6 = -6$

$-6 = -6$

2.029

$\frac{2}{3}x = 12$

$\frac{3}{2} \cdot \frac{2}{3}x = \frac{3}{2} \cdot 12$

$x = 18$

Check:

$\frac{2}{3} \cdot 18 = 12$

$12 = 12$

2.030

$8 = 3b - 1$

$\underline{1 = \quad\quad 1}$

$9 = 3b$

$\frac{9}{3} = \frac{3b}{3}$

$3 = b$

Check:

$8 = 3 \cdot 3 - 1$

$8 = 9 - 1$

$8 = 8$

2.031

$6.5 + 2x = 0.5$

$\underline{-6.5 \quad\quad = -6.5}$

$2x = -6$

$\frac{2x}{2} \quad \frac{-6}{2}$

$x = -3$

Check:

$6.5 + 2(-3) = 0.5$

$6.5 - 6 = 0.5$

$0.5 = 0.5$

2.032

$7k = k - 18$

$\underline{-k = -k}$

$6k = -18$

$\frac{6k}{6} = -\frac{18}{6}$

$k = -3$

2.032 cont.

Check:

$7(-3) = -3 - 18$

$-21 = -21$

2.033

$x - 2x = 3x - 27$

$-x = 3x - 27$

$\underline{-3x = -3x}$

$-4x = -27$

$\frac{-4x}{-4} = \frac{-27}{-4}$

$x = \frac{27}{4}$ or $6\frac{3}{4}$

Check:

$\frac{27}{4} - 2 \cdot \frac{27}{4} = 3 \cdot \frac{27}{4} - 27$

$\frac{27}{4} - \frac{54}{4} = \frac{81}{4} - \frac{108}{4}$

$-\frac{27}{4} = -\frac{27}{4}$

2.034

$\frac{3x}{2} - \frac{3}{4} = \frac{x}{2} + \frac{1}{4}$

$4(\frac{3x}{2} - \frac{3}{4} = \frac{x}{2} + \frac{1}{4})$

$6x - 3 = 2x + 1$

$\underline{3 = 3}$

$6x = 2x + 4$

$\underline{-2x = -2x}$

$4x = 4$

$\frac{4x}{4} = \frac{4}{4}$

$x = 1$

Check:

$\frac{3 \cdot 1}{2} - \frac{3}{4} = \frac{1}{2} + \frac{1}{4}$

$\frac{6}{4} - \frac{3}{4} = \frac{2}{4} + \frac{1}{4}$

$\frac{3}{4} = \frac{3}{4}$

2.035 $5(2x - 6) = 2x + 2$

$$10x - 30 = 2x + 2$$
$$\underline{30 = 30}$$
$$10x = 2x + 32$$
$$\underline{-2x = -2x}$$
$$8x = 32$$
$$\frac{8x}{8} = \frac{32}{8}$$
$$x = 4$$

Check:

$$5(2 \cdot 4 - 6) = 2 \cdot 4 + 2$$
$$5(8 - 6) = 8 + 2$$
$$5(2) = 8 + 2$$
$$10 = 10$$

SELF TEST 3

3.01 c

3.02 f

3.03 g

3.04 h

3.05 a

3.06 k

3.07 1

3.08 j

3.09 e

3.010 d

3.011 m

3.012 b

3.013 b

3.014 a

3.015 c

3.016 a

3.017 d

3.018 b

3.019 c

3.020 b

3.021 c

3.022 c

3.023 b

3.024 Three more than four times a number is one less than twice the number.

3.025

$$\frac{1}{5}a - 2 = 8$$
$$\underline{2 = 2}$$
$$\frac{1}{5}a = 10$$
$$5 \cdot \frac{1}{5}a = 5 \cdot 10$$
$$a = 50$$

Check:

$$\frac{1}{5} \cdot 50 - 2 = 8$$
$$10 - 2 = 8$$
$$8 = 8$$

3.026

$$4x + 2x - 5 = -3x + 11 + 5x$$
$$6x - 5 = 2x + 11$$
$$\underline{5 = 5}$$
$$6x = 2x + 16$$
$$\underline{-2x = -2x}$$
$$4x = 16$$
$$\frac{4x}{4} = \frac{16}{4}$$
$$x = 4$$

3.027 $2n - 5 = 15$

3.028
$$\frac{1}{2}n - 8 = 12$$
$$8 = 8$$
$$\frac{1}{2}n = 20$$
$$2 \cdot \frac{1}{2}n = 2 \cdot 20$$
$$n = 40$$
Check:
$$\frac{1}{2} \cdot 40 - 8 = 12$$
$$20 - 8 = 12$$
$$12 = 12$$

3.029 Let x = one number

$2x$ = another number
$$x + 2x = 42$$
$$3x = 42$$
$$\frac{3x}{3} = \frac{42}{3}$$
$$x = 14$$
$$2x = 28$$
The numbers are 14 and 28.

3.030 Let c = the cost of each pound

$7c$ = \$3.43 or 343¢
$$\frac{7c}{7} \quad \frac{343}{7}$$
c = 49¢ or \$0.49

3.031
$$3x - 1 = 2x + 3$$
$$1 = 1$$
$$3x = 2x + 4$$
$$-2x = -2x$$
$$x = 4$$
Check:
$$3 \cdot 4 - 1 = 2 \cdot 4 + 3$$
$$12 - 1 = 8 + 3$$
$$11 = 11$$

3.032
$$5a - 2a + 6 = a - 2$$
$$3a + 6 = a - 2$$
$$-6 = -6$$
$$3a = a - 8$$
$$-a = -a$$
$$2a = -8$$
$$\frac{2a}{2} = \frac{-8}{2}$$
$$a = -4$$
Check:
$$5(-4) - 2(-4) + 6 = -4 - 2$$
$$-20 + 8 + 6 = -6$$
$$-6 = -6$$

3.033 Let n = first integer

$n + 1$ = 2nd consecutive integer
$$n + n + 1 = 15$$
$$2n + 1 = 15$$
$$-1 = -1$$
$$2n = 14$$
$$\frac{2n}{2} = \frac{14}{2}$$
$$n = 7$$
$$n + 1 = 8$$
The integers are 7 and 8.

3.034 Let n = first even integer

$n + 2$ = 2nd consecutive even integer
$$n + n + 2 = 26$$
$$2n + 2 = 26$$
$$-2 = -2$$
$$2n = 24$$
$$\frac{2n}{2} = \frac{24}{2}$$
$$n = 12$$
$$n + 2 = 14$$
The integers are 12 and 14.

3.035 -9, -8

3.036
$$8(x - 2y - a) =$$
$$8 \cdot x - 8 \cdot 2y - 8 \cdot a =$$
$$8x - 16y - 8a$$

3.037 Let a = Susan's age

$a + 8 = 24$

$-8 = -8$

$a = 16$

Susan is 16 yrs. old.

3.038 Let d = distance traveled in 23 hrs.

$\frac{400}{7} = \frac{d}{23}$

$7 \cdot d = 400 \cdot 23$

$7d = 9{,}200$

$\frac{7d}{7} = \frac{9{,}200}{7}$

$d = 1{,}314.3$ miles

SELF TEST 1

1.01 u

1.02 f

1.03 n

1.04 s

1.05 b

1.06 o

1.07 r

1.08 t

1.09 i

1.010 d

1.011 j

1.012 k

1.013 g

1.014 a

1.015 c

1.016 1

1.017 e

1.018 p

1.019 q

1.020 h

1.021 four hundred twelve

1.022 1,000 + 400 + 60 + 2

1.023 600

1.024 279

1.025 1,889

1.026 1,115,416

1.027 456

1.028 4,200

1.029 $9 \cdot 141 = 1{,}269¢ = \$12.69$

1.030 $1_5, 2_5, 3_5, 4_5, 10_5, 11_5, 12_5$

1.031 $1111_2 = 2^3 + 2^2 + 2^1 + 1 =$
$8 + 4 + 2 + 1 = 15$

1.032 3243_5

1.033 $A \cap B = \{2, 4\}$

1.034 $24 = 2 \cdot 2 \cdot 2 \cdot 3 = 2^3 \cdot 3$

1.035 $18 = 2 \cdot 3 \cdot 3$
$27 = 3 \cdot 3 \cdot 3$
$\text{GCF} = 3 \cdot 3 = 9$

1.036 $8 = 2 \cdot 2 \cdot 2$
$10 = 2 \cdot 5$
$\text{LCM} = 2 \cdot 2 \cdot 2 \cdot 5 = 40$

1.037 $4\frac{2}{3} = \frac{3 \cdot 4 + 2}{3} = \frac{14}{3}$

1.038 $\frac{28}{72} = \frac{4 \cdot 7}{4 \cdot 18} = \frac{7}{18}$

1.039 $\frac{1}{4}, \frac{1}{2}, \frac{2}{3}, 1, \frac{4}{3} = 1\frac{1}{3}, \frac{12}{7} = 1\frac{5}{7}, 4\frac{1}{2}$

1.040 c

1.041 b

1.042 a
$\frac{5 \text{ yds.}}{12 \text{ ft.}} = \frac{5}{12 \div 3} = \frac{5}{4} = 5{:}4$

1.043 d
$\frac{8}{9} = \frac{x}{27}$
$9 \cdot x = 8 \cdot 27$
$9x = 216$
$\frac{9x}{9} = \frac{216}{9}$
$x = 24$

1.044 a
$4{,}000 = 4 \cdot 1{,}000 = 4 \cdot 10^3$

1.045 b

1.046 b

1.047 c

1.048 d

1.049 a

SELF TEST 2

2.01 c

2.02 d

2.03 b
$f(n) = 3n + 4 =$
$3 \cdot 6 + 4 =$
$18 + 4 = 22$

2.04 c
$2 = 2$
$3 = 3$
$8 = 2 \cdot 2 \cdot 2$
$\text{LCD} = 2 \cdot 2 \cdot 2 \cdot 3 = 24$

2.05 a
$\text{LCD} = 17$
$14 = 13\frac{17}{17}$
$-2\frac{9}{17} = -2\frac{9}{17}$
$11\frac{8}{17}$

2.06 d

2.07 b
spread = 10 – 1 = 9

2.08 a

2.09 c

2.010 a
$34 = \% \cdot 85$
$\% \cdot 85 = 34$
$\% = 34 \div 85$
$\% = 0.4$
$0.4 = 40\%$

2.011 d

2.012 q

2.013 j

2.014 b

2.015 h

2 016 s

2.017 p

2.018 t

2.019 r

2.020 f

2.021 g

2.022 l

2.023 a

2.024 k

2.025 n

2.026 c

2.027 u

2.028 i

2.029 o

2.030 e

2.031 $\text{LCD} = 4 \cdot 3 = 12$
$\frac{1}{4} = \frac{3}{12}$
$\frac{2}{3} = \frac{8}{12}$
$\frac{11}{12}$

2.032 LCD = 4 • 5 = 20

$$1\frac{3}{4} = 1\frac{15}{20}$$
$$3\frac{4}{5} = 3\frac{16}{20}$$
$$4\frac{31}{20} = 5\frac{11}{20}$$

2.033 LCD = 7 • 4 = 28

$$\frac{5}{7} = \frac{20}{28}$$
$$\frac{1}{4} = \frac{7}{28}$$
$$\frac{13}{28}$$

2.034 LCD = 3 • 2 = 6

$$3\frac{1}{3} = 3\frac{2}{6} = 2\frac{8}{6}$$
$$2\frac{1}{2} = 2\frac{3}{6} = 2\frac{3}{6}$$
$$\frac{5}{6}$$

2.035 $\frac{8}{15}$

2.036 $2\frac{2}{3} \cdot 3\frac{1}{5} = \frac{8}{3} \cdot \frac{16}{5} = \frac{128}{15}$ or $8\frac{8}{15}$

2.037 $\frac{5}{6} \div \frac{1}{4} = \frac{5}{\cancel{6}_3} \cdot \frac{\cancel{4}^2}{1} = \frac{10}{3}$ or $3\frac{1}{3}$

2.038 $2\frac{5}{7} \div 3\frac{1}{6} = \frac{19}{7} \div \frac{19}{6} = \frac{\cancel{19}^1}{7} \cdot \frac{6}{\cancel{19}_1} = \frac{6}{7}$

2.039 LCD = 4 • 3 = 12

$$4\frac{1}{4} = 4\frac{3}{12}$$
$$8\frac{1}{3} = 8\frac{4}{12}$$
$$7\frac{2}{3} = 7\frac{8}{12}$$
$$9\frac{15}{12} = 20\frac{3}{12} = 20\frac{1}{4}$$

2.040 68.351

2.041 87.89

2.042

```
   4.16
   3.24
  -----
   1664
   832
  1248
 -------
 13.4784
```

2 043

```
        5.4
0.3.)1.6.2
     1 5
     ---
       12
       12
       --
        0
```

2.044 To change a decimal to a percent, move the decimal point to the right two places and add the percent sign.

a. 0.51 = 51%
b. 0.06 = 6%
c. 1.61 = 161%

2.045 a. $\frac{1}{2} = 0.5 = 50\%$

b. $\frac{1}{4} = 0.25 = 25\%$

c. $\frac{3}{5} = 0.6 = 60\%$

2.046 16% • 214 = N
0.16 • 214 = N
34.24 = N

2.047 8 = % • 40
% • 40 = 8
% = 8 ÷ 40
$\% = \frac{1}{5}$
$\frac{1}{5} = 20\%$

2.048 13% • B = 52
B = 52 ÷ 13%
B = 52 ÷ 0.13
B = 400

2.049 1, 1, 6, 6, 6, 6, 6, 7, 8, 9, 9, 10, 10, 11

a. mean = $\frac{96}{14} = 6\frac{6}{7}$ or 6.9

b. median = $\frac{6+7}{2} = \frac{13}{2} = 6\frac{1}{2}$ or 6.5

c. mode = 6

2.050 (red) = $\frac{2}{10} = \frac{1}{5}$

(white) = $\frac{8}{10} = \frac{4}{5}$

SELF TEST 3

3.01 n

3.02 h

3.03 a

3.04 t

3.05 e

3.06 d

3.07 s

3.08 m

3.09 c

3.010 g

3.011 i

3.012 r

3.013 q

3.014 j

3.015 k

3.016 o

3.017 u

3.018 p

3 .019 l

3.020 b

3.021 -6, 9, 0, -14, 96

3.022 0

3.023 12

3.024 $3\frac{1}{4} \cdot 4\frac{2}{3} = \frac{13}{\not{4}_2} \cdot \frac{\not{14}^7}{3} = \frac{91}{6}$ or $15\frac{1}{6}$

3.025 300 + 80 + 6

3.026 2 + (-1) + 6 + (-19) =
2 – 1 + 6 – 19 = -12

3.027 $1_5, 2_5, 3_5, 4_5, 10_5, 11_5$

3.028 To change a decimal to a percent, move the decimal point two places to the right and add the percent sign.
$\frac{1}{2} = 0.5 = 50\%$
0.06 = 6%
1.6 = 160%
0.002 = 0.2%

3.029 $4^2 = 4 \cdot 4 = 16$
$3^3 = 3 \cdot 3 \cdot 3 = 27$
$2^5 = 2 \cdot 2 \cdot 2 \cdot 2 \cdot 2 = 32$
$5^4 = 5 \cdot 5 \cdot 5 \cdot 5 = 625$

3.030 $A = \pi \cdot r^2$
$A = 3.14 \cdot 8.1^2$
$A = 3.14 \cdot 65.61$
$A = 206.02$ sq. cm. = 206.0 sq. cm.

3.031 $V = l \cdot w \cdot h$
$V = 14 \cdot 8 \cdot 6$
$V = 672$ cu. in.

3.032 $-pq + 4r - 3s$

3.033 50,000

3.034 $(x + 3) \cdot (y - 4) =$
$x(y - 4) + 3(y - 4) =$
$xy - 4x + 3y - 12$

3.035 $2x - 4 = x + 1$
$\quad\;\; 4 = \quad 4$
$2x \quad = x + 5$
$-x \quad = -x$
$x \quad = \quad 5$

3.036 $1101_2 = 2^3 + 2^2 + 1 =$
$8 + 4 + 1 = 13$

3.037 $4n + 2 = n - 1$

3.038 57,000

3.039 One less than twice a number is three more than the number.

3.040 Let x = 1st integer
$x + 1$ = 2nd consecutive integer
$x + 2$ = 3rd consecutive integer
$x + 3$ = 4th consecutive integer

$$x + x + 1 + x + 2 + x + 3 = 18$$
$$4x + 6 = 18$$
$$-6 = -6$$
$$4x = 12$$
$$\frac{4}{4} = \frac{12}{4}$$
$$x = 3$$
$$x + 1 = 4$$
$$x + 2 = 5$$
$$x + 3 = 6$$

3.041 b

$$3x - 1 = 11$$
$$1 = 1$$
$$3x = 12$$
$$\frac{3x}{3} = \frac{12}{3}$$
$$x = 4$$

3.042 c

3.043 c

3.044 b

3.045 c

3.046 a

3.047 c

3.048 a

$12 = 2 \cdot (2 \cdot 3)$

$18 = (2 \cdot 3) \cdot 3$

$30 = (2 \cdot 3) \cdot 5$

$GCF = 2 \cdot 3 = 6$

3.049 a

$$n + 11 = 87$$
$$-11 = -11$$
$$n = 76$$

3.050 c

$$4x = 96$$
$$\frac{4x}{4} = \frac{96}{4}$$
$$x = 24$$

1. k
2. i
3. j
4. f
5. d
6. h
7. e
8. b
9. c
10. g
11. five thousand, four hundred sixty-one
12. 15,876
13. 1,900
14. 2
15. 6
16. 1,257
17. 10,918
18.
```
  1,284
  x 397
  8988
11556
3852
509,748
```
19.
```
      248
23)5,704
   46
   110
    92
    184
    184
      0
```
20.
```
      73 = 73 36/38 = 73 18/19
38)2,810
   266
    150
    114
     36
```
$73 = 73\frac{36}{38} = 73\frac{18}{19}$

21.
```
  7,346
   x 86
  44076
 58768
631,756
```
22. 3,975
23. 14,557
24. c
25. c
26. b
27. d
28. a
29. a
30.
```
  37,513
– 31,466
   6,047 miles
```
31.
```
$ 85
  146
   18
 x 12
$261
```
32.
```
 14
x 2
$28
```
33. 200 + 70 + 5

1. s
2. r
3. a
4. n
5. o
6. e
7. g
8. f
9. c
10. b
11. j
12. h
13. numerals
14. Roman
15. $11101_2 = (1 \times 2^4) + (1 \times 2^3)$ thousand
$+ (1 \times 2^2) + (1 \times 1)$
$= 16 + 8 + 4 + 1$
$= 29$
16. yes
17. commutative
18. The sum of the digits
$= 8 + 6 + 4 + 2 + 3 + 4 + 1$
$= 28$. 28 is not divisible by 3, so
8,642,341 is not divisible by 3
19. $72 = 2^3 \times 3^2$
$84 = 2^2 \times 3 \times 7$
$GCF = 2^2 \times 3 = 4 \times 3 = 12$
20. $24 = 2^3 \times 3$
$36 = 2^2 \times 3^2$
$LCD = 2^3 \times 3^2 = 8 \times 9$
$= 72$
21. domain
22. $110010_2 = (1 \times 2^5) + (1 \times 2^4)$
$+ (1 \times 2^1) = 32$
$+ 16 + 2 = 50.$
23. $10100_2 = (1 \times 2^4) + (1 \times 2^2)$
$= 16 + 4 = 20$
24. $111101_2 = (1 \times 2^4) + (1 \times 1)$
$+ (1 \times 2^2) + (1 \times 1)$
$=16 + 8\ 4\ 1$
$=29$
25. $1000001_2 = (1 \times 26) + 1 \times 1)$
$= 64 + 1 = 65$
26. five million, five hundred
five thousand, fifty
27. fifty thousand, five hundred, five
28. five hundred thousand, fifty-five
29. fifty-five million, five thousand
30. {1, 2, 3, 4, 5}
31. {2, 3, 4}
32. {3, 4}
33. {3, 4, 5}

1. c

2. j

3. a

4. g

5. b

6. f

7. d

8. k

9. h

10. e

11. $\frac{5}{7}$

12. $40 \div 8 = 5$
$\frac{7}{8} = \frac{7 \times 5}{8 \times 5} = \frac{35}{40}$

13. $\frac{75}{15} = 75 \div 15 = 5$

14. $5\frac{2}{3} = \frac{5 \times 3 + 2}{3} \quad \frac{15 + 2}{3} = \frac{17}{3}$

15. $3\frac{27}{100}$

16. 0.75

17. Drop the % sign and move the decimal point two places to the left. 3% = .03

18. Drop the % sign and write 100 as the denominator. 20% = $\frac{20}{100} = \frac{20 \div 20}{100 \div 20} = \frac{1}{5}$

19. $\frac{75}{175} = \frac{75 \div 25}{175 \div 25} = \frac{3}{7}$ or 3:7

20. 10^5

21. 1dm = 100 mm, so
15 km = 15 x 100 = 1,500 mm

22. Move the decimal point to the right two places and sdd the % sign. 0.0016 = 0.16%

23. Drop the % sign and move the decimal point two placces to the left. 427% = 4.27

24. $\frac{4}{8} = \frac{4 \div 4}{8 \div 4} = \frac{1}{2}$

25. 1:50 + 12:?
50 x 12 = 1 x ?
600 = ?
600 km

26. 0.50

27. $\frac{42}{70} = \frac{42 \div 14}{70 \div 14} = \frac{3}{5}$

28. $\frac{15}{4} = 15 \div 4 = 3\frac{3}{4}$

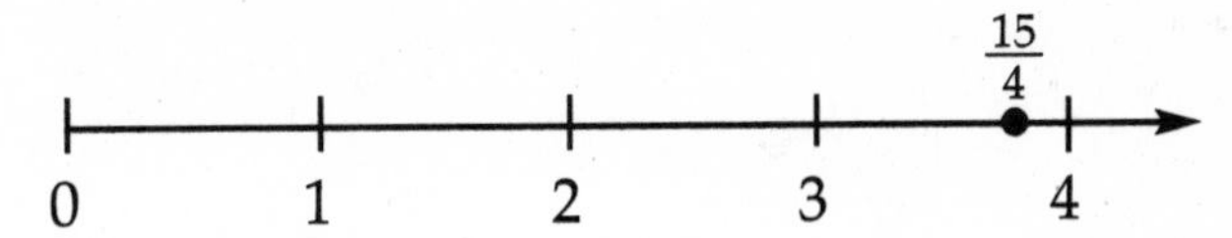

29. Examples:
$\frac{5}{4} = 1\frac{1}{4} = 1.25$
$\frac{5}{4} = 1\frac{1}{4}$
$\frac{5}{4} = 1.25 = 125\%$
$\frac{5}{4} = \frac{5 \times 2}{4 \times 2} = \frac{10}{8}$
$\frac{5}{4} = \frac{5 \times 3}{4 \times 3} = \frac{15}{12}$
$\frac{5}{4} = \frac{5 \times 5}{4 \times 5} = \frac{25}{20}$
$\frac{5}{4} = \frac{5 \times 6}{4 \times 6} = \frac{30}{24}$
$\frac{5}{4} = \frac{5 \times 7}{4 \times 7} = \frac{35}{28}$
$\frac{5}{4} = \frac{5 \times 8}{4 \times 8} = \frac{40}{32}$
$\frac{5}{4} = \frac{5 \times 10}{4 \times 10} = \frac{50}{40}$

30. b

31. b

32. c

33. b

34. a
0.006 = 0.6%

35. b

36. d

1. a. equal
   b. different

2. larger

3. a. 75,643.35
   b. 75,643.4
   c. 75,600
   d. 76,000
   e. 80,000

4. $20.00
   – 15.72
   $4.28 (count off 1 point if $ sign is not used)

5.
$$\begin{array}{r} 100 = 99\frac{5}{5} \\ -98\frac{3}{5} = 98\frac{3}{5} \\ \hline 1\frac{2}{5} \text{ degrees} \end{array}$$

6. $\frac{2}{6} = \frac{1}{3}$

7. $\frac{3}{7}$

8.
$$\begin{array}{r} 7\frac{1}{5} = 7\frac{16}{80} \\ +11\frac{11}{16} = 11\frac{55}{80} \\ \hline 18\frac{71}{80} \end{array}$$

9.
$$\begin{array}{r} 4\frac{1}{5} = 3\frac{6}{5} \\ -2\frac{4}{5} = 2\frac{4}{5} \\ \hline 1\frac{2}{5} \end{array}$$

10.
$$\begin{array}{r} 14\frac{5}{6} = 14\frac{20}{24} \\ +17\frac{7}{8} = 17\frac{21}{24} \\ \hline 31\frac{41}{24} = 32\frac{17}{24} \end{array}$$

11.
$$\begin{array}{r} 6\frac{1}{7} = 6\frac{5}{35} = 5\frac{40}{35} \\ -2\frac{4}{5} = 2\frac{28}{35} = 2\frac{28}{35} \\ \hline 3\frac{12}{35} \end{array}$$

12. 90.98

13. 17,544.053

14. 1,048.753

15. 10 = 2 x 5
    6 = 2 x 3
    3 = 3
    LCD = 2 x 3 x 5 = 30

$$\begin{array}{r} 4\frac{7}{10} = 4\frac{21}{30} \\ 17\frac{5}{6} = 17\frac{25}{30} \\ +27\frac{2}{3} = 27\frac{20}{30} \\ \hline 48\frac{66}{30} = 50\frac{6}{30} = 50\frac{1}{5} \end{array}$$

16. 3 = 3
    5 = 5
    6 = 2 x 3
    LCD = 2 x 3 x 5 = 30

$$\begin{array}{r} 16\frac{1}{3} = 16\frac{10}{30} \\ +14\frac{1}{5} = 14\frac{6}{30} \\ \hline 30\frac{16}{30} \end{array} \qquad \begin{array}{r} 30\frac{16}{30} = 29\frac{46}{30} \\ -29\frac{5}{6} = 29\frac{25}{30} \\ \hline \frac{21}{30} = \frac{7}{10} \end{array}$$

17. 7,452.34
    + 563.3
    8,015.640
    –6,942.543
    1,073.097

18. 55.32
    + 67.7
    123.02
    –119.9
    3.12

19. 0.0022

20. 13.3375

1. a. multiply the numerators
   b. multiply the denominators

2. $\frac{\not{3}^{1}}{8} \times \frac{5}{\not{6}_{2}} = \frac{5}{16}$

3. $\frac{2}{_{1}\not{3}} \times \frac{\not{6}^{2}}{7} = \frac{4}{7}$

4. $\frac{\not{5}^{1}}{8} \times \frac{1}{\not{5}_{1}} = \frac{1}{8}$

5. $\frac{\not{4}^{2}}{9} \times \frac{1}{\not{6}_{3}} = \frac{2}{27}$

6. $\frac{10}{3} \div 10 =$

   $\frac{\not{10}^{1}}{3} \times \frac{1}{\not{10}_{1}} = \frac{1}{3}$

7. $\frac{25}{6} \div \frac{7}{3} =$

   $\frac{25}{_{2}\not{6}} \times \frac{\not{3}^{1}}{7} = \frac{25}{14} = 1\frac{11}{14}$

8. $\frac{11}{4} \div \frac{4}{3} =$

   $\frac{11}{4} \times \frac{3}{4} = \frac{33}{16} = 2\frac{1}{16}$

9. $\frac{26}{5} \div \frac{11}{3} =$

   $\frac{26}{5} \times \frac{3}{11} = \frac{78}{55} = 1\frac{23}{55}$

10.
$$\begin{array}{r} 347.3 \\ \underline{\times\ 0.014} \\ 13892 \\ \underline{3473} \\ 4.8622 \end{array}$$

11. 0.09

12.
$$\begin{array}{r} 571.1 \\ \underline{\times\ 0.0043} \\ 17133 \\ \underline{22844} \\ 245573 \end{array}$$

13.
$$\begin{array}{r} 2.4 \\ \underline{\times\ 0.024} \\ 96 \\ \underline{48} \\ 0.0576 \end{array}$$

14.
$$\begin{array}{r} 1.44 \\ 45\overline{)6480} \\ \underline{45} \\ 198 \\ \underline{180} \\ 180 \\ \underline{180} \\ 0 \end{array}$$

15.
$$\begin{array}{r} 4{,}407 \\ 161\overline{)709527} \\ \underline{644} \\ 655 \\ \underline{644} \\ 1127 \\ \underline{1127} \\ 0 \end{array}$$

16.
$$\begin{array}{r} 64.34 \\ 3413\overline{)219592.42} \\ \underline{20478} \\ 14812 \\ \underline{13652} \\ 11604 \\ \underline{10239} \\ 13652 \\ \underline{13652} \\ 0 \end{array}$$

17. $150 - 120 = 30$

    $\frac{30}{120} = \frac{1}{4} = 0.25 = 25\%$ increase

18. $14 - 9.8 = 4.2$

    $\frac{4.2}{14} = 0.3 = 30\%$ decrease

19. $0.22 \times 4.3 = N$

$$\begin{array}{r} 0.22 \\ \underline{\times\ 4.3} \\ 66 \\ \underline{88} \\ 0.946 = N \end{array}$$

20. $R \times 25 = 4.5$

$R = 4.5 \div 25$

$$\begin{array}{r} 0.18 \\ 250\overline{)45.00} \\ \underline{250} \\ 2000 \\ \underline{2000} \\ 0 \end{array}$$

$R = 18\%$

21. $0.0003 \times 75 = N$

$0.0225 = N$

22. $0.27 \times B = 17.037$

$B = 17.037 \div 0.27$

$$\begin{array}{r} 63.1 \\ 27\overline{)1703.7} \\ \underline{162} \\ 83 \\ \underline{81} \\ 27 \\ \underline{27} \\ 0 \end{array}$$

$B = 63.1$

23. $0.265 \times B = 0.12455$

$B = 0.12455 \div 0.265$

$$\begin{array}{r} 0.47 \\ 265\overline{)124.55} \\ \underline{1060} \\ 1855 \\ \underline{1855} \\ 0 \end{array}$$

$B = 0.47$

24. $R \times 25 = 0.07$

$R = 0.07 \div 25$

$$\begin{array}{r} 0.0028 \\ 2500\overline{)7.0000} \\ \underline{5000} \\ 20000 \\ \underline{20000} \\ 0 \end{array}$$

$R = 0.28\%$

1. $f(n) = 4 \div 2 + 5$
   $= 2 + 5$
   $= 7$
2. $f(n) = 6 \div 2 + 5$
   $= 3 + 5$
   $= 8$
3. $f(n) = 8 \div 2 + 5$
   $= 4 + 5$
   $= 9$
4. $f(n) = 10 \div 2 + 5$
   $= 5 + 5$
   $= 10$
5. $f(n) = 12 \div 2 + 5$
   $= 6 + 5$
   $= 11$

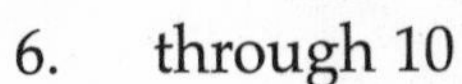

6. through 10

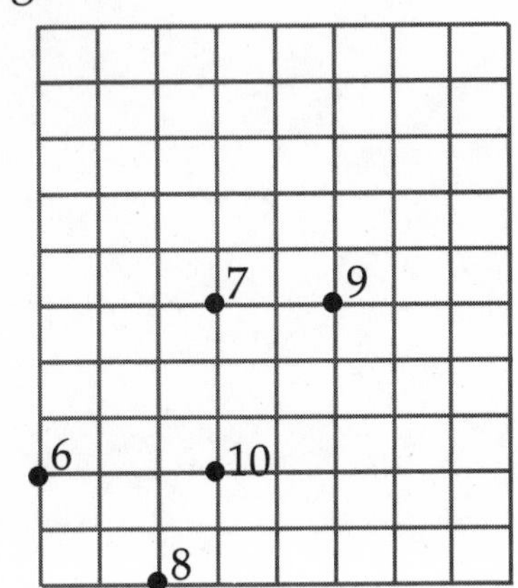

11. 4
12. 5
13. 6
14. $f(n) = n \div 2 + 1$
15.

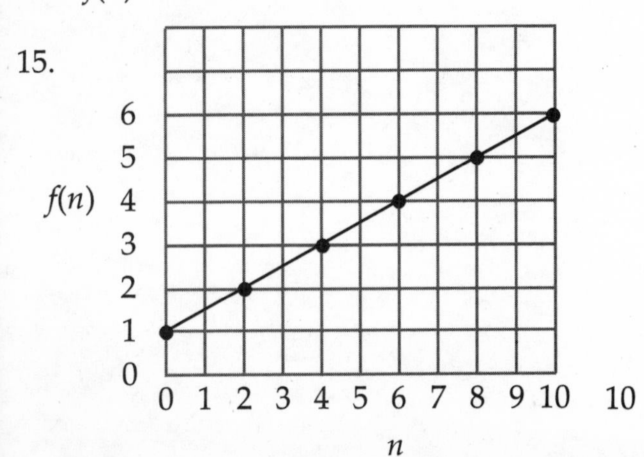

16.

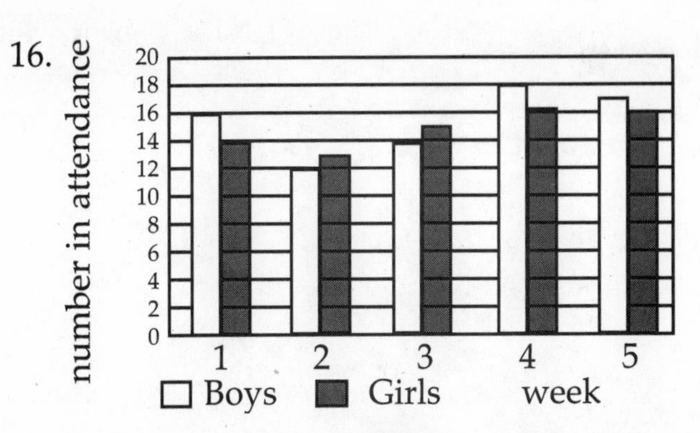

17.

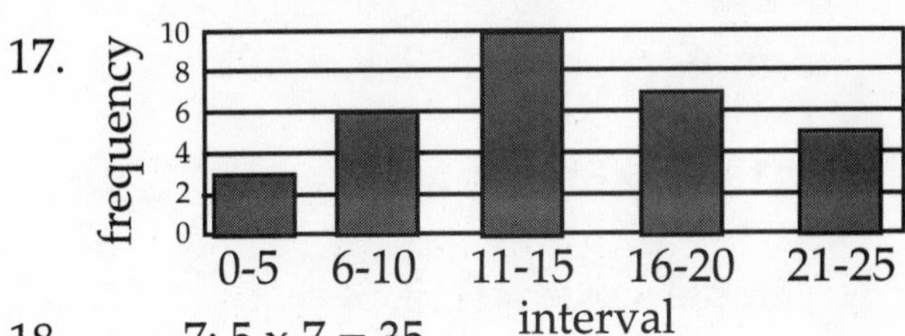

18. 7; 5 x 7 = 35
19. 9; 15 x 9 = 135
20. 18; 25 x 18 = 450
21. 10; 35 x 10 = 350
22. 6; 45 x 6 = 270

18. through 22. totals = 50; 1,240

23. $\frac{1,240}{50} = 24\frac{4}{5} \rightarrow 25$
24. The twenty-fifth and twenty-sixth numbers are in the 21-30 interval; therefore, the median interval is 21-30.
25. mode interval = 21-30
26.

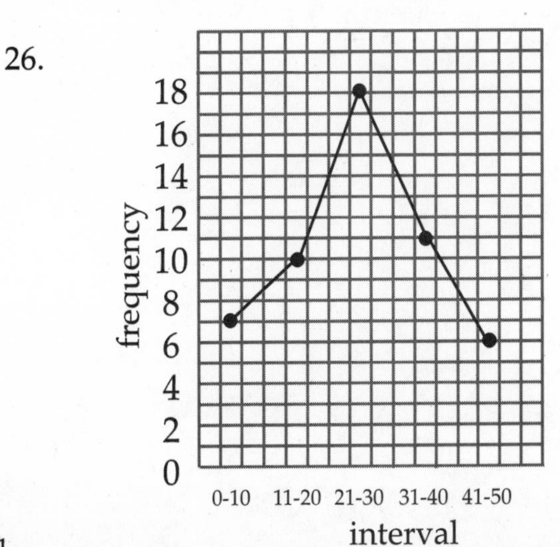

27. $\frac{3}{10}$

28. $\frac{3}{10} \times \frac{2}{9} = \frac{{}^{1}\cancel{3}}{{}_{5}\cancel{10}} \times \frac{\cancel{2}^{1}}{\cancel{9}_{3}} = \frac{1}{15}$

29. $\frac{2}{10} = \frac{1}{5}$

30. $\frac{2}{10} \times \frac{3}{9} = \frac{1}{5} \times \frac{1}{3} = \frac{1}{15}$

31. $\frac{1}{10}$

32. sample space = {6, 7, 8, 9, 10}

$P(1 \text{ ball} > 5) = \frac{5}{10} = \frac{1}{2}$

33. sample space = {1, 2, 3}

$P(1 \text{ ball} < 4) = \frac{3}{10}$

34. $\frac{1}{10} + \frac{1}{10} = \frac{2}{10} = \frac{1}{5}$

35. sample space (even number) =

{2, 4, 6, 8, 10}

$P(\text{even number}) = \frac{5}{10} = \frac{1}{2}$

36. $P(\text{even}) = \frac{5}{10} = \frac{1}{2}$

$P(\text{odd}) = \frac{5}{9}$

$P(\text{even, then odd}) = \frac{1}{2} \times \frac{5}{9} = \frac{5}{18}$

37. $\frac{3}{10} \times \frac{2}{9} \times \frac{5}{8} = \frac{{}^{1}\cancel{3}}{{}_{1}\cancel{10}} \times \frac{\cancel{2}^{1}}{\cancel{9}_{3}} \times \frac{\cancel{5}^{1}}{8} = \frac{1}{24}$

38. $\frac{1}{10} \times \frac{1}{9} \times \frac{1}{8} = \frac{1}{720}$

1. 31
2. -5
3. 0, 1, 2, 3
4. -2, 0, 2, 4
5. $5 > 2$
6. $-11 > -200$
7.

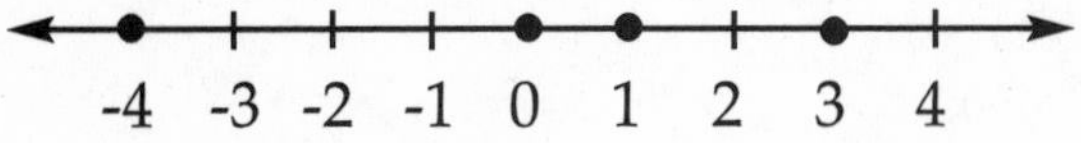

8.

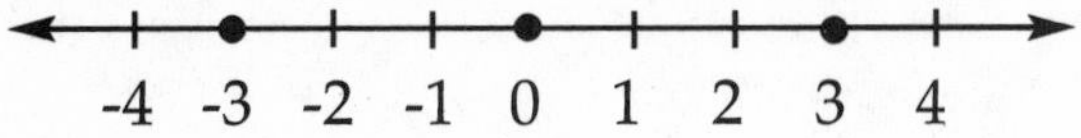

9. (-3, -5)
10. (0, 2)
11. 22
12. -15
13. 40
14. -24
15. -72
16. 8
17. -125
18. $\frac{1}{5^3} = \frac{1}{125}$
19. -5
20. 0
21.

$$-7 - 3 + (-2) =$$
$$\underbrace{-7 + (-3)} + (-2) =$$
$$\underbrace{-10 \quad + (-2)} =$$
$$-12$$

22.

$$\underbrace{4 \cdot (-7)} - \underbrace{3 \cdot (-2)} =$$
$$-28 - (-6) =$$
$$\underbrace{-28 + 6} =$$
$$-22$$

23.

$$4 \cdot 5^2 =$$
$$4 \cdot 25 =$$
$$100$$

24.

$$5^0 + (-1)^2 =$$
$$1 + 1 = 2$$

25.

m	3	0
$n = 7 - m$	$7 - 3$	$7 - 0$
n	4	7

n	-1
$m = 7 - n$	$7 - (-1)$ $7 + 1$
m	8

(3, 4), (0, 7), (8, -1)

26.

r	-3	0	5
$-7r$	$-7 \cdot (-3)$	$-7 \cdot 0$	$-7 \cdot 5$
t	21	0	-35

(-3, 21), (0, 0), (5, -35)

27.

a	-2	-1	0
$a - 2$	$-2 - 2$	$-1 - 2$	$0 - 2$
b	-4	-3	-2

1	2	3	4
$1 - 2$	$2 - 2$	$3 - 2$	$4 - 2$
-1	0	1	2

(-2, -4), (-1, -3), (0, -2)

(1, -1), (2, 0), (3, 1), (4, 2)

27. cont.

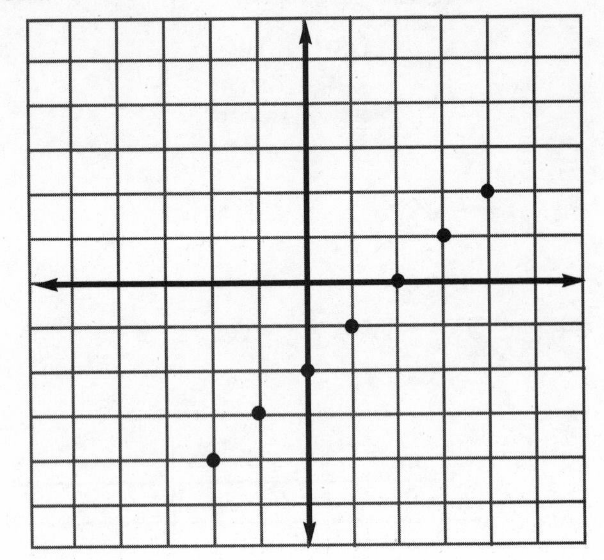

28.

x	-2	-1	0	1	2
$2x^2 - 3$	$2\cdot(-2)^2 - 3$	$2\ (-1)^2 - 3$	$2 \cdot 0^2 - 3$	$2\ 1^2 - 3$	$2\ 2^2 - 3$
	$2 \cdot 4 - 3$	$2\ 1 - 3$	$0 - 3$	$2\ 1 - 3$	$2\ 4 - 3$
	$8 - 3$	$2 - 3$		$2 - 3$	$8 - 3$
y	5	-1	-3	-1	5

(-2, 5), (-1, -1), (0, -3), (1, -1),
(2, 5)

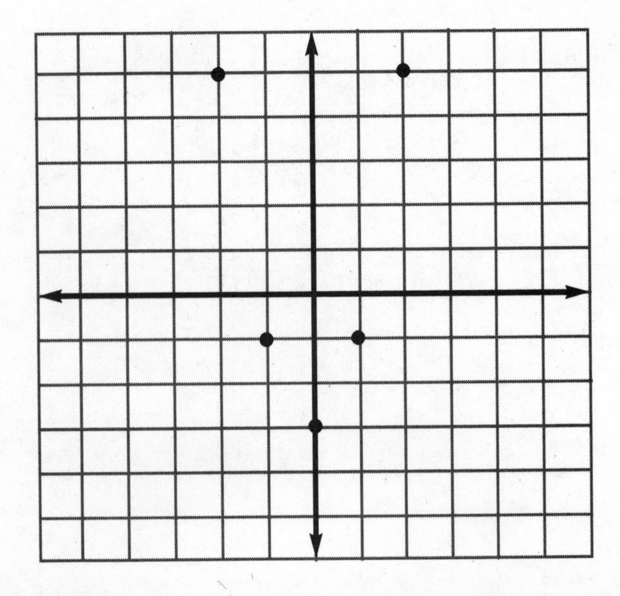

1. Arithmetic method:

$$\begin{array}{l} \;\;\;1\;\;8.\;7\;\;0 \rightarrow 18.7 \\ \sqrt{3\;\;50.00\;\;00} \end{array}$$

$$\begin{array}{lll} & 1 & \\ 2\underline{8} & 2\,50 & \\ & \underline{2\,24} & \\ 36\underline{7} & 26\,00 & \\ & \underline{25\,69} & \\ 374\underline{0} & 31\,00 & \\ & \underline{0} & \\ & 31\,00 & \end{array}$$

Divide-and-average method:
$18^2 = 324$ is too small.
$19^2 = 361$ is too large.
Guess 18.5.

$$\begin{array}{r} 18.91 \\ 18.5\overline{)350.0.00} \\ \underline{185} \\ 1650 \\ \underline{1480} \\ 1700 \\ \underline{1665} \\ 350 \\ \underline{185} \\ 165 \end{array}$$

$$\frac{18.5 + 18.91}{2} = \frac{37.41}{2} = 18.705$$

Round to 18.7

2. $A = l \cdot w$
$= 14 \cdot 12$
$= 168 \text{ ft.}^2$

3. $C = \pi \cdot d$
$= 3.14 \cdot 12$
$= 37.68$ ft

4. $r = \frac{1}{2} \cdot d = \frac{1}{2} \cdot 12 = 6$ ft.
$A = \pi \cdot r^2$
$= 3.14 \cdot 6^2$
$= 3.14 \cdot 36$
$= 113.04 \text{ ft.}^2$

5. $A = \frac{1}{2} \cdot b \cdot h$
$= \frac{1}{2} \cdot 12 \cdot 8$
$= 6 \cdot 8$
$= 48 \text{ m}^2$

6. a. parallelogram
b. $A = b \cdot h$
$= 18 \cdot 7$
$= 126 \text{ mm}^2$

7. $a = 18$
$b = 12$
$c =$ unknown

$18^2 + 12^2 = c^2$
$324 + 144 = c^2$
$468 = c^2$
$\sqrt{c^2} = \sqrt{468}$
$c \doteq 21.63 \doteq 21.6$ yds.

8. $A = \frac{1}{2} \cdot b \cdot h$
$= \frac{1}{2} \cdot 12 \cdot 18$
$= 6 \cdot 18$
$= 108 \text{ yd.}^2$

9. $504 \text{ ft.}^2 \cdot \frac{1 \text{ yd.}^2}{9 \text{ ft.}^2} =$

$\frac{504 \text{ yd.}^2}{9} = 56 \text{ yd.}^2$

10. a. trapezoid
b. $A = \frac{1}{2} \cdot h \cdot (b_1 + b_2)$
$= \frac{1}{2} \cdot 8 \cdot (24 + 12)$
$= \frac{1}{2} \cdot 8 \cdot 36$
$= 4.36$
$= 144 \text{ cm}^2$

11. a. $A = l \cdot w \cdot h$
$= 14 \cdot 8 \cdot 10$
$= 112 \cdot 10$
$= 1{,}120 \text{ in.}^3$

b. area if top:
$14 \cdot 8 = 112 \text{ in.}^2$
area of bottom:
$14 \cdot 8 = 112 \text{ in.}^2$
area of back side:
$10 \cdot 8 = 80 \text{ in}^2$
area of front side:
$10 \cdot 8 = 80 \text{ in.}^2$
area of left side:
$14 \cdot 10 = 140 \text{ in.}^2$
area of right side:
$14 \cdot 10 = \underline{140 \text{ in.}^2}$
total $S.A. = 664 \text{ in.}^2$

12. $V = \frac{1}{2} \cdot b_{\Delta} \cdot h_{\Delta} \cdot h_{p}$
$= \frac{1}{2} \cdot 30 \cdot 25 \cdot 50$
$= 15 \cdot 25 \cdot 50$
$= 375 \cdot 50$
$= 18{,}750 \text{ ft.}^3$

13. *S.A.* of each rectangle $= l \cdot w$ or $= b \cdot h$
S.A. of 2 rectangles $= 2 \cdot l \cdot w$ or $2 \cdot b \cdot h$
$= 2 \cdot 50 \cdot 30$
$= 100 \cdot 30$
$= 3{,}000 \text{ ft.}^2$

14. $V = \frac{1}{3} \cdot B \cdot h$
$= \frac{1}{3} \cdot s^2 \cdot h$
$= \frac{1}{3} \cdot 20^2 \cdot 16$
$= \frac{1}{3} \cdot 400 \cdot 16$
$= \frac{1}{3} \cdot 6{,}400$
$= 2{,}133.33 \text{ ft.}^3$

15. $V = \pi \cdot r^2 \cdot h$
$= 3.14 \cdot 3^2 \cdot 11$
$= 3.14 \cdot 9 \cdot 11$
$= 28.26 \cdot 11$
$= 310.86 \text{ ft.}^3$

16. $TA = 2 \cdot \pi \cdot r \;(r + h)$
$= 2 \cdot 3.14 \cdot 3 \cdot (3 + 11)$
$= 2 \cdot 3.14 \cdot 3 \cdot 14$
$= 6.28 \cdot 3 \cdot 14$
$= 18.84 \cdot 14$
$= 263.76 \text{ ft.}^2$

17. $V = \frac{1}{3} \cdot \pi \cdot r^2 \cdot h$
$= \frac{1}{3} \cdot 3.14 \cdot 3^2 \cdot 18$
$= \frac{1}{3} \cdot 3.14 \cdot 9 \cdot 18$
$= 3.14 \cdot 3 \cdot 18$
$= 9.42 \cdot 18$
$= 169.56 \text{ cm.}^3$

18. $V = \frac{4}{3} \cdot \pi \cdot r^3$
$= \frac{4}{3} \cdot 3.14 \cdot 40^3$
$= \frac{4}{3} \cdot 3.14 \cdot 64{,}000$
$= \frac{4}{3} \cdot 200{,}960$
$= \frac{803{,}840}{3}$
$= 267{,}946.67 \text{ ft.}^3$

19. $S = 4 \cdot \pi \cdot r^2$
$= 4 \cdot 3.14 \cdot 40^2$
$= 4 \cdot 3.14 \cdot 1{,}600$
$= 12.56 \cdot 1{,}600$
$= 20{,}096 \text{ ft.}^2$

1. g
2. a
3. k
4. f
5. h
6. e
7. j
8. c
9. l
10. m
11. i
12. b
13. c
14. d
15. a
16. c
17. b

$\frac{1}{5}a = 6$

$5 \cdot \frac{1}{5}a = 5 \cdot 6$

$a = 30$

18. a
19. b
20. b

$2n - 1 = 11$

$\underline{\quad 1 = 1}$

$2n \quad = 12$

$\frac{2n}{2} = \frac{12}{2}$

$n = 6$

21. b
22. $4 \cdot a + a \cdot c + a \cdot a = 4a + ac + a^2$
23. $8 \cdot 7 = 56$ and $7 \cdot 8 = 56$:
    56 = 56, so the statement is true.
24. $8a + 4a - 6a - 14a + 5a = 12$

$-3a = 12$

$-\frac{1}{3}(-3a) = -\frac{1}{3}(12)$

$a = -4$

25. $14xy - 6x - 7xy + 8x - 6xy =$

$14xy - 7xy - 6xy - 6x + 8x =$

$xy + 2x$

26. $3x - 6 = 2x - 9$

$\underline{\quad 6 = \quad 6}$

$3x \quad = 2x - 3$

$\underline{-2x \quad = -2x}$

$x = \quad -3$

Check:

$3(-3) - 6 = 2(-3) - 9$

$-9 - 6 = -6 - 9$

$-15 = -15$

27. $8h = 2h + 24$

$\underline{-2h = -2h}$

$6h = 24$

$\frac{6h}{6} = \frac{24}{6}$

$h = 4$

Check:

$8(4) = 2(4) + 24$

$32 = 8 + 24$

$32 = 32$

28. $\frac{1}{2}a - 3 = 7$

$\underline{\quad 3 = 3}$

$\frac{1}{2}a = 10$

$2 \cdot \frac{1}{2}a = 2 \cdot 10$

$a = 20$

Check:

$\frac{1}{2} \cdot 20 - 3 = 7$

$10 - 3 = 7$

$7 = 7$

29. $A = \frac{1}{2}h(a + b)$

$A = \frac{1}{2} \cdot 6(5 + 7)$

$A = 3(12)$

$A = 36$

30. Three less than eight times a number is one more than twice the number.

31. Let $3n$ = one number

n = another number

$3n + n = 52$

$4n = 52$

$\frac{4n}{4} = \frac{52}{4}$

$n = 13$

$3n = 39$

The numbers are 13 and 39.

32. Let n = 1st integer

$n + 1$ = 2nd consecutive integer

$n + n + 1 = 21$

$2n + 1 = 21$

$-1 = -1$

$2n = 20$

$\frac{2n}{2} = \frac{20}{2}$

$n = 10$

$n + 1 = 11$

The integers are 10 and 11.

33. Let x = number of revolutions

$300 = \frac{x}{15}$

$15 \cdot 300 = \frac{15x}{15}$

$4500 = x$

The engine will make 4,500 revolutions in 15 minutes.

34. Let a = Ann's Aunt's age

$\frac{1}{5}a$ = Ann's age

$\frac{1}{5}a = 14$

$5 \cdot \frac{1}{5}a = 5 \cdot 14$

$a = 70$

Ann's Aunt's age is 70 yrs. old.

1. r
2. a
3. p
4. q
5. e
6. s
7. m
8. l
9. f
10. g
11. b
12. d
13. h
14. i
15. j
16. k
17. n
18. c
19. o
20. t
21. 2,387
22.

$$\begin{array}{r} 5.116 \\ \underline{0.813} \\ 15348 \\ 5116 \\ \underline{40928} \\ 4.159308 \end{array}$$

23.

$$0.26\overline{)4836.00} = 18{,}600.$$

$$\begin{array}{r} \underline{26} \\ 223 \\ \underline{208} \\ 156 \\ 156 \\ 0 \end{array}$$

24.

$$\begin{array}{rcl} 2n-3 & = & 9-n \\ 3 & = & 3 \\ \hline 2n & = & 12-n \\ n & = & n \\ \hline 3n & = & 12 \\ \frac{3n}{3} & = & \frac{12}{3} \\ n & = & 4 \end{array}$$

Check

$2 \cdot 4 - 3 = 9 - 4$

$8 - 3 = 5$

$5 = 5$

25.

$$\begin{array}{rcl} x + 2x - 6 & = & x + 8 \\ 3x - 6 & = & x + 8 \\ 6 & = & 6 \\ \hline 3x & = & x + 14 \\ -x & = & -x \\ \hline 2x & = & 14 \\ \frac{2x}{2} & = & \frac{14}{2} \\ x & = & 7 \end{array}$$

Check:

$$\begin{array}{rcl} 7 + 2 \cdot 7 - 6 & = & 7 + 8 \\ 7 + 14 - 6 & = & 15 \\ 15 & = & 15 \end{array}$$

26.

6 dozen = $6 \cdot 12 = 72$

$\frac{1}{39} = \frac{72}{x}$

$1 \cdot x = 39 \cdot 72$

$x = 2{,}808¢ = \$28.08$

27.

$1_5,\ 2_5,\ 3_5,\ 4_5,\ 10_5,\ 11_5,\ 12_5,$
$13_5,\ 14_5,\ 20_5$

28.

$75 = 3 \cdot 5 \cdot 5$ (5 • 5 circled)

$50 = 2 \cdot 5 \cdot 5$ (5 • 5 circled)

GCF = $5 \cdot 5 = 25$

29. $2\frac{3}{5} \cdot 4\frac{1}{4} = \frac{13}{5} \cdot \frac{17}{4} = \frac{221}{20}$ or $11\frac{1}{20}$

30. $A \cup B = \{A, B, C, 1, 2, 3\}$

31. $48 = 2 \cdot 2 \cdot 2 \cdot 2 \cdot 3 = 2^4 \cdot 3$

32. To change a decimal to a per cent, move the decimal point to the right two places and add the percent sign.
$0.05 = 5\%$
$0.001 = 0.1\%$
$\frac{1}{4} = 0.25 = 25\%$
$1.2 = 120\%$
$3.5 + 350\%$

33. $1\frac{1}{3} \div 5 = \frac{4}{3} \div \frac{5}{1} = \frac{4}{3} \cdot \frac{1}{5} = \frac{4}{15}$

34. $C = 2 \cdot \pi \cdot r$
$C = 2 \cdot 3.14 \cdot 14.6$
$C = 91.688 = 91.7$ cm

35. $(3a - 4d)(4x + 3y) =$
$3a(4x + 3y) - 4d(4x + 3y) =$
$12ax + 9ay - 16dx - 12dy$

36. $\frac{14}{7} = 2$
$\frac{0}{2} = 0$
$\frac{0}{2}, \frac{2}{3}, \frac{4}{5}, \frac{7}{8}, 1, \frac{14}{7}, 2\frac{3}{4}, 8$

37. c

38. d
$14 = 2 \cdot 7$
$24 = 2 \cdot 2 \cdot 2 \cdot 3$
$LCD = 2 \cdot 2 \cdot 2 \cdot 3 \cdot 7 = 168$

39. b
$45\% \cdot B = 180$
$B = 180 \div 45\%$
$B = 180 \div 0.45$
$B = 400$

40. b

41. d

42. c
4, 5, 5, 5, (6) 6, 7, 7, 8